Computer-Mediated Communication for Linguistics and Literacy: Technology and Natural Language Education

Adams B. Bodomo
University of Hong Kong, Hong Kong

Information Science REFERENCE
INFORMATION SCIENCE REFERENCE
Hershey · New York

Director of Editorial Content: Kristin Klinger
Senior Managing Editor: Jamie Snavely
Assistant Managing Editor: Michael Brehm
Publishing Assistant: Sean Woznicki
Cover Design: Lisa Tosheff
Printed at: Yurchak Printing Inc.

Published in the United States of America by
Information Science Reference (an imprint of IGI Global)
701 E. Chocolate Avenue,
Hershey PA 17033
Tel: 717-533-8845
Fax: 717-533-8661
E-mail: cust@igi-global.com
Web site: http://www.igi-global.com/reference

Library of Congress Cataloging-in-Publication Data

Bodomo, Adams.
Computer-mediated communication for linguistics and literacy : technology and natural language education / by Adams B. Bodomo.
p. cm.
Includes bibliographical references and index.
Summary: "This book investigates the way humans communicate through the medium of information technology gadgets, focusing on the linguistic, literacy and educational aspects of computer-mediated communication"--Provided by publisher.

ISBN 978-1-60566-868-0 (hardcover) -- ISBN 978-1-60566-869-7 (ebook) 1. Telematics. 2. Education--Computer network resources. I. Title.
TK5105.6.B63 2009
384.3--dc22

2009013749

British Cataloguing in Publication Data
A Cataloguing in Publication record for this book is available from the British Library.

Table of Contents

Foreword

In the name of progress, we have perhaps adopted computer technology at such a pace that few of us can slow down enough to ponder whether it is technology that is driving us, or whether it is that we have clear purpose in guiding where technology is heading. There is no better example than the interaction between human languages and technology. Dr. Adams Bodomo has been in an ideal position to observe, collect data, experiment, and analyze the phenomenon in one of the most dynamic places on earth, from a social and linguistic perspective: Hong Kong SAR, China. This book is the culmination of his research work on the relationship between human language and information technology there. To help appreciate its significance, let me explain my viewpoint on the role of technology in the Information Age.

The Lana Project (1971-1976) by Duane Rumbaugh, Sue Savage-Rumbaugh, and William Fields had the clear purpose of investigating the ability of chimpanzees to acquire language. The application of information technology resulted in the development of a computer-based language training system, as summarized on www.greatapetrust.org:

Lana is a female chimpanzee born in 1970 at the Yerkes National Primate Research Center. Her name derives from the LANguage Analogue (LANA) project, which sought to develop a computer-based language training system in an effort to investigate the ability of chimpanzees to acquire language. Lana joined the research as a subject when she was two and a half years old. The research was the first to interface a keyboard with a chimpanzee. At that time, it was believed that only humans could use symbols. Lana demonstrated that she could discriminate between lexigrams and associate them with ideas. As she progressed, she would sequence words and use them grammatically, later starting to create novel utterances in response to unplanned events that affected her life. For example, Lana would request that the research technician refill her computer vending device when it was empty of treats, or request an item she had seen outside her room that the computer had no facility to provide to her. Lana exhibited language learning, and her experimental accomplishments were extraordinary. Equally important to her legacy is the lexigram keyboard, developed by Duane Rumbaugh, which has served as the primary communicative interface for ape language research at Decatur, Georgia for the last several decades. This keyboard is composed of three panels with approximately 384 noniconic arbitrary symbols. When the apes depress a key, the word represented there is spoken by a digital voice and the lexigram is displayed on a video screen.

This is a classic example of successful application of technology to serve a clearly stated purpose in computer-mediated communication (CMC). It starts with what one wishes to accomplish, to be followed by how technology can help. By contrast, the development of many of the information communication technology (ICT) in common use today are driven by space-age trends of miniaturization and speed, rather than any linguistic or literacy concerns. Designers strive mainly to package more processing power into ever-shrinking devices. It is then little surprise that the need to type on tiny keyboards and read on even smaller screens, often while one is in motion, leads to the distortion of standard languages. The merit and legitimacy of such emerging forms of language is subject of debate. However, whether viewed as evolution or deformation, one can hardly argue that they are by design and on purpose.

Meaningful purpose is the domain of education. With clear standards for what one wishes to accomplish, the question can then be asked on how CMC can help. Views on literacy can be orthodox (adherence to standard languages) or liberal (open to new language standards), but should not be allowed to drift, even in the guise of multiculturalism and diversity. In the perhaps unique and peculiar case of Hong Kong, the politically-correct and complacent claim of bi-literacy and tri-lingualism [两文三语] may have degenerated into a euphemism for deficiency in all. In the past, citizens motivated by conforming to colonial rule maintained language standards that afforded them the relative competitive advantage of knowing better English than their mainland compatriots, and better Chinese than Westerners. With sinking standards in English and the native Cantonese not being official Chinese, Hong Kong runs the risk of being marginalized in this age of globalization. Is code-mixing (the prevalent blending of Cantonese and English) the culprit? Is this predicament exacerbated by ICT-driven colloquialism? The fact that we have the technology to communicate with Lana the chimpanzee, does not mean that we all have to use CMC that way, do we? Can educators rise to the occasion to set clear purpose to improve language standards, with the help of CMC technology? For anyone intrigued by or having a stake in these critical issues, this book by Dr. Adams Bodomo provides indispensible insight with background data and analysis for constructive discourse and deliberation.

Professor James K. Ho
Department of Information & Decision Sciences
University of Illinois at Chicago
USA

James K. Ho *is Professor of Information & Decision Sciences at the University of Illinois at Chicago. A native of Hong Kong, he graduated from Columbia University in 1970 and obtained his PhD from Stanford University in 1974. He has published widely in academic and professional journals and authored four books, including* Prosperity in the Information Age—Creating Value with Technology, from Mailrooms to Boardrooms *(1994) and* Cyber Tigers: How Companies in Asia can Prosper from E-Commerce *(2000). His recent research interests focus on data mining and visualization with applications in topological analysis of online auction markets, technology diffusion, investment climates, and customer relations management. In Computer Mediated Communication, he did pioneering work in multilingual Web interfaces to promote e-commerce among small and medium size enterprises, which culminated in the APEC Multilingual International Trade Project. He is on the editorial board of the* Journal of Computer Mediated Communication.

Preface

INTRODUCTION: CMC

Computer-mediated Communication (CMC) is an amazingly multi- and inter-disciplinary subject area that spans fields as diverse as computer science, information technology, communication studies, linguistics, literacy, education, business, ethics, and law. Given this vast subject-matter it would be practically impossible for any single volume to cover all aspects of CMC to any appreciable depth. There is thus the need to focus on one or the other of these component disciplines.

THE BOOK'S FOCUS

Within this wider, interdisciplinary arena, this book, titled, *Computer-mediated Communication for Linguistics and Literacy: Technology and Natural Language Education*, occupies an important position. It has a clear focus on the linguistic, literacy and educational aspects of CMC. The book investigates the way humans communicate through the medium of information technology gadgets. Based on extensive research on how we use natural languages like English and Chinese in media such as emails, MSN, and mobile phones, the book outlines new forms and ways of speaking, reading, and writing in an age in which there is a pervasive presence of communication technologies in offices and homes. This interaction between human language and technology has created new forms and uses of language and literacy the study of which has given birth to this exciting new field of Computer-mediated Communication (CMC) that we are about to delve into.

Alternative names for this focus on the linguistic and literacy aspects of the field of CMC include Human Language Technology (HLT), Language and Computation, or even Internet Linguistics. CMC, HLT and their allied fields have as a subject-matter the theoretical study of this interaction between natural languages and communications technology but also a study of the practical implications and applications of how computers and other communications technologies can be used to analyze and process natural languages for the design of communication systems and devices like automatic translation devices, mobile phones, voice recognition devices and all kinds of computer systems involving a human language component. This book takes a closer look at many of these issues and guides the reader through the way language is used in various media of technology and the implications this has for learning to speak, read, and write languages.

FURTHER ON THE ROLE OF CMC

The era in which we live has often been referred to as the Information Age or even the Knowledge Society. In this age and society we constantly witness a massive explosion of new types and styles of communications gadgets such as computers of all types (including desktops and laptops), PDAs, mobile phones, Blackberries, and so forth; and media like the Internet, emails, ICQ, chat, text messaging and others. We can expect that new ones will be invented.

In this book, I will discuss how these different types and styles of information and communication technologies impact the way we communicate. In particular, I will examine the nature and uses of language in what may be described as a revolutionary environment for information and communication. As these communication technologies bring us closer together, we are forming a global village, and indeed, this era has also been referred to as an era of globalization. We will find that in this global village our everyday communications may involve not just one language, but several languages. In fact, it is hardly ever sufficient to operate in only one language anymore. Bilingualism/multilingualism is now an essential part of this information communication revolution. I will examine how these combinations of technologies and languages affect the nature and structure of each language and whether this leads to changes in language structure and language use.

Not only are these technologies creating an environment of multiplicity of languages, they also have an impact on the way we read, write, and process information. I will examine the different ways in which we are called upon to speak, read, and write in these different technological environments. We will notice that we are called upon to juggle different ways of reading and writing in these different environments. Literacy is no longer a mono-modal practice but very much multi-modal; various kinds of literacies are needed to function well in the information society.

This ecology of languages, literacies, and technologies will be shown to be very dynamic, and in this ever-changing ecology many issues about human nature will be discussed. In particular, an examination of these interrelationships helps us understand some important aspects of our culture and society. Throughout the book I will examine some consequences for learning and knowledge acquisition in this complex ecology, and attempt to show how we can harness these languages, literacies, and technologies to improve education for our future generations.

DISTINCTIVE ASPECTS OF THE BOOK

Case Studies

One of the most distinctive aspects of this book is that most chapters incorporate a case study to anchor the issue being discussed within the sphere of actual empirical data. This is possible because it is a book that is the outcome of funded competitive research projects as diverse as *Linguistic Features of Mobile Phone Communication, The Use of Computers in Teaching Languages and Linguistics*, and *Communication in the Age of Information Technology: New Forms of Language and their Educational Implications*. It is thus based on current research results and discussions that would lead the reader to valuable insights to the sort of issues that experts in the field are currently grappling with.

Addressing Everyday Issues

While the book is grounded on top-notch research, it still addresses everyday issues. The main theme and purpose of this book is to analyze and explain everyday linguistic and cultural dynamics triggered by new communication technologies. This is an important theme and a cultural issue that has the potential to impact most, if not all, citizens of the 21st century society. Books such as this are thus needed by not just only academics and other experts but also the man on the street to help explicate these linguistic, cultural, and communication dynamics surrounding our everyday lives.

More Than One Medium

A third aspect of the book is that it touches on many media. There are a number of valuable books on this important theme of CMC but my book is distinguished from them in the sense that it treats contemporary themes surrounding language and literacy dynamics in more than one medium. Many books on CMC either concentrate on email, or some single internet system or the other but my book treats all these linguistic, literacy, and communication innovations on email, MSN, mobile phone texting, and even video-based CMC media like YouTube and online games. All these media are treated with data based on actual university-level academic research by myself, my students, and other members of an informal research group that I head: The Linguistic Theory and Technology Group (LTTG).

READERSHIP

Each chapter in this book takes up comprehensive treatments of the frequent themes on the interaction between languages, literacies, and technologies; it is based on facts and figures gained from actual research projects on how the youth use language in the new media, it is international in scope, and it does a rigorous survey of the literature in the area. Therefore this book will be useful for undergraduates, postgraduates, and scholars in computer-mediated communication fields as diverse as Linguistics, Literacy, Education, Computer Science, Information Science, and Human Language Technology. It would also be of interest to the general public.

ORGANIZATION OF THE BOOK

The book comprises 12 chapters. Chapters 1 to 4 may be regarded as constituting Section 1 of the book, covering the foundational aspects. Chapters 5 to 8 may be considered as Section 2, the technological and linguistic empirical base of our study, where I focus on the various CMC technological environments like email, msn, and mobile phone one at a time. Chapters 9 to 11 may be seen as focusing on the educational and pedagogical aspects of the link between ICT and language. Chapter 12, the final chapter ties together many of the issues discussed, and points to emerging themes and emerging CMC tools and media that will surely shape the future of CMC.

In Chapter 1, *Definitions and Basic Conceptual Notions*, I introduce fundamental conceptual terms, such as communication, language, literacy, computer science, IT, and of course computer-mediated communication. I do this by providing short definitions and explanations of these key topics that constitute the subject matter of the book.

In Chapter 2, titled, *Is Technology Changing the Way We Communicate*? I present the main theme or thesis of the book in the form of a discussion. I present different positions, indicating how different scholars approach these issues. On the one hand are works of scholars like Adams (1986), Baron (1984) and Crystal (2001), and on the other side are those like Kress (1998) and Luke (2000).

In Chapter 3, titled *Digital Literacy: Reading in the Age of ICT*, I focus on the changing patterns of reading in the environment of new technologies. I first show how technology is affecting the way we read, both in terms of the choice of media and in terms of the strategies we employ to read. More importantly I focus on the surveys of reader preferences, as it is evident that given a wide choice of media that technology provides us, we have developed preferences. Preference surveys have become an important way of measuring changes in reading communication due to changes in technology.

In Chapter 4, *TeLCU: A Model for Technology Conditioned Language and Literacy Change*, I shall build on this discussion by proposing a model in which we can capture and conceptualize these perceived new forms of language and new ways of speaking. Indeed, I go further as to present a particular study of the way some bilingual speakers of Chinese and English create new forms of language through mobile phone texting (though the concept of mobile phone texting is taken up more comprehensively in chapter seven). I shall present and discuss concrete examples throughout the chapter.

In Chapter 5, *Insights from an MSN Corpus*, I look closely at one type of Computer-mediated Communication, Microsoft Network (MSN) instant messaging. MSN instant communication is quite popular among the youth in most parts of the world. In Hong Kong, it has fast replaced I-seek-you (ICQ) and QQ (in mainland China) as one of the primary instant communication tools among the youth.

In Chapter 6, *Insights from Mobile Phone Voice Communcation*, I will be looking particularly at the linguistic features of this kind of communication. More specifically, Hong Kong being a trilingual and biliterate society, we will see how participants juggle these languages within the medium of the mobile phone instant messaging and what linguistic features are thus produced. As is usual with most chapters our observations are based on a case study. A distinctive feature of this chapter from some others is that I present and describe a comprehensive corpus as an example of an actual CMC exchange. This will give us insights into actual communication sessions of young people who use mobile phone in Hong Kong and worldwide.

Chapter 7, *The Grammar of Mobile Phone Written Language*, is dedicated to the grammatical analyses of language produced in the environment of mobile phone texting or communication through Short Message Service (SMS). SMS has quickly emerged as a frequent daily linguistic, literacy, or general communicative practice in which two or more people exchange messages by coding and decoding texts received and sent from their cell phones. In this chapter, I focus on the relationship between communications technology and language change with evidence from the peculiar nature of the grammar of mobile phone texting.

Chapter 8 is titled, *New Forms of Reading and Writing: Cell Phone Novels*. In this chapter I discuss a radical consequence of the new forms of reading and writing on the literacy and literary world that would have been discussed in earlier chapters: the cell phone or mobile phone novel. Instant messaging, such as text messaging, is taking the idea of a novel to new heights. While the traditional novel as we know it, is a product of, presumably, many years of laborious writing and imagination by a single individual, the author, and delivered to the reading public as a carefully edited and published product by editors and publishers, the "novel" as produced through computer-mediated instant messaging is a radically different sub-genre or even a different genre altogether. It is interactive and instantaneous. In this chapter I present the outlines of the cell phone novel, including how it started, and what its main features are.

Beginning with Chapter 9, *New Languages, New Literacies, and the School Curriculum*, I focus on the educational applications of linguistic practices within CMC media for the next three chapters. In previous chapters I would have undertaken a quite detailed analysis of the features of language and literacy practices within the medium of communication technologies, including email, MSN, mobile phone speech, and mobile phone texting. This has given us the opportunity to examine carefully new or peculiar linguistic structures, and new ways of communicating and expressing oneself within these new technological environments. Of course, it is not everybody who is comfortable with these new ways of using whatever language it is by experimenting with these new structures and strategies of encoding such as shortening words and explicitly disregarding standard grammatical encoding. In particular, parents and teachers have been quite critical about these new languages and new literacies produced within the context of new communication technologies by their children and students. To address these concerns, I draw on a case study based on a funded project whose aim was to do a focused and careful analysis of email, ICQ and mobile phone practices among young people in Hong Kong with a view to examining if there are any consequences of these language and literacy practices on the way they learn and use language in the classroom.

In Chapter 10, *Educational Technologies (WebCT): Creating Constructivist and Interactive Learning Communities*, I continue this theme of education by discussing how to take advantage of youth interests in these new technologies that have been discussed at length in previous chapters. I show how we can turn them into learning and pedagogical technologies, leading to a model for language learning using these media – the Conversational Learning Community. Two main concepts, constructivism (Bodomo 2005b, 2007) and interactivity (Bodomo 2006, 2008, will be highlighted as important concepts in the area of using learning technologies to create good pedagogical environments for teaching issues of linguistics and literacy, and for that matter any other subject. The empirical basis for this chapter is a series of studies undertaken in the context of a teaching development project at the University of Hong Kong led by me, the author. I will be particularly concerned with how we can achieve interactivity in the learning environment I term Conversational Learning Community (CLC). CLC is based on a constructivist theory I term Conversational Learning Theory (CLT).

In Chapter 11, the penultimate chapter of the book titled, *Evaluating Learning Technologies*, I continue with our discussion of how we can take advantage of these youth interests and practices with ICTs for enhancing learning and teaching by actually evolving ways to evaluate these communication and learning environments. As with most chapters in the book, I focus on a case study as a way to give an in-depth study to the subject matter. In this case, interactivity is the subject matter. Interactivity, discussed at length in the previous chapters, leading to the creation of a new learning theory, the Conversational Learning Theory, and a new learning model, the Conversational Learning Community is even further emphasized here in terms of how to evaluate it.

In Chapter 12, *Computer-Mediated Communication: Emerging Media and Themes*, I point to emerging media and new themes that will shape the future of CMC. About every month a new technology is introduced by one company or the other that has the potential to impact the discipline in profound ways. So much so that it is hard for a research academic to pretend that at any one moment in time one can capture a representative snap-shot of the discipline that can stand the test of time. New technologies - new media, new issues - new themes are constantly emerging. In previous chapters, I would have focused mostly on what may be termed text-based Computer-Mediated Communication, which may be defined as interaction and transfer of information through the medium of the computer and related digital devices mainly in the written word. One emerging trend to be pointed to in this chapter, however, is that a new theme has emerged from text-based CMC to video-based CMC. Video-based Computer-Mediated Communication may be defined as interaction and transfer of information through the medium of the computer and related digital devices mainly in the form of dynamic image streams. Most contemporary social networking tools like Facebook and YouTube are implemented with video-based CMC. Of course, Video CMC still contains the written word, but the written word is mainly meant to just express talk around the main issue, the Video event. Young users of the internet have radically moved away from communication through the plain written word to communication in the medium of video clips and voice-image interactions through video-based media such as Facebook, YouTube, video games, and skype. It is this emerging paradigm shift from text-based to video-based CMC that this last chapter addresses.

SUMMARY

Taken together then, the chapters in this book address many aspects of the vast field of Computer-mediated Communication. I do this with a particular focus on linguistic or natural language and literacy aspects of the technology that is the basis of CMC. Each of the 12 chapters reveals some new facts and insights, or outlines the various positions surrounding some discussion or the other and then takes a position based on the results that I have obtained from my research. It is hoped that students and scholars of computer-mediated communication will find in these pages facts, figures, insights, theoretical positions, and practical solutions that can only make our field even more exciting.

REFERENCES

Adams, A. (1986). Language awareness and information technology. *Curriculum and Teaching, 11*(2), 69-76

Baron, N. S. (1984). Computer mediated communication as a force in language change. *Visible Language, 18*(2), 118-141.

Bodomo, A. B. (2005a). Constructing knowledge through online bulletin board discussions. In C. Howard et al. (Eds.), *Encyclopedia of Distance Learning*. Hershey, PA, USA: IGI Global.

Bodomo, A. B. (2006). Interactivity in web-based learning. *International Journal of Web-Based Learning and Teaching Technologies, 1*, 18-30. Hershey, PA, USA: IGI Global.

Bodomo, A. B. (2007). Constructivism in higher education: The case of online bulletin board discussions in a linguistics course. In S. Mohamad & N. J. Tuah (Eds.), *Application of Computing and Information Technology in Higher Learning*. Asaihl (pp. 1-22). University of Brunei.

Bodomo, A. B. (2008). Instructional interactivity in a web-based learning community. In L. Esnault (Ed.), *Web-Based Education and Pedagogical Technologies: Solution for Learning* Applications (pp. 131-146). Hershey, PA, USA: IGI Global.

Crystal, D. (2001). *Language and the Internet*. Cambridge: Cambridge University Press.

Kress G. (1998). Visual and verbal mode of representation in electronically mediated communication: The potentials of new forms of text. In Snyder (Ed.), *Page to screen: Taking literacy into the electronic era* (pp. 53-79). London/New York: Routledge.

Luke, C. (2000). Cyber-schooling and technological change: multiliteracies for new times. In B. Cope & M. Kalantzis (Eds.), *Multiliteracies: Literacy Learning and the Design of Social Futures* (pp. 69-91). Australia: Macmillan.

Acknowledgment

This book is the outcome of many years of researching issues and teaching courses in Linguistics, Language and Literacy, Language and IT, Cognitive Science, and Human Language Technology mainly at the University of Hong Kong but also in other places in the US and mainland China where I have held visiting positions. I am very grateful to my colleagues, my numerous students and teaching/research assistants for discussing many of these issues with me and for helping me gather data and analyze and construct corpora mainly on the way the youth in Hong Kong and other parts of the world speak, read, and write using Chinese and English on ICT gadgets.

While the book is based on many of my research findings in Hong Kong, most of it was finally written up during a sabbatical year in California, USA. I have been fortunate to discuss these issues at Stanford University where I served as Visiting Scholar in 2006/07. I also thank colleagues and students at Santa Clara University in the heart of Silicon Valley where I was fortunate to test out the draft of this book in the form of a seminar course in the Spring Quarter of 2007. In particular, I am grateful to Prof Pedro Ramos for the interest he showed in my book project and for the many suggestions he gave me.

Some of the facts, figures, screenshots, and analyses contained in this book are based on some of my journal papers published in various avenues. I thank my co-authors and the editors of these publications for permission to include or refer to some of the articles in writing this book. In particular, I thank the editors of *Literacy and Numeracy Studies*, *Reading Matrix*, *Global E-journal of Open Flexible and Distance Education*, and *IGI Global* journals and books where some of my single or joint articles have appeared. I also thank Peter Sidorko and Lucinda Wong of the HKU Library for processing my request for permission to use their library screenshots.

I am also particularly grateful to my past and current research assistants and students: Sophia Lee, Carmen Lee, Jovita Lam, Olivia Lam, Sally Mok, Charles Marfo, Pan Yanhong, Lucille Hu, Lemon Lau, Xiong Jiajuan, Frances Lung, Yee Wai Man, Cindy Fu, Zhou Zhou Tiffani, Erin Eng, Maha Alfreih, and Leonard Chow. Since SMS, MSN, Facebook, Online gaming, YouTube and other CMC environments are mainly patronized by the youth, it was not always easy for a man in his middle ages like me to get into all these networks and participate with a view to mining data and facts. Fortunately, I was able to rely on your insights and efforts in obtaining data by participating in these environments for distributing my questionnaire or for your own class projects and thesis write-ups. Thanks for helping me to format my writing or for researching one piece of evidence or the other at various stages of this project. I am especially grateful to Lucille Hu who spent many hours, in the final stages of the writing, formatting and putting together all the chapters into a single volume.

I also wish to record my gratitude to two anonymous reviewers and the editors and other editorial staff of IGI Global for very important suggestions for reformulations that lead to better presentations and analyses and a better structure of the book. In particular I am grateful to Rebecca Beistline and Tyler Heath the Assistant Development Editors who handled the book writing process at various stages.

Finally, I thank my family for unfailing support, as usual.

Adams B. Bodomo
University of Hong Kong, Hong Kong
March 2009

Chapter 1
Definitions and Basic Conceptual Notions

INTRODUCTION

Given the multi- and inter-disciplinary nature of computer-mediated communication in an era that is often referred to as the Information Age or even the Knowledge Society, it is necessary to carefully delineate the key terms and concepts of this complex area of study in this first chapter. In this age and society we constantly witness a massive explosion of new types and styles of communications gadgets such as computers of all types (including desktops and laptops), PDAs, mobile phones, Blackberries, etc; and media like the internet, emails, ICQ, chat, text messaging and others. We can expect that new ones will be invented. All these types and styles of communication gadgets lead to new terms and concepts that can potentially have differing interpretations from context to context.

DOI: 10.4018/978-1-60566-868-0.ch001

In this book, *Computer-Mediated Communication for Linguistics and Literacy: Technology and Natural Language Education*, I will discuss how these different types and styles of information and communication technologies impact the way we communicate. In particular, I will examine the nature and uses of language in what may be described as a revolutionary environment for information and communication. As these communication technologies bring us closer together, we are forming a global village, and indeed, this era has also been referred to as an era of globalization. We will find that in this global village our everyday communications may involve not just one language, but several languages. In fact, it is hardly ever sufficient to operate in only one language anymore. Bilingualism/multilingualism is now an essential part of this information communication revolution. I will examine how these combinations of technologies and languages affect the nature and structure of each language and whether this leads to changes in language structure and language use.

Not only are these technologies creating an environment of multiplicity of languages, they also have an impact on the way we read, write, and process information. I will examine the different ways in which we are called upon to speak, read, and write in these different technological environments. We will notice that we are called upon to juggle different ways of reading and writing in these different environments. Literacy is no longer a mono-modal practice but very much multi-modal; various kinds of literacies are needed to function well in the information society.

This ecology of languages, literacies, and technologies will be shown to be very dynamic, and in this ever-changing ecology many issues about human nature will be discussed. In particular, an examination of these interrelationships helps us understand some important aspects of our culture and society. Throughout the book I will examine some consequences for learning and knowledge acquisition in this complex ecology, and attempt to show how we can harness these languages, literacies, and technologies to improve education for our future generations.

Outline of the Chapter

Towards addressing all these issues, this introductory chapter is organized as follows. The next part will outline the objectives of the book. Following from this I will then address important foundational concepts such as language, linguistics, literacy, and computer-mediated communication before examining their interrelations in the next part. The discussion concludes with an attempt to put together all the ideas introduced in this chapter.

OBJECTIVES OF THE BOOK

A first objective of the book is to enable readers to analyze and reflect on the interrelationships between languages, literacies, and technologies. Such analysis and reflection are critical to understanding how human language, literacy as a knowledge base, and information technology coalesce to constitute the building blocks of the emerging field of computer-mediated communication.

A second objective is to become acquainted with a variety of arguments surrounding language use, literacy practices, and general communication habits within the context of ICTs. These various arguments and "debates" such as whether or not we get new forms of language or whether or not language finds different uses with the evolution of new technology, among others, will be taken up in different chapters.

A third objective is to consider questions on the extent to which these new technologies and practices can be harnessed and incorporated into the school curriculum for educational gains. The internet as a product of the age of IT, for instance, was not invented with clear educational goals in mind. But it and allied technologies have fast become important pedagogical tools. I will examine the theories underlying learning and consider how to use these technologies to develop adequate learning environments for language education, and indeed how to evaluate these learning environments.

A fourth objective is to enable users of this book to get acquainted with academic research and writing on new ways of communication in the information age. Students and other readers who use this book can start developing the skills to write academic papers based on the discussions, arguments and theoretical positions taken on various issues.

In sum, then, a major objective of this book is to explore the ways in which language and linguistics, on the one hand, and information communications technology, on the other are related. One would want to know if there are causal and deterministic relations between these entities. Do they have a common subject matter? Do changes in one influence the other? Can we use knowledge in one to enhance understanding of the other? A basic analysis of each of these, especially simple working definitions and explanations, should help us to tease out the relationships.

With these stated objectives, I now look at the various foundational concepts of the book.

FOUNDATIONAL CONCEPTS

Language and Linguistics

Language, as is widely known among linguists and other language scientists, is a system of codes for communication. In this respect we need to recognize that there are various kinds of languages, including human or natural language, animal language, artificial language and other systems of communication, and in this book I would be more focused on the relationship between human language and information communications technology.

Linguistics is the science that studies the nature, structure, and functions of language. This science of language seeks to answer questions including how it is that language is acquired, how it is processed in the brain, the structure of language, in general, and individual languages, in particular, and how this linguistic structure is used as a tool for human-to-human and human-to-machine communication. Linguistics is sometimes referred to as the most scientific of the humanities disciplines and the most artistic of the science disciplines, thus testifying to its vast interdisciplinary area. Indeed, Linguistics, as a major area of academic studies, has a vast scope, spanning the humanities, social sciences, and natural science. Each of the chapters in this book touches on some aspect of the structure of language in general and particular human languages such as English and Chinese, and how these structures are used in each of the sociotechnological environments that we will see, including the popular social networking systems like MSN, YouTube, and Facebook. In this sense then this book covers linguistic phenomena that span entire areas of the humanities, social science, and natural science.

Information Communications Technology

The turn of the century has witnessed a massive presence of computers in our everyday life. Computers are now ubiquitous artifacts in homes and offices. The historical evolution of CMC is in itself a topic worthy of a book length publication and while it is not the focus in this book, it is important to note that the events that have led to what is a massive presence of computers in this century have engendered a number of terms to describe the disciplines that deal with computers and cognate tools. Computer Science (CS) is the de facto term that represents the traditional disciplines with computers – their architecture and their functions - as central concerns. Information Technology (IT) is an alternative term for Computer Science, and is historically more recent than the term Computer Science but it is much more than that since it actually lays emphasis on the information processing aspects, rather than on just the raw technology per se. In recent times, the term Information Technology (IT) has been extended to include 'communications', with varying nomenclatures such as Information and Communications Technology (ICT) or just simply Information Communications Technology (ICT). This extension underscores the communicative aspects of information processing; in other words, the discipline has grown, historically, from just emphasis on the science and technology of computer architecture to the information processing aspects, especially those that deal with meaningful transfers of information between entities on different locations (Shortis 2001). So ICT is simply defined as the discipline which deals with the use of technologies to communicate and to process information.

In this sense, the tools involved go beyond computers to include other digital tools such as mobile phones, PDAs, and Blackberries. This emphasis on computers in facilitating communication between entities, especially with a massive explosion of the internet and allied tools like email, MSN, and blogs has led to newer terms such as Computer-Mediated Communication (CMC).

Computer-Mediated Communication (CMC)

The term Computer-mediated communication (CMC) has been chosen as the title of our book mainly to stress the point that this and allied terms underlie the evolution over time and space of newer and interdisciplinary areas of study involving how human languages and computers/IT intermingle to serve society's needs. These newer areas of study include Computational Linguistics, Human Language Technology, and IT-based learning and teaching. There are even terms like Internet linguistics or e-linguistics to express the fact that so much language and linguistic analysis is facilitated by the internet and other electronic resources both as tools and as repositories of these analyses.

The term, CMC itself has been used in various ways by various authors such as Baron (1998), December (1996), and Thurlow, Lengel and Tomic (2004); it thus necessarily has a varied number of definitions. Baron (1998) simply sees it as "a domain of information exchange via computer". (p142). December (1996)'s definition of CMC, along with Baron's, is one of the earliest and it is as follow:

Internet-based, computer-mediated communication involves information exchange that takes place on the global, cooperative collection of networks using the TCP/IP protocol suite and the client-server model for data communication. Messages may undergo a range of time and distribution manipulations and encode a variety of media types. The resulting information content exchanged can involve a wide range of symbols people use for communication. (p24)

Our Own Definition of CMC

For us *CMC is defined as the coding and decoding of linguistic and other symbolic systems between sender and receiver for information processing in multiple formats through the medium of the computer and allied technologies such as PDAs, mobile phones, and blackberries; and through media like the internet, email, chat systems, text messaging, YouTube, Skype, and many more to be invented.* As is seen, the term *computer* itself is no longer limited to desktop and laptop devices but generalizes onto smaller but even more powerful gadgets like palmtops, mobile phones, and PDAs, all with internet connectivity. We can therefore comfortably extend the term *computer-mediated communication* (CMC) to *Internet-mediated communication* (IMC) which involves *the processing of linguistic and other symbolic systems through the internet and allied technologies by interaction between sender(s) and receiver(s).*

Of course, in the book, as seen from the stated objectives above, the focus on CMC is to enable us analyze the use of natural languages on the internet and the new forms of language that are being produced. I will also analyze the educational implications and applications of natural language in the context of CMC.

Another aspect I highlight in my treatment of CMC that is not often distinguished in earlier works is this: rather than focus on distinguishing between synchronous (real-time communication) CMC and asynchronous CMC (delayed communication) I instead concentrate on a distinction between text-based CMC and video-based CMC. This is because the distinction between types of CMC based on synchrony of communication is getting blurred by developments in technology. For instance, email has often been listed as an asynchronous CMC but developments in broad-band and internet connectivity make it possible for email to be used by two people separated by vast oceans to engage in a fairly reasonable real-time communication with each other, just as is done with most other real-time chat systems such as MSN chatting.

However, with regards to text-based and video-based CMC, even though one can talk of "mixed mode CMC", there is a fundamental distinction in the kind of activities that accompany either. Text-based CMC involves communication partners transmitting information between each other mainly through the medium of the written word along with other symbolic systems such as numbers and emotional icons. However, video-based CMC involves primary communication through the medium of moving images. Text-communication may accompany the video-communcation process but this is secondary and is meant to serve as *talk around the image*, so to speak.

Having outlined definitions and discussions about the foundational terms, it is now time to discuss how they inter-relate and are tied together into one whole theme about communication.

INTER-RELATING THE FOUNDATIONAL CONCEPTS

Communication

It can be deduced from the above that once we define language and process this concept in terms of a system of communication, we immediately recognize a clear relation, link and, possibly, common subject matter between language and information communications technology. Communication, including mass communication as discussed by authors such as McQuail (2005), is one major feature that glues the concepts of Language, Linguistics, CS, IT, ICT and, finally, CMC together. All are knowledge bases, disciplines and tools that are in one way or the other concerned with the subject matter of facilitating communication between entities. A possible methodology then in understanding the workings and interrelationships is to embark on a sustained examination of their architecture or internal logic and how this internal logic facilitates communication.

Literacy

Another way in which we can find a relationship among the terms is to seek to understand what is necessary to have a basic understanding of these tools, i.e. how we can understand and help others to understand how to code and decode the information or communicative messages processed by each of these tools? This brings us into the realm of literacy, an area of study dealing with the ability to code and decode linguistic and other symbolic systems for communication and information processing (Bodomo 2000). Literacy, like communication, is a subject matter of all these knowledge bases. To function well in a society characterized by a heavy information overload, individuals must have basic ideas about these knowledge bases. The area of study that can facilitate this has commensurate terms relating these bodies of knowledge, such as basic literacy, computer literacy, information literacy, IT-literacy, electronic literacy (Warschauer 1998, 2006), etc. All these types of literacy involve, not just general language study, but also the development of literacy for specific purposes and for people with learning disabilities (Macia, Cervera and Ramos, 2006). It would seem that a second methodological perspective towards a sustained study of these knowledge bases and their interrelations is a sustained examination of these concepts in the field of literacy studies.

CONCLUSION

In the age of IT, at least, two kinds of knowledge would be crucial for individuals to function optimally in society: linguistic and IT skills. Therefore a study of these two kinds of knowledge and how they relate to each other and are used is a worthwhile enterprise. Thus this book on Computer-mediated Communication dealing with language, literacy, and IT, is very relevant for understanding developments in the Information Age. There are many differing opinions about the cause and effect relationships between language and technology as a whole, and I will look at this issue more extensively in the next chapter. Indeed, in later sections of the book I will discuss a particular model for capturing the dynamic relationships between language and (information) technology.

REFERENCES

Baron, N. S. (1998). Letters by phone or speech by other means: the linguistics of email. Language & Communication , 18, 133–170. doi:10.1016/S0271-5309(98)00005-6

Bodomo, A. B. (2000). Lecture notes on Language and Literacy. *Computer literacy and language educational technology*. The University of Hong Kong. URL: http://ecourse.hku.hk:8900/public/LING2011

December, J. (1996). Units of analysis for internet communication. Journal of Computer-Mediated Communication , 46(1), 14–38.

Macia, E. A., Cervera, A. S., & Ramos, C. R. (Eds.). (2006). *Information technology in languages for specific purposes: Issues and prospects.* New York: Springer.

McQuail, D. (2005). *Mcquail's Mass Communication Theory*. 5th ed. London: SAGE Publications.

Shortis, T. (2001). *The Language of ICT: Information and Communication Technology.* London and New York: Routledge.

Thurlow, C., Lengel, L., & Tomic, A. (2004). Computer mediated communication: Social interaction and the internet. London: SAGE Publications.

Warschauer, M. (1998). Electronic literacies: Language, culture and power in online education. Mahwah, NJ: Lawrence Erlbaum Associates.

Warschauer, M. (2006). Laptops and literacy: learning in the wireless classroom: Teacher's College, Columbia University.

Chapter 2
Is Technology Changing the Way We Communicate?

INTRODUCTION

In this and the next chapter, I will discuss to what extent technology is changing the way we speak, read, and write. The present chapter takes the form of a debate or discussion where we look at different positions on the issue. This issue is taken up early on in the book because its consequences come up again and again throughout the book, as it is one of the main themes of the book. In chapter three I do a particular study of reader preferences vis-à-vis electronic and e-books and their traditional hard copy counterparts. The chapter is structured as follows. In the first part, I outline and define more clearly the aims, objectives and theoretical positions that constitute the discussion in the chapter. Following this I then take up the question whether there is a causal relation between new forms of language and

DOI: 10.4018/978-1-60566-869-7.ch002

new technologies, and discusses whether the internet in particular and other ICT tools are changing forms and uses of language. After that I then provide a list of the most frequent shortenings as an example of the new forms of linguistic expressions that emerge through the use of the internet, championed mostly by young people. The chapter concludes with a summary of the issues discussed in it.

ICT FEATURES AND TOOLS

We have already defined and discussed ICT in Chapter I. I will now outline the main features and tools of ICTs that will enable us better understand the roles they place in bringing about linguistic and communicative innovations.

Main Features of Information Communication Technology (ICT)

In this section of the chapter I outline the main features of ICT, based on which I show later on that these features distinguish ICTs from previous communications technologies which is the basis for the new forms and new uses of language that we observe in text-based CMC. These are multimedia integration, flexibility of use, connectivity and interactivity (Blurton 1999, Bodomo 2000, Shortis 2001).

Multimedia integration underscores the fact that ICTs enable us to combine various types of media for the purpose of communication. It is impossible, for instance, to seamlessly integrate text and images (drawings), and even voice features into a piece of writing or lecture presentation far more than would have been the case without computers and allied technologies.

Flexibility of use underlies not only the fact that it is easy to do this integration, but that one can have many options in a communication or information exchange situation. For instance, one can choose to do real-time communication or delayed communication using the same ICT tools, such as email.

Connectivity deals with the ability of ICTs to bring together people who may be living oceans apart. For instance, it is now possible for academics in Asia and North America to collaborate reasonably easily on writing an article for publication. This involves a lot of communication which would have been virtually impossible or too slow prior to the advent of ICTs.

Interactivity is the ultimate feature of ICTs. Because of all the above features participants in communicative situations using ICTs are more interactive, that is, they can control the communication situation in terms of back and forth exchange of information and in terms of how they control the content of the information. In a learning situation for instance, we may distinguish between learner-teacher interaction, learner-learner interaction, and learner resource interaction, all of which would be more enhanced with the use of ICTs than would otherwise be the case, even to the point of making learning from a distance look like a physical, on campus classroom learning. Interactivity and its role in creating learning environments will be discussed in-depth in chapters X and XI.

All these four features lead to an incredibly generous choice of modes of communication that are at the disposal of ICT users. These modes of communication include one-to-one, one-to-many, many-to-one, and many-to-many communication scenarios. One-to-one communication may involve a simple mobile phone or MSN chatting between two people, one on each side; one-to-many communication may involve a blogger, a kind of online diary, sharing his or her life experiences with his or her readers; many-to-one communication may involve an interview situation where a candidate is linked to a group of interviewers; and many-to-many communication could for instance involve an online chat situation or a situation where two classrooms are interacting across vast oceans.

ICT Tools

ICT tools may be divided into two broad types. These include internet technologies and digital ICT tools though it must be quickly stated that this broad division is getting very blurred with improvements in technology whereby integrated technologies are the order of the day.

Internet technologies include internet media such as email, instant messaging like MSN, YouTube, Internet video gamming, weblogs (blogs), learning systems such as webCT, and all other kinds of chatting systems.

Digital ICTs tools include hardware such as computers of all types, software connected to computers, writing tools, PDAs, mobile phones, blackberries, and other tools yet to be invented.

As mentioned above, there is a tendency to blur this division with the building of digital ICT tools integrated with internet technologies. For instance, blackberries and i-phones are both digital ICTs with integrated internet technologies, making very easy to access the internet on the go as easily as on a desk top computer.

CAN NEW TECHNOLOGIES CHANGE OUR LANGUAGE?

The main aim of the chapter is to explore primary issues concerning the possible *causal* relationship between language and new technologies. Two relevant questions in this regard are: is there a causal relationship between new forms of language and new technologies; and are the internet and other ICT tools changing forms and uses of language? I focus on the first question here beginning with the idea of change.

The idea of 'change' as used involves both change in linguistic forms and change in use of language. Another aspect is to look at change as *modification of existing forms and uses* and change as *emergence of novel, original creations* of language. To give two examples, change as modification involves abbreviations such writing the second person singular pronoun in English, *You* as *U*. Change as novel creations involve the creation of novel forms or entirely new words that are directly linked with the information revolution, such as *YouTube*(noun), *to (you)tube* (verb) and a *(you)tuber* (noun). Change in the way in which we use language may involve the pervasive use of emoticons to express our feelings when we write. Emoticons are a salient feature of what has often been described in the literature as visual language (Horn 1999), an emergent international language that blends words and images, including the emoticons that I have focused on here, to convey complex ideas in a communication situation.

With the idea of change now clarified, let us discuss the various positions taken by various writers on the causal relationship between language and new technologies in answer to the question whether technology influences change in language or otherwise.

While not directly arguing against the idea that there is no causal relationship between the emergence of new technology, in general, and the new ways in which language is used, Kress (1998:53), for instance, observes that '...when we look at the far-reaching and deep changes in forms of communication which characterize the present-email and its changing forms of language, for instance, it is tempting to attribute these changes to some technological innovation but erroneous to do so'. Luke, C. (2000:83) also takes a similar position, believing that new forms of literacy practices do not simply emerge with technological change. Rather, 'technologies always emerge as products of specific cultural practices, literate traditions, and the interests and desires of those groups who design and name them'.

Yet there are other writers in the field who argue that there is indeed a significant causal relationship between communications technology and new language and literacy practices or more specifically the evolution of new ways of using language. As Adams (1996) puts it, 'the new technologies are themselves dramatically changing the nature of the language we use'. Such an approach is also supported by Baron (1984: 139) which concludes that '[n]o one in the computer industry has any hidden agenda for using hardware or software development to alter human language. Yet technology can indeed drive linguistic and social change'.

Indeed, new practices of language and literacy may be attributed to a set of unique properties in new communications technology.

This is quite an important debate in the field CMC and constitutes a major theme in this book: whether or not the internet and other CMC tools have causal effects on language and literacy in terms of changing their forms and uses, and the consequences of these for linguistic analysis and for general and language education. In Chapter IV, these views will once again be highlighted and then used to develop a specific model of technology-conditioned language change process.

Readers may indeed be more specific and just concentrate on debating the following topic: "Do you think that the Internet is changing the ways we use our language? If so, give evidence of some new forms and uses of language. If not, explain why not."

Indeed, this and related issues were the topic of a radio discussion by David Crystal, writer of the book, Language and the Internet (Crystal 2006). Crystal describes these language changes, both in form and use, as leading to a new form of language which he refers to as "Netspeak": Language and the Internet, *The Connection,* Boston University and wbur. 4th January, 2002 http://archives.theconnection.org/archive/2002/01/0104b.shtml (retrieved: November 2008, website not accessible on March 7, 2009)

A further allied discussion to the above debate topics is whether the use of the internet has consequences for our choice of language. In other words are we obliged to use one particular language more on the Internet? This was indeed an issue in the early years of the internet when English was the only language used, since the internet began in the US, an English speaking country.

Some demographic data would tend to support this view (Source: Global Internet Statistics (by Language) http://global-reach.biz/globstats/index.php3). The internet population as at September 2003 comprised 35.6% native English speakers. However, the percentage is steadily dropping (cf. 55% in 1998, 36.5% in 2002) as a result of the internet becoming more and more multilingual and thus compatible with many languages. Chinese is in second place after English, followed by Japanese, Spanish, German and Korean:

2nd Chinese 12.2%
3rd Japanese 9.5%
4th Spanish 8.0%
5th German 7.0%
6th Korean 4.0%

In terms of statistics about languages *on-screen* (as of June 2000), over 60% (over 2 billion) of web pages were published in English, with other languages in the following order:

2nd Japanese 5.85%
3rd German 5.77%
4th Chinese 3.87%

5th French 3.00%

Why English Dominates the Situation?

A further discussion point would be why English, in particular, dominates the CMC situation. First, as mentioned above, the internet began in the US; second, English is still regarded as an international language for global information exchange; third it has a fair amount of native speakers and many second language speakers worldwide, such that even native speakers of other languages still have the option to use English as well as their language on the internet. A fourth reason for the dominance of English on the internet is that there are quite substantial character inputting and encoding problems, which leads to difficulties in using other languages, especially those that do not have an alphabetical writing system.

Some Solutions

There have been several attempts to find a solution to this problem and thus make the internet more multilingual. To solve character inputting and decoding problems, attempts have been made to develop codes compatible for most languages, known as the Unicode system. Second there are also some inputting systems in place for converting alphabetical writing into other orthographical systems such as Chinese and Japanese characters. A third way is to develop multilingual websites, situations in which the same information is written in several languages on the same website. A fourth way is to develop what is known as Real-time multilingual translators for on the spot translation of websites. This has not been too successful but two of the most efficient ones are: Systran (http://www.systransoft.com/Papers/ppr_alta.htm) and AltaVista Babel Fish (http://world.altavista.com/) (retrieved March 7, 2009)

ACRONYMY: SHORTENINGS AS AN EXAMPLE OF NEW LINGUISTIC FORMS

Besides older ways of forming acronyms, such as *ASAP* (as soon as possible), *BTW (by the way),* acronymy, the process of creating acronyms, on the internet comprises many new methods, some of which include the following:

Acronyms of sentences
GTG (got to go), BRB (be right back), LOL (laughing out loud)
Letter homophones
U (you), R (are)

Number Homophones (near-homophones)
88 (bye-bye), 99 (nite nite)
Combination of letter and number homophones
B4 (before)
Reduction of individual words
tml (tomorrow), coz (because), gd nite (good night)
Combination of letter initials and letter homophones
TTUL (talk to you later), IC (I see)

In preparation for deciphering acronyms in later chapters, readers may try to identify the following short forms that have appeared in some of the CMC texts that I will be discussing.

HAND
CUL8R
IMHO
RUOK
GR8
TTYL

CONCLUSION

In this chapter, have I presented the basic features and tools of CMC and then used that as the basis to discuss the possible causal relationship between ICTs and language change. I have in so doing examined different views on the causal relationship between language and new technologies, along with allied issues such as why English has been a dominant language on the internet and what is being done to make the internet more multilingual or user-friendly for other languages. I concluded by suggesting and discussing shortenings as one of the pieces of evidence of new forms of language and literacy, as a result of the use of ICTs. In the next chapter, i discuss some quite profound changes in the way we read as a result of our interaction with CMC tools and the production of electronic materials such as e-books and e-journals.

REFERENCES

Adams, A. (1996). Language awareness and information technology. *Curriculum and Teaching*, *11*(2), 69–76.

Baron, N. S. (1984). Computer mediated communication as a force in language change. *Visible Language, 18*(2), 118–141.

Blurton, C. (1999). *New directions of ICT-use in education.* UNESCO's World Communication and Information Report 1999. on-line available at http://www.unesco.org/education/educprog/lwf/dl/edict.pdf

Bodomo, A. B. (2000). *Lecture notes on Language and Literacy. Computer literacy and language educational technology.* The University of Hong Kong. URL: http://ecourse.hku.hk:8900/public/LING2011

Crystal, D. (2006). *Language and the Internet.* Cambridge, UK: CUP.

Global Internet Statistics. (by Language) http://global-reach.biz/globstats/index.php3)

Horn, R. (1999). *Visual Language: Global communication in the 21st Century.* Macrovu Inc.

Kress, G. (1998). Visual and verbal mode of representation in electronically mediated communication: the potentials of new forms of text. In I. Synder (Ed.), *Page to Screen: Taking Literacy into the Electronic Era* (pp. 53-79). London and New York: Routledge.

Luke, C. (2000). Cyber-schooling and technological change: multiliteracies for new times. In B. Cope & M. Kalantzis (Eds.), *Multiliteracies: Literacy Learning and the Design of Social Futures* (pp. 69-91). Australia: Macmillan.

Shortis, T. (2001). *The Language of ICT: Information and Communication Technology.* London and New York: Rouledge.

Chapter 3
Digital Literacy
Reading in the Age of ICT

INTRODUCTION[1]

Digital literacy is an important aspect of Computer Mediated Communication (CMC) and in this chapter I focus on particular elements of it, including new ways of reading and processing written language, with implications of how literacy is embedded in language. Reading in the 21st Century is no longer confined to reading print books. Electronic versions of many print materials have been made available on the World Wide Web. Some of the advantages of using electronic or digital materials include their relative accessibility and flexibility in terms of time and space. Electronic materials also allow the inclusion of multi-media elements like sound and video clips, which cannot be presented in print books. These and other factors have contributed to growing acquisition rates of e-books and e-journals in academic libraries around

DOI: 10.4018/978-1-60566-868-0.ch003

the world. As urban areas all over the world become more and more crowded, the save-space feature of digital libraries is an even more attractive reason why users of library facilities in academic and public libraries are being encouraged to read electronic versions of publications.

While libraries are encouraging the use of digital resources, a salient issue that one has to take into consideration is the actual acceptability of these materials to users. Are digital materials preferable to print materials, or are print books used more often? Will digital resources replace print ones in the future? Will people still continue to read books in the 21st century? In this chapter I will begin to explore user preferences for print and digital resources in academic libraries from these perspectives. I will do this through a case study of students at the University of Hong Kong, but with occasional comparisons with, and references to, other studies and surveys in other parts of the world, such as in the US (University of Texas at Austin and California State University library surveys) and in Malaysia (Ismail and Zainab, 2005). I would like to show that, however laudable the advantages of electronic books may be over physical print books, user preferences in every situation must be carefully analyzed before a paradigm shift from print to digital library resources can be successfully effected.

The chapter is organized as follows. The first part discusses the nature of digital libraries in relation to the concept of digital literacy, which then leads to the study of user preferences presented in the following section. I next describe the major findings drawn from a survey of a group of undergraduate students and in-depth interviews with eight students and two librarians. Following from this, I highlight the various problems we have observed from the survey results with regard to digital library acceptability. Possible solutions and measures are also proposed. In the concluding part, I summarize the key issues of the chapter.

DIGITAL LITERACY AND DIGITAL LIBRARIES[2]

In this age of information technology, the traditional notion of 'literacy', which is understood as 'the ability to read and write' is far from sufficient. In order to adequately characterize literacy in the 21st century, the definition has to capture the relationship between literacy and new technologies. One of such relationships is the ability to process electronic information. In this section, I discuss the concept of digital literacy and its related concepts in the literature. The relationship between digital literacy and our present study of digital library or digital libraries is also discussed.

Re-Conceptualizing 'Literacy'

We are living in an era of rapid technological changes. The concept of literacy has to be refined in order to respond to the needs and demands of the information society. Gee (1990, 1996), for instance, speaks of the 'plurality' of literacy, focusing on the coinage, 'literacies', which emphasizes that literacy is a dynamic concept which changes over time and space. Bodomo (2000) also recognizes the problem in the traditional notion of literacy and redefines literacy as 'the ability to code and decode linguistic and other symbolic systems for communication and information processing'. Bodomo and Lee (2002) further proposes a 'technology-sensitive' approach to literacy in their study of changing forms of language in the information age, which is an instantiation of context-sensitive approaches to literacy studies. It is argued that the concept of context-sensitivity may be rather broad, and to really understand the impact of ICTs on literacy, the specification of a technology-sensitive approach is necessary. Lee (2002) provides an in-depth study of digital literacy practices in Hong Kong, especially the way Hong Kong youth use email and ICQ, an instant messaging programme which allows users to 'chat' online.

Gilster's (1997) conceptualization of digital literacy is one of the various forms of new literacies that have emerged under the impact of new communication tools. The concept is defined as 'the ability to understand and use information in multiple formats from a wide range of sources when it is presented via computers' (Gilster 1997:33). This characterization may be compared with works within the framework of critical literacy, since they have also recognized the multiplicity of skills involved in information processing. They look into the idea of multiliteracies (e.g. Fairclough 2000, Kress 1998, 2000, Luke 2000). Kress even speaks of 'multimodality' in his discussion of the processing of multimedia materials.

Other 'new' literacies being generated in the context of this technological dynamism include computer literacy, visual literacy, information literacy, and IT literacy. All of these concepts have become prevalent in present-day society, especially in the domain of education, due to the increasing use of ICTs in pedagogy. In this chapter, I discuss the concept of digital literacy in connection with the rise of digital materials in academic libraries.

Digital Literacy and Reading Digital Resources

Information and skills are indispensable concepts in the characterization of literacy in a knowledge-based society. Digital literacy is concerned with 'the ability to understand information', and to evaluate and integrate information in multiple formats that the computer can deliver' (Gilster, in Pool 1997). The emergence of e-book technology and its related reading practices would be a good manifestation of digital literacy. The increasing availability of this technology has attracted the attention of many university libraries for at least two reasons: (i) electronic materials are convenient in the sense that students can have remote access to these materials; and (ii) e-books can save space.

The experience of reading a book published online, however, differs quite dramatically from reading a print book (Lewenstein, 2000). The skills involved are certainly more complex. As Gilster (1997) acknowledges, '[w]e read books but we browse the web'. Interaction between the 'reader' and the e-book is no longer static. The reader has become more 'active' in the process of reading by clicking and browsing through Web pages and hyperlinks. Holmqvist, et al., (2003: 657) note the following:

Net paper readers have been shown to read deeper into articles than newspaper readers. It has also been claimed that newspaper readers rather scan than read newspapers. Do these findings mean that net paper readers read proportionally more than newspaper readers? This paper presents results showing that in fact net paper readers scan more and read less than newspaper readers. We furthermore investigate whether this result can be explained by the differences in layout, navigation structure and purpose of reading between the two media.

The question, then, is if reading electronic books requires such a new reading experience or even a new kind of literacy, are students of this generation who have been trained with print books ready to fully accept such new technology? In other words, are they digitally literate enough to 'master' e-book reading? What are their habits of using such materials? And more significantly, what do they actually prefer? These are the questions that this chapter is concerned with, based on a case study among students at the University of Hong Kong. The study aims to investigate university students' preferences for print and digital materials. The study eventually proposes possible measures in aid of a potential paradigm shift from print to digital libraries.

Previous Studies on Digital Libraries

Some surveys conducted by academic libraries in other parts of the world have already looked into issues that center around digital libraries or the implementation of e-book technology.

The libraries at the University of Texas at Austin conducted an informal survey[3] in 1999. The survey question was: 'If you were to choose between the library's receiving an electronic subscription or a paper subscription to a given journal, which would you prefer?' Of the 788 respondents, 627 (80%) preferred e-journal access, while only 162 (20%) preferred print. Most of those who chose electronic materials stated that e-journals can save resources and are more accessible. The capability of keyword search was also another reason for the preference for electronic materials.

A similar or related survey was carried out at the California State University[4]. This survey asked users to comment on the NetLibrary system. Interestingly, of the 211 library users who responded, 57% preferred print book version if they had used both the print and e-book version of a book, though no reasons were specified. The conclusion of the survey report was: 'While print books are currently being used at a higher rate than e-books, there is a clear indication that e-books are becoming accepted and used, especially in certain subject disciplines. The use of e-books is likely to increase over time as more materials are available and as users become familiar with them. The use of print books is expected to remain steady.' While accepting the fact that more electronic materials will be available, whether print books will remain popular or not is still unclear. In this book, the suggestion is made that while we are not sure about the future of print books, with the rise of electronic materials, a careful study of user preferences would help libraries acquire appropriate electronic resources and thus encourage higher usage rates. We also propose measures to sustain or enhance e-book access rates.

THE CASE STUDY

The context in which this study is situated is the University of Hong Kong (HKU)'s rich set of digital library collections. To the best of our knowledge, academic libraries in Hong Kong and many parts of the world have not yet conducted any large-scale surveys or studies on user preferences for digital library resources, especially among students. Such surveys are long overdue. This study would like to address this issue with specific reference to the libraries at HKU. It also highlights the relationship between students' level of digital literacy and their preferences for e-materials.

Electronic Resources at HKU Libraries

Since 1997, HKU libraries have been acquiring web-based electronic resources. These electronic resources mainly include e-journals, e-books, and e-newspapers. The University has subscribed to most of the key online research databases like ProQuest, LEXIS-NEXIS Academic Universe, and ERL Database. Over the years, there has been a significant increase in the number of electronic materials in the libraries to the extent that one can safely state that HKU has one of the richest digital collections in the Asia-Pacific region.

Figure 1. HKU e-journals access homepages

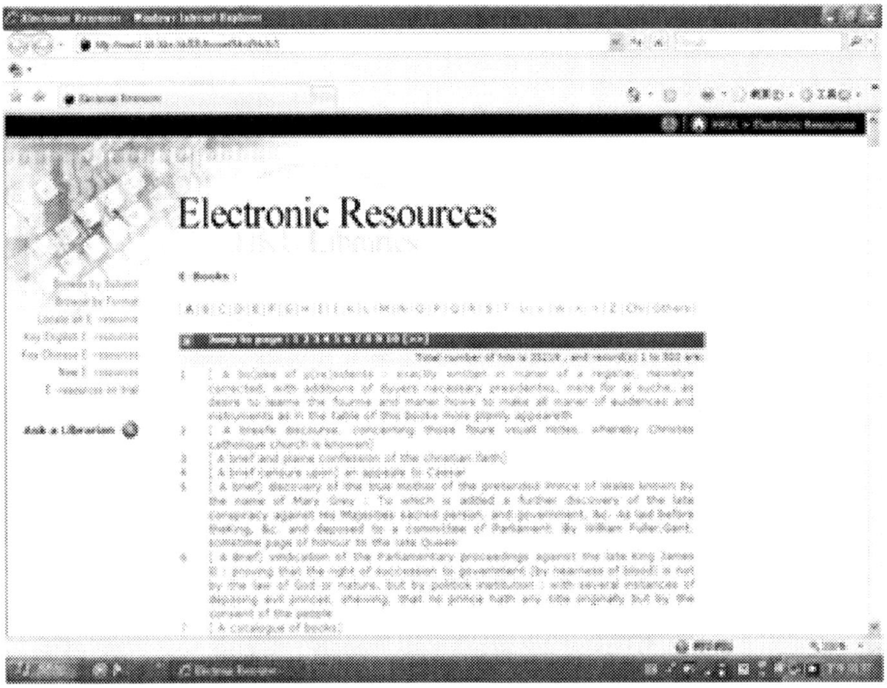

From June 2000 to June 2002, there was an increase of over 57,000 new titles in the electronic library collections, of which over 3,000 were e-journals and over 30,000 were e-books[5]. Overall, according to the libraries' Annual Report, the HKU libraries have acquired a total of 13,755 e-journals and 63,961 e-books in the 2001/2002 academic year. With these figures, it is obvious that the University strives to enhance the accessibility of library information by encouraging library users to access electronic resources. The actual access rates of digital materials, however, are yet to be explored. This chapter, therefore, highlights this issue and attempts to look into users' preferences for, as well as their habits of, using digital libraries.

Methodology

The data for this chapter come mainly from a library user preference survey conducted in February 2003 (Lam 2003). In this study, both quantitative and qualitative data were collected. First, a questionnaire survey was conducted. Some 35 undergraduates of HKU participated in the questionnaire survey. They were asked questions in three major areas: (i) self-evaluation of level of computer literacy, (ii) their habits of using print and electronic materials, and (iii) their preferences for print

Figure 2. HKU e-books access homepages

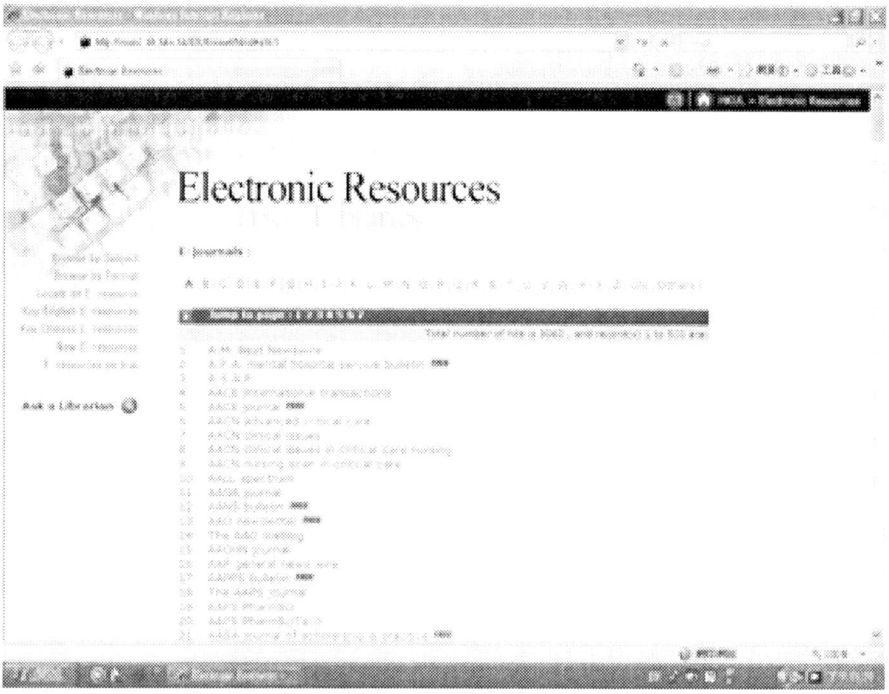

and digital library resources. Possible correlations of these aspects are discussed later in the chapter.

To better understand the reasons behind the users' preferences, eight of the respondents who participated in the questionnaire survey were also interviewed (Bodomo, Lam, and Lee, 2003). The qualitative data serve as further evidence and explanation for the results drawn from the questionnaire survey.

Further still, two staff members of the HKU libraries were interviewed. The idea behind interviewing the librarians was to obtain a fuller picture of the current situation of acquiring and using digital library resources in HKU.

FINDINGS

In this section, some of the most significant findings of the study, which lead to a more substantial discussion of the observations in a later section, are presented. These findings are presented according to students' (i) level of computer literacy, (ii) habits of using library resources, and (iii) preferences for print and digital library resources. Results from the questionnaire survey are first presented, and are then

Figure 3. Do you think you are highly proficient in using computers?

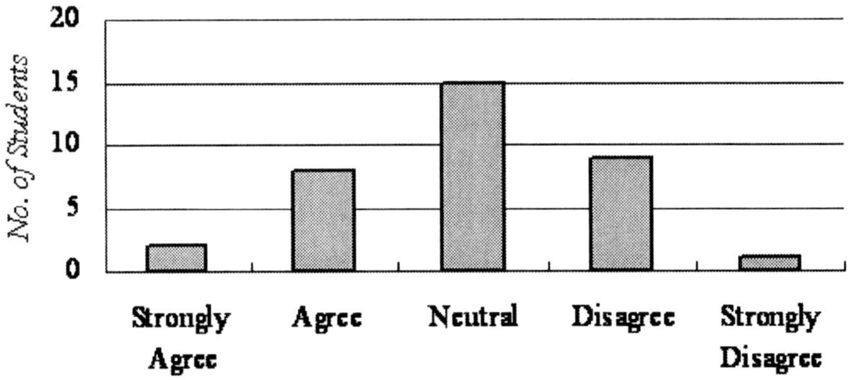

followed by a summary of the interview results.

Self-Evaluation of Level of Computer Literacy

We have already discussed the concept of digital literacy and we have also recognized that reading electronic materials requires a different set of skills. Therefore, we asked students to evaluate their own level of computer literacy, which might or might not directly affect their preferences for digital library resources. The question was 'Do you think you are highly proficient in using computers?' The result is shown in Figure 3.

It is found that, of the 35 respondents, only 10 agreed or strongly agreed that they were computer literate. Most of the respondents were not sure about their levels. The rest disagreed with the fact that they are proficient computer users. Although this is a self-evaluation and may not accurately reflect the actual level of computer literacy, we may conclude that more than half of the subjects were not completely confident in using the computer.

Habits of Using Print and Electronic Library Resources

The survey also looks into students' habits of using print and digital library resources. Respondents were asked to indicate, in the past three months, how often they used (i) print materials and online materials, (ii) physical libraries and digital libraries, and (iii) e-books and e-journals. These are summarized in the following charts (Figures 4-6). Figure 4 shows that more than half of the respondents never

Figure 4. Habits of using print and electronic books

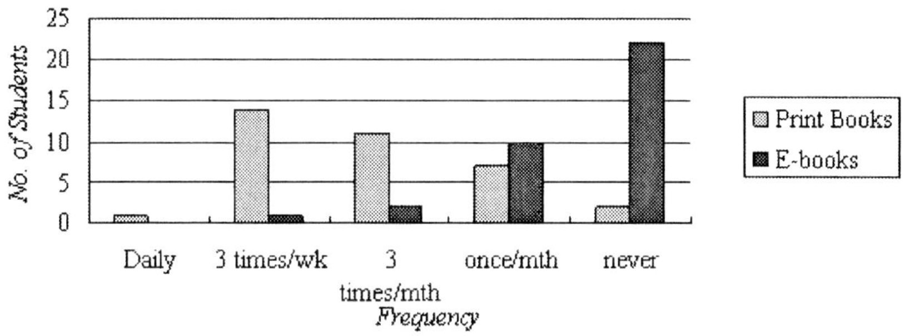

used an e-book in the past three months.

Though none of the respondents 'visited' any of the libraries daily, physical libraries were used more frequently, as shown in Figure 5. Of the 35 respondents, 26 went to a physical library more than once a month, while only 14 used digital libraries more than once a month. 10 of them even claimed that they never used digital libraries in the past three months.

We also compared students' habits of using print journals and e-journals. As illustrated in Figure 6, the frequencies of using these two types of journals were quite similar, though print journals were used slightly more often.

In sum, online materials were used less often than physical ones. The most significant evidence is found in the comparison between print books and e-books. More than 20 (60%) respondents never used one in the past three months prior to the survey. However, only two of them never used print books during the specified period of time.

Figure 6. Habits of using print journals and e-journals

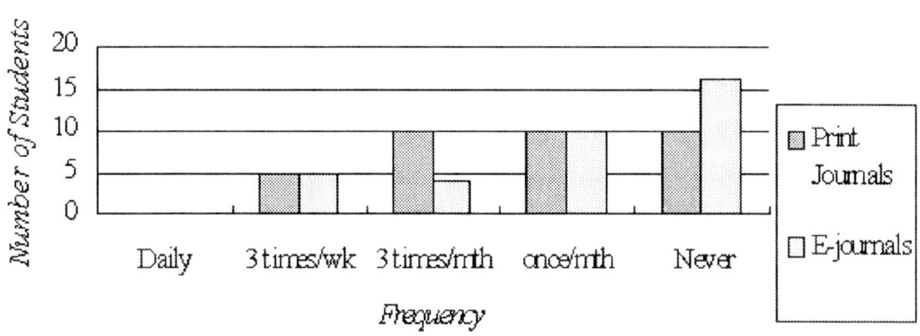

Figure 5. Habits of using physical libraries and digital libraries

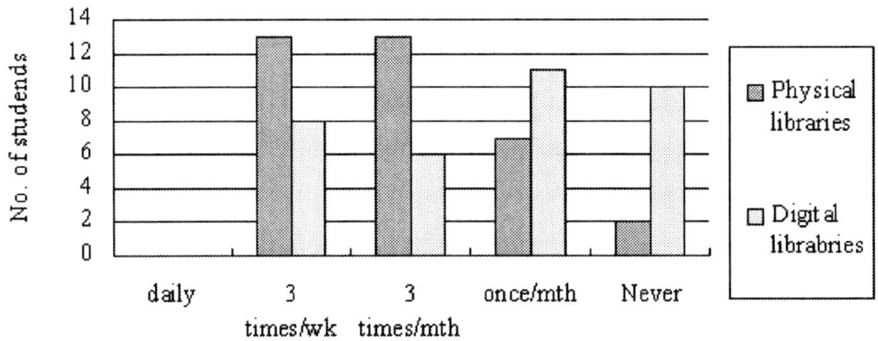

Preferences for Print and Electronic Library Resources

As shown in Figure 7, 27 of the 35 respondents (approximately 77%) preferred print materials, while only three had a preference for digital resources. The results seem to correspond to the respondents' habits of using these two types of materials. The previous section has shown that students used physical print resources more often than digital ones, which reflects their preferences for these materials.

There are three main reasons for the preferences, as can be seen in Figure 8. The majority who preferred print materials said that using print resources was a habit they had developed. Others thought that using print materials was more convenient

Figure 7. Preferences for print materials and e-materials

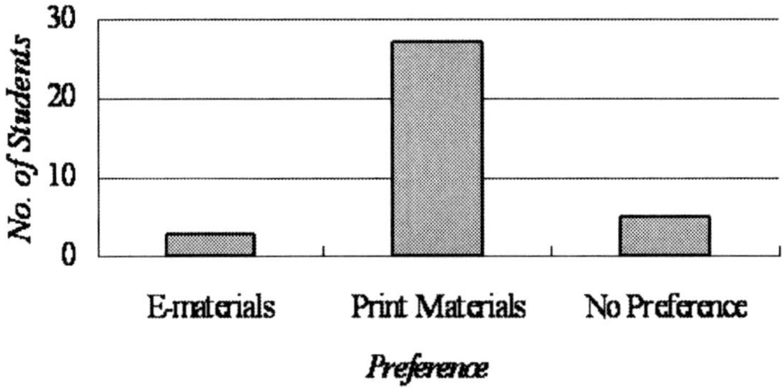

Figure 8. Reasons for choosing print materials

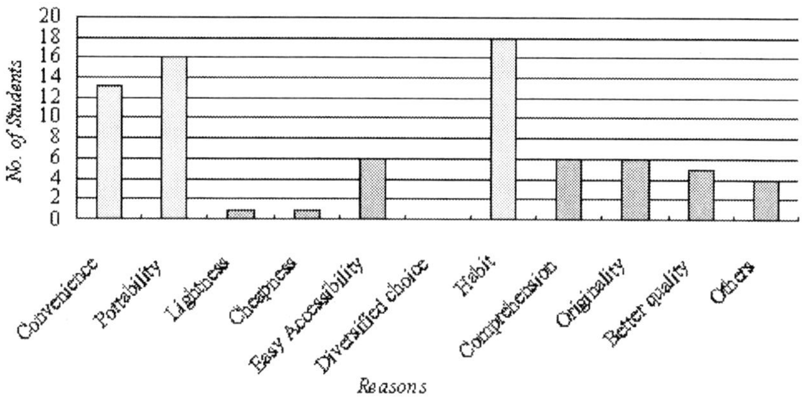

as print books are actually more portable. In cases where respondents answered 'others', four of them found that reading physical materials is better for the eyes.

Interview Results

This section reports on results drawn from in-depth interviews of eight of the respondents who participated in the survey (Lam 2003). They were mainly asked to comment on the salient issue of whether print materials would be replaced by digital ones.

In general, the students recognized that digital libraries were very convenient since they did not need to actually go to libraries and could still read and download books or journals from home. They, however, did not think that digital libraries would entirely replace physical libraries. Print books still have their advantages. Students indeed enjoyed the experience of reading with the physical presence of a bound book.

The interviewees raised three main issues regarding a potential paradigm shift from print to digital libraries.

First of all, some of them were concerned with the technical problems in using electronic resources. They doubted if the computer systems could always preserve all the materials. There might be unpredictable failures or bugs in the digital library system, the Internet server, or even their own computers. There would be no guarantee that users can always retrieve whatever they want. But with print materials, unless a book is lost or checked out, the book would always exist without any technical concern.

The second consideration is the issue of copyright. Students are not familiar with the copyright regulations governing digital materials. They are not sure to what extent they can make copies of digital resources.

The subjects also pointed out that some people in Hong Kong neither have a computer nor any knowledge of computer usage at all, and thus they do not have access to resources like digital libraries. This shift from print to digital literacy would take place only if computers and the Internet are widely used.

Views from the Librarians[6]

So far, we have only looked at views from students i.e. from a user perspective, which shows that there is a general preference for print materials and only a cautious acceptance of digital resources. However, we should not neglect views from the people who actually administer the acquisition and management of electronic resources. We, therefore, interviewed two librarians at the HKU libraries. In the interview, they were asked to provide general information as well as their own opinions of the acquisition and usage of digital resources at HKU.

Unlike the views from the student respondents, these two librarians enthusiastically supported the acquisition of electronic resources in the libraries. They both thought that digital resources are highly accessible from the users' perspective i.e. they can get the materials without actually visiting the library. Another attractive advantage, from the library administrators' perspective, is that e-books and e-journals do not take up much storage space in the physical libraries. This is very important for a densely populated city like Hong Kong, where space considerations pose many more challenges than in comparable cities in Europe and North America.

When asked their opinions of the future of libraries, they said that the number of electronic resources would continue to rise, given the save-space advantage of e-materials. One of the interviewees even predicted that e-journals would eventually replace all the print journals at HKU libraries.

PROBLEMS OBSERVED AND MEASURES PROPOSED

The findings presented above underlie three potential problems in using electronic materials in Hong Kong: (i) the low user preference for digital resources, (ii) the low access rates of digital libraries, and (iii) a possible communication gap between library users and library administrators.

Low Preference for Digital Resources

The study showed that print materials are still highly preferred, though the amount of electronic materials acquired in the libraries is on the rise. As compared to the results of user preference surveys conducted in other parts of the world, Hong Kong students seem to have a relatively low preference for digital resources. This may be attributed to two main reasons: (i) lack of promotion done by libraries and other authorities, (ii) low awareness of using digital resources i.e. students have not developed a 'habit' of using books published through 'unconventional' media. In other words, students are not 'digitally ready' to read e-books. 'Being digitally literate is multidimensional and interactive' (Gilster 1997). As has been discussed, special 'acts' like clicking and hyperlinking are involved in processing electronic information. Students are still more used to the experience and literacy practices of reading bound and print books.

Apparently, the access rates of digital resources are still low, as demonstrated in Figures 4-6 in the previous section. Students consider print books more portable than digital ones, since a print book can be carried to places while reading a digital book requires a networked computer or other special devices. Although various portable e-book readers are available on the market, this technology is not very popular in Hong Kong. Having spelt out these issues, let us now propose some possible measures to tackle the problems raised in the study.

A Disregard for User Preferences?

Based on the data collected from students and librarians, we noticed that there exists a gap between user and provider expectations towards digital libraries. The findings clearly indicate that, while the libraries hold a positive view for digital resources, most of the students, the users, actually still prefer print materials, hence the title of the chapter, some students still read (print) books in the 21st Century. We have thus realized that a successful implementation of the idea of digital libraries should take into account the actual acceptability and user preferences for different types of materials.

Enhancing Access Rates of Digital Libraries

In order to address the issues discussed above, some measures have to be taken. Here, we propose the following solutions to the problems we have observed regarding the implementation of the concept of digital libraries in Hong Kong and beyond.

Promoting the Advantages of Digital Materials

One of the most significant findings of this study is that librarians appreciate the potential values brought by digital resources while students are still quite skeptical about using these rather 'unconventional' resources. As such, one of the major tasks for library administrators and e-book developers is to pay careful attention to the needs of their users. They should emphasize the advantages of electronic materials over the print. Once users come round to realizing that reading e-books is an equally enjoyable experience, they might begin to develop a positive attitude towards reading electronic materials.

Raising Awareness of the Rich Collection of Electronic Materials

Even though there is a huge collection of electronic resources in the libraries, it may not be surprising that not many users know the actual coverage of electronic resources. This may have contributed to the low access rates of electronic resources. Therefore, we suggest that academic libraries should emphasize the dissemination of information about digital materials, such as announcing new additions of e-journals and e-books much more regularly, in order to draw potential users' attention to the availability of these vital resources.

Taking Full Advantage of Electronic Media in Publishing E-Books

Some e-book versions of print books today are not exact duplicates of the original print versions. Images like photos or charts and book covers may be missing. It is hoped that developers of electronic resources would do their best to keep all contents of the print versions in the electronic versions, or better still, incorporate multimedia elements which can only be done on the electronic version.

Enhancing E-Book Technologies

Portability is one of the major criteria that users consider in evaluating digital materials. Although we recognize that electronic materials possess a certain level of flexibility and they save storage space, using e-books and e-journals is not necessarily more convenient than using print books. A number of restrictions are often imposed on electronic materials, just as they are imposed on print books. For instance, e-journals and e-books in academic libraries can only be accessed by authorized users within the registered computer network or server. If users wish to use e-books or e-journals from home, they must first connect their computers to the registered network. Such inconvenience may discourage the acceptance of electronic resources to a certain extent. Portable e-book readers like eBOOKMAN® and AlphaBook have been launched in recent years. It is hoped that such technologies will soon be more widespread and become a solution to the problem of portability.

Encouraging the Use of Electronic Books in Other Education Sectors

Most of today's e-books are reference works or academic materials for people at the tertiary level. If, however, students could begin using e-books at an early age, they would develop an early awareness of the advantages, or at least, the 'existence' of electronic materials. We suggest that the provision of electronic materials should not only be confined to university libraries. Other sectors like primary and secondary schools should also enjoy the benefits of electronic materials, especially since space in school libraries would be very limited in a place like Hong Kong. Of course, due to the high subscription fees of electronic materials, not many schools can actually afford the implementation of such technologies. It is hoped that funding bodies would continue to support not just tertiary academic libraries, but also primary and secondary school libraries.

Clarifying Copyright Policies Governing Electronic Resources

Based on information provided by one of the librarians interviewed, copyright laws of electronic materials are still unclear to library users and even to professional librarians. This has also contributed to the low user preference for electronic resources. We, therefore, suggest that copyright laws which govern the use of electronic information be made clearer to library users as well as library administrators. The extent to which users can copy, share, disseminate, and use the materials should be stated. Relevant guidelines can be disseminated in the form of seminars, pamphlets, or perhaps some guidelines on the Internet.

CONCLUDING REMARKS: THE FUTURE OF PRINT RESOURCES

From the literature reviews and the surveys done in this chapter, one may conclude that reading electronic materials is quite different from reading print materials. Electronic resources seem likely to replace most print materials, especially journals, but uncertainty still remains. Some students still prefer print books in the 21st

Figure 9. HKU Libraries User Preference Survey 20087

Overall Preference for Print vs Electronic

In terms of the overall preference for print versus electronic resources, 65.85% of respondents prefer reading journal online while 21.93% prefer print and 12.22% has no preference between print and electronic journals; whereas 60.6% of respondents prefer reading the printed book, prefer 29.39% the electronic book and 10.01% has no preference.

Material Format	No Preference	% No Preference	Online	% Online	Print	% Print	Total No. of Respondents
Journals	249	12.22%	1,342	65.85%	447	21.93%	2,038
Books	204	10.01%	599	29.39%	1235	60.60%	2,038

With the growth in electronic resource provision, we have witnessed a growth in preference for online materials. For journals, this year's 65.85% preference for electronic represents a marked increase on last year's 59.05%, and for books this year's 29.39% is also increased over last year's 22.68%.

century! The paradigm shift from print to digital libraries, if this were the target at all, would depend on measures taken to encourage a positive attitude of users of digital resources. The fact is that most students of this early part of the 21st Century are still more accustomed to print books. Schools in Hong Kong and other parts of the world adopt more print literacy practices than electronic ones, i.e. students are trained to be more 'print literate' than 'digitally literate'. It is thus understandable that there is a long way from full acceptability. In another 10 years' time, when e-book technologies would have become more developed and widespread, one may not be surprised to see a paradigm shift, where more students prefer e-books to print books. Indeed, the University of Hong Kong library has started an annual user preference survey (Sidorko, 2004), which is beginning to show that more and more readers are tending to favour electronic journals over print journals, as can be seen in the excerpts of figures and commentary (Figure 9) from the survey located at the following website: http://lib.hku.hk/survey2008/ (The University of Hong Kong Libraries (2008) accessed: March 25, 2009).

Future studies should take into consideration user preferences across disciplines, levels of education and even across different cultures and economic zones of the world. It is believed that students of the humanities may access electronic resources less often than those in the hard sciences. Of course, this needs to be supported by further evidence. Other user preference studies could, for instance, include e-book reading habits of undergraduate vis-à-vis research postgraduate students, user preferences for e-books vis-à-vis e-journals, and user preferences for reference materials vis-à-vis fiction. It would also be interesting to see how students and young readers in developing economies, such as in Asia and Africa fare alongside readers in developed economies in Hong Kong, Japan and many parts of Europe and North America, where resources and cultures vary in terms of broad access to the internet infrastructure. A step in this direction is a study in Malaysia by Ismail and Zainab (2005) but we need more of such studies.

Having focused on the concept of reading literacies, with particular reference to digital literacy reading practices and preference studies, I will in the next chapter return to the issue of how language users actually speak and write in the medium of new communication technologies, and the new linguistic forms and uses that are the result of this new kind of communication. I will develop a particular model to capture the changing forms and uses of language in the medium of new communications technologies.

REFERENCES

Bodomo, A. B. (2000). Lecture notes on Language and Literacy. *Computer literacy and language educational technology*. The University of Hong Kong. Available at http://ecourse.hku.hk:8900/public/LING2011

Bodomo, A. B.,, & Carmen,, K. M. Lee. (2002). Changing forms of language and literacy: Technobabble and mobile phone communication. *Literacy and Numeracy Studies*, *12*(1), 23–44.

Bodomo, A. B., Lam, M. L., & Carmen, K. M. Lee. (2003). Some students still read books in the 21st century: A study of user preferences for print and electronic libraries. *Reading Matrix. 3* (3). Special Issue on Reading and Technology. [electronic version] http://www.readingmatrix.com/articles/bodomo_lam_lee/article.pdf

Fairclough, N. (2000). Multiliteracies and language: Orders of discourse and intertextuality. In B. Cope and M. Kalantzis (Eds.), *Multiliteracies* (pp: 162-182). Australia: Macmillan.

Gee, J. P. (1996). [1990] *Sociolinguistics and literacies* (2nd ed.). London: Taylor and Francis.

Gilster, P. (1997). *Digital literacy*. New York: John Wiley & Sons.

Holmqvist, K. Jana Holsanova, Maria Barthelson, and Daniel Lundqvist. (2003). Reading or scanning? A study of newspaper and net paper reading. In Radach, R., Jukka Hyon and Heiner Deubel (eds) *The mind's eye: Cognitive and applied aspects of eye movement research* Elsevier Science Ltd.

Ismail and Zainab. (2005). *Patterns of e-book use among undergraduates of Malaysia* [electronic version] http://72.14.253.104/search?q=cache:5lll8LOZ3KAJ:majlis.fsktm.um.edu.my/document.aspx%3FFileName%3D327.pdf+bodomo&hl=en&gl=us&ct=clnk&cd=296

Kress, G. (1998). Visual and verbal mode of representation in electronically mediated communication: The potentials of new forms of text. In I. Synder (ed.) *Page to screen: Taking literacy into the electronic era* (pp. 53-79). London and New York: Routledge.

Kress, G. (2000). Design and transformation: New theories of meaning. In B. Cope and M. Kalantzis (Eds.), *Multiliteracies* (pp: 153-162). Australia: Macmillan.

Lam, M. L. (2003). *Digital literacy and the rise of digital libraries: A perceptual learning style study of user preferences for print and digital materials of Hong Kong students*. Unpublished undergraduate dissertation. Hong Kong: University of Hong Kong.

Lee, C. K. M. (2002). *Chinese and English computer-mediated communication in the context of new literacy studies*. Unpublished Master's dissertation. Hong Kong: University of Hong Kong.

Lewenstein, M., et al. (2000). *Stanford-Polyner Project*. Project website:http://www.poynter.org/eyetrack2000

Luke, C. (2000). Cyber-schooling and technological change: Multiliteracies for new times. In B. Cope and M. Kalantzis (Eds.), *Multiliteracies* (pp: 69-91). Australia: Macmillan.

Pool, C. R. (1997). A new digital literacy: A conversation with Paul Gilster. *Educational Leadership. 55*(3). [electronic version] http://www.ascd.org/readingroom/edlead/9711/pool.html

Sidorko, P. E. (2004, November 21-24). The remaking of the librarian: Average customer review 3 out of 5. 2004 Asian Library and Information Conference (ALIC), Bangkok, Thailand. Retrieved March, 24, 2009, from http://hub.hku.hk/browse?type=author&order=ASC&rpp=20&value=Sidorko%2C+Peter+E

The University of Hong Kong Libraries. (2008). 2008 user survey report. Hong Kong: The University of Hong Kong Libraries (pp. 38-41). Retrieved March 25, 2009, from http://lib.hku.hk/survey2008/User%20survey%20report2008_for%20Webpage.doc

The University of Hong Kong Libraries Annual Report 01-02. (2002). Hong Kong: University of Hong Kong.

ENDNOTES

1. The data and analyses contained in this chapter first appeared in the journal *Reading Matrix* as Bodomo, Lam and Lee (2003). I thank Carmen Lee and Jovita Lam, my former students for collaborative research in the early 2000s on new ways of reading in the age of IT.
2. I thank a reviewer for drawing my attention to the idea that the term "digital library" has a specific meaning for computer scientists, with many of them considering digital libraries as 'social spaces' offering much more than mere access to information. In this book the use of the term simply refers to digital collections, as we have discussed throughout chapter 3. I sometimes also use the term "digital library collection" to emphasize this aspect of stocking up digital resources in major academic libraries.
3. The results of this survey are available online at: http://www.lib.utexas.edu/admin/cird/surveyresponse.html
4. The California State University NetLibrary Service Survey Results: http://www.calstate.edu/SEIR/eBK_FINAL_RPT/eBk_Final_RPt_Svy_Result_1.pdf
5. Thanks to staff members at the Reference Department of the HKU libraries for providing statistical data of the university's acquisition of electronic resources from 2000 to 2002! It has to be stressed that the definition of an electronic material has changed slightly over the past few years.
6. Grateful acknowledgement to two librarians at HKU, Mr. David Palmer, Systems Librarian, and Ms. Gayle Chan, Collection Development Librarian, for the invaluable information they provided in interviews conducted on June 12th 2003!
7. Thanks to Peter Sidorko, Deputy Librarian, for drawing my attention to these regular biennial User Preference Surveys that began in 2004.

Chapter 4
TeLCU
A Model for Technology-Conditioned Language and Literacy Change

INTRODUCTION[1]

In line with our theme of investigating the relationship between new communications technology and the way we process (i.e. speak, read, and write) natural language, in chapter 2 I outlined different views about language forms and language use in the context of new technologies, and in chapter 3, I looked at new ways of reading and accessing reading materials with the advent of new computer-mediated communication platforms that promote the production of e-materials. We found in chapter 2 that there were two main views about the relationship between CMC technology and language. One was that CMC technologies cannot actually change language and that whatever transformations we observe are part of a larger social transformation; indeed that technology itself is part of social transformation. The second view was

DOI: 10.4018/978-1-60566-868-0.ch004

that technology actually has a causal effect on language structures and use, leading to the idea that new forms of language and new ways of using and processing language arise from the introduction of new communications technologies.

In this chapter I shall build on this discussion by proposing a model in which we can capture and conceptualize these perceived new forms of language and new ways of speaking. Indeed, I go further as to present a particular study of the way some bilingual speakers of Chinese and English create new forms of language through mobile phone texting (though the concept of mobile phone texting is taken up more comprehensively in chapter seven). I shall present and discuss concrete examples throughout the chapter.

The information age is characterized by a rapid introduction of new tools and media of communication. These new tools and media include digital products such as computers, the internet and the World Wide Web (WWW), means of computer-mediated communication like email (electronic mail) and ICQ (I-Seek-You), and mobile phones. A natural issue that arises then is a study of the relationship between these new tools and human language for which they serve as media of transmission. As we encountered in chapter 2, answers to a set of questions are necessary to comprehend the issues involved. Does the introduction of these media influence the way we understand, analyze, and use language in its various forms? If so, what is the nature of this influence? Does it involve different uses of existing elements of language or does the prevalence of these media lead to new forms of language and new ways of using language?

This chapter aims to address these questions and propose some answers. In examining these questions, based on observation and analysis of issues of language, literacy, and communications technology, I propose a model called Technology-conditioned approach to Language Change and Use (TeLCU). This approach projects the view that there is a causal relationship between the emergence of new tools and media of communication and the creation of new forms of language and literacy. New tools and media of communication demand the creation of new forms and ways of communication. These new forms compete with existing forms and ways of communication, leading to changes in the way we use language in its various forms, including spoken and written forms.

A potential anti-thesis to TeLCU is that there is little or no causal relationship between the prevalence of new media of communication and changes in the forms of language and ways in which we use human language. While not directly arguing against the idea that there is no causal relationship between the emergence of new technology, in general, and the new ways in which language is used, Kress (1998:53), for instance, observes that '...when we look at the far-reaching and deep changes in forms of communication which characterize the present-day email and its changing forms of language, for instance, it is tempting to attribute these changes to some technological innovation but erroneous to do so'. Luke, C. (2000:83) also takes a similar position, believing that new forms of literacy practices do not simply emerge with technological change. Rather, 'technologies always emerge as products of specific cultural practices, literate traditions, and the interests and desires of those groups who design and name them'.

In this chapter, I build on this fruitful discussion in the literature on the relationship between new technologies and the way language is used within these technologies and argue that there is indeed a significant causal relationship between communications technology and new language and literacy practices or more specifically the evolution of new ways of using language. As Adams (1996) puts it, 'the new technologies are themselves dramatically changing the nature of the language we use'. Such an approach is also supported by Baron (1984: 139) which concludes that '[n]o one in the computer industry has any hidden agenda for using hardware or software development to alter human language. Yet technology can indeed drive linguistic and social change'.

Indeed, new practices of language and literacy may be attributed to a set of unique properties in new communications technology. Modern digital communications technology is characterized by flexibility, connectivity, and interactivity (Blurton 1999) that traditional forms of technology like radio and TV lack. In other words, it is possible to have many-to-many, many-to-one, one-to-many, and one-to-one modes of communication with modern digital information communications technology (ICT). These features of digital ICTs enable them to have a more pervasive influence on forms and uses of language.

Such an approach to the dynamics of language and literacy has exciting consequences on literacy research, given the rapidity with which new digital ICTs are emerging. TeLCU provides the practitioner in the field of language and literacy research with a framework in which to undertake a critical approach to definitions and theoretical perspectives in the field. Such a critical approach would be commensurate with perspectives of literacy in the theoretical framework of critical literacy (Luke, A. 1997, Luke, C. 2000)

This chapter is organized as follows. I first outline some of the theoretical perspectives within which I develop my model of TeLCU. I briefly explain my views and reviews of the relationship between language and literacy and show how TeLCU fits into the paradigm of New Literacy Studies (NLS) (Street 1984, Barton 1994, Gee 1996 [1990], Barton, Hamilton, and Ivanic (2000). Next, I examine some new forms and uses of language under topics like technobabble, acronymy, punctuation, ellipsis, and emoticonymy. I also point to some creative ways in which language is used through the mobile phone. Following from this is a brief discussion of new digital literacies with a focus on how these constitute evidence for the TeLCU model of new forms of language and literacy. Finally, I conclude the chapter by summarizing and discussing the consequences of this model for future research in language and literacy studies.

THEORETICAL PERSPECTIVES

I believe that there is an inter-connection between language and literacy. Indeed, one may even claim that it is hard to examine these concepts as separate entities. Nevertheless, since they are not one and the same thing, one must attempt to establish the exact nature of the relationship between these closely related and almost identical entities. To understand any relationship, one needs to examine the features of each entity in this relationship. This section will first discuss different conceptualizations of language and literacy. It will then attempt to make a connection between the two concepts. I will also present the model of TeLCU and show how it fits into the NLS framework.

Views of Language and Definitions of Literacy

Language may be analyzed from several dimensions, such as from the perspective of structural or formal linguistics, in terms of functional approaches or sociolinguistics, and in terms of cognitive approaches, etc. Another approach, which this chapter is based on, is called the constructivist approach to language. A constructivist model investigates language in three major aspects: history, acquisition, and use of language. With this approach, the study of language is not the study of a system as a set of structures. It is, instead, the study of 'a dynamic social activity which serves people's purposes' (Barton 1994: 54).

Literacy is a contentious concept and, as such, there is a vast array of definitions and ways of conceptualizing it in the literature on literacy studies (e.g. Goody and Watt 1968, Scribner and Cole 1981, Street 1984, etc), without there ever being a universal consensus on what it really means. A broad and commonly accepted, but hardly sufficient, definition of literacy is 'the ability to read and write'. Indeed, this has been the original and basic definition of the concept of literacy but the concept has evolved to include much more than just issues of reading and writing. Some might see literacy as a set of skills, with the development of functional literacy in our society. The concept can even be extended to different types of literacies, e.g. political literacy, vernacular literacy, information literacy, etc. From a linguistic point of view, however, no definition of literacy is adequate unless it addresses the issue of coding a linguistic system. The evidence for this assumption is that in the definitions of literacy in almost all of the world's languages (Table 1), literacy is

Table 1. 'Literacy' in various languages

European Languages	**Asian Languages**
French: Alphabétisation Alphabetization 'literacy'	***Japanese***: Mon-mou Sentence- blindness 'literacy'
Norwegian: Lese-og skirive-opplaering Reading and writing training 'literacy'	***Chinese***: 讀寫能力 'literacy' *duk6 se2 nang4 lik6* Read write ability 識字能力 *Sik1 zi6 nang4 lik6* Know character ability ***Chinese***: 文盲 'literacy' *Man4 Maang4* Written language blindness
African Languages	
Dagaare: *Gán-b@ng yɛ@lɛ@* Book-know-matter 'literacy'	
Twi: *Engoma nim* Book-know 'literacy'	

often conceptualized in terms of dealing with the ability or inability of processing written language.

Bodomo (2000), and Bodomo and Lee (2001, 2002) put forward the above ideas and defines literacy as 'the ability to code and decode linguistic and other symbolic systems for communication and information processing'. An obvious relationship between language and literacy can be inferred from this definition.

Conceptualizing the Interrelationships between Language and Literacy

The relationship between language and literacy can be examined by dealing with topics like how literacy is embedded in language, the use of literacy in multilingual communities, the relationship between spoken and written language, as well as the relationship between language acquisition and literacy acquisition.

Literacy is embedded in language. This relationship can be explored through the study of how language may be used in different ways. Language itself forms part of a discourse. Literacy is embedded in discourses which structure our knowledge. Much of learning is learning to participate in new discourses. Discourses will then result in texts, which are bound to each other through intertextuality. Barton (1994: 53) concludes that 'the study of literacy, as of all language, is the study of practices and the study of texts'. People are then positioned by different ways of using text and discourse.

Another way of characterizing the linguistic basis of literacy is to study the use of literacy practices in a multilingual set-up. 'Being literate is always expressed in a particular language.'(Barton 1994: 69). Different literacies are associated with different languages and their associated practices. For example, in Hong Kong, there are code-mixing and code-switching phenomena, as well as a biliterate trilingual language policy. Thus, the meaning of 'being literate' differs in different languages in different societies and their associated practices. Literacy-related issues are also discussed in a number of cross-cultural studies of literacy such as Heath (1983), Street (1984), and Baynham (1993).

Literacy can also be explained through the relationship between spoken and written language. These two media of language are important in the study of literacy and its related events and practices. In the study of literacy events, the written text is always a starting point but it can be extended to the uses of spoken language which are often integrated in the use of writing (Heath 1983, Barton and Hamilton 2000).

To sum up, there is an intertwining relationship between language and literacy. Literacy, in its basic conceptualization as the knowledge required to code and decode a linguistic system, is embedded in language through different ways of using text and discourse. The meaning of literacy also differs in different linguistic settings. Moreover, speech and writing are intermingled in literacy events and we have to study both of them when analyzing the relationship between language and literacy.

Technology-Conditioned Approach to Language Change and Use (TeLCU)

Having outlined the interrelationship between language and literacy, which is highly subject to social patterns, I then propose a model which is called Technology-conditioned approach to Language Change and Use (TeLCU). I suggest that there is a causal relationship between the emergence of new tools and media of communication in our society and the use of language.

New technologies often require new forms of language and literacy to express new concepts that emerge along with these new media and tools. New media of communication, then, can lead to changes in the way people use their language. In fact, works like Lakoff (1982) and Ong (1982) have discussed the impact of oral media like TV and radio on people's use of language.

The information age is characterized by the rapid advancement of technology, with the introduction of new tools and media that constitute computer-mediated communication (CMC). Among these, CMC or Information Communication Technologies (ICTs) constitute the type of technology that this book is concerned with. ICTs, according to Blurton (1999), are characterized by flexibility, connectivity, and interactivity, which are different from passive, one-way media such as TV and radio. These distinctive features of ICT tools and media allow pervasive changes in language forms and uses. Apart from these, I believe that the popularity of a particular tool or media can also be a major factor in the discussion of how new ICTs

Figure 1. TeLCU – the relationship between new ICTs and new forms of language and literacy

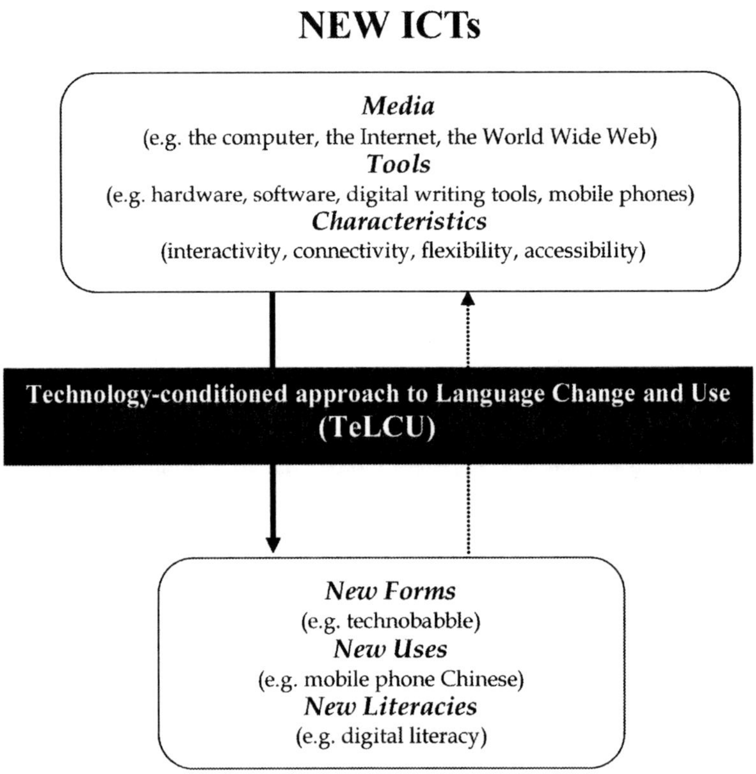

introduce new forms and uses of language. One might expect that the products of the TeLCU model, i.e. new forms of language and literacy, are gradable. That is, the more of the above features a new technology carries, the more likely new forms of language and literacy will be introduced and the more widespread these new forms will be. Figure 1 is a diagram which illustrates the cause-and-effect relationship between new digital ICTs and new forms of language and literacy.

In this model, I show that new ICTs include all the tools and media along with a bundle of features, which act as the 'inputs' of this model. These features undergo the main process, TeLCU, and new forms of language and literacy are then generated. This is indicated with the downward arrow in the model. However, I do not exclude the possibility of a 'reverse' process, i.e. the possibility where new forms of language and literacy reinforce changes in new technologies. Hence, the upward arrow of the diagram illustrates this process. This process, nevertheless, cannot exist without TeLCU i.e. new ICTs foster changes in language and literacy before new forms of language and literacy can have impact on ICTs.

TeLCU can also fit into the model of New Literacy Studies (NLS). NLS is a theoretical framework in which literacy is defined in terms of social practices (Street 1984, Barton 1994, Gee 1996, Barton, Hamilton, and Ivanic< 2000). It is a social approach to literacy that explains literacy in terms of different social contexts. Thus, the meaning of literacy varies from situation to situation. This ideological approach to literacy is in contrast with the traditional 'autonomous' approach (Street 1984) of regarding literacy as a set of cognitive skills (Ong 1982). As such, TeLCU itself allows new forms of language and literacy in which the use of these new forms can be explained by the dynamic nature of social contexts and practices.

Having introduced the theoretical framework of TeLCU, I shall discuss in the following sections some of the new forms of language and literacy under the influence of new ICTs.

NEW FORMS OF LANGUAGE

One of the major 'products' of TeLCU is the emergence of new forms of language and their associated practices. As Halliday (1985:82) predicts:

When new demands are made on language... [it] changes in response to them. ... [W]e are making language work for us in ways it never had to do before, it will have to become a different language in order to cope.

Digital ICT tools and media then present these 'new demands' which promote new language forms and uses. In this section, I shall focus on the new forms and ways of using language in two types of communication tools: the networked computer and the mobile phone.

Technobabble

Crystal (1997: 382) notes that ' [a]s society develops new facets, so language is devised to express them. In recent times, whole new areas of expression have emerged, in relation to such domains as computing, broadcasting, commercial advertising, and popular music.' In the area of ICTs, we have what is called Netspeak (Crystal 2001), or Technobabble (Barry 1991). Several definitions of technobabble have been found. The Online Oxford English Dictionary defines technobabble as: 'colloq.(orig. U.S.) [after psychobabble], outlandish or pretentious (pseudo-) scientific jargon'. Barry (1991: xiii) defines technobabble as 'the gamut of languages that describe the area subsumed under the rubric of high technology: computers, biotechnology, aerospace, robotics, and the like.' and is 'a form of communication among people in the rapidly advancing computer and other high-technology industries'.

Here I define the concept as *a language characterized by the pervasive use of technical jargon, or a speech or a piece of writing characterized by the pervasive and extreme injection of technical jargon to the extent that the language is barely comprehensible to non-specialist speakers of the particular language.*

Technobabble is of many kinds, according to their uses and settings. It is used and exists in communicative events in the physical world and also in the cyberworld. The following categories of technobabble have been identified:

Technobabble for Real World Communication

One of the purposes of technobabble is to describe computer technology (Barry 1991, Crystal 2001), such as what is called e-terminologies, in which the prefix e- denotes an all-electronic operation over a network - usually over the Internet, always in the Internet protocols.3 Below are some examples of e-terminologies which are relevant to language and literacy (Table 2).

E-terminologies constitute only one of the many instances of technical jargon in the ICT industry. The prefix e- seems to be highly productive and can simply be attached to any existing noun (e.g. text, book, classroom, and library...etc). This is like a new trend or convention of word formation in ICT technobabble - whenever something is computerized, it is prefixed by an 'e' 4.

Technobabble for Online Communication

Another level of technobabble is the kind of language used in computer-mediated communication (CMC) such as email and ICQ. When CMC is used as a medium for text-based message exchanges, some features are adopted according to different contexts of communication. For instance, in an interpersonal 'chat' situation through systems like ICQ, in order to preserve the 'real-time' effect of face-to-face conversation and to create a conversation-like environment in online communication, users have developed a number of new ways to express themselves in the messages. These novel forms of language may only be comprehensible within the so-called cyber- or online- community. From the perspective of the NLS, one must recognize these changing forms of literacy practices that exist in a computer-mediated communication environment (de Pourbaix 2000, Tusting, Ivanic, and Wilson 2000). I here discuss the various forms under the topics of acronymy, punctuation, and emoticonymy.

Table 2. Some e-terms which are associated with language and literacy

E-term	Meaning
e-book and e-book reader	- an e-book is a book presented in electronic form to be read primarily on a screen. It may provide interactivity through dynamic links, quizzes, or simulations. An e-book reader is a device, which may be in the form of a simulated book with two foldout screens, for viewing e-books. [Source: Reading Online, www.readingonline.org Posted October 1999. Published simultaneously in the Journal of Adolescent & Adult Literacy]
e-classroom	- an online learning environment
e-course	- a computerized course which makes uses of the internet for teaching and learning
e-dictionary	- a dictionary which is made available online such as the online Oxford English Dictionary (http://www.oed.com).
e-glossary	(1) glossary of technology/internet –related terms (2) an online glossary of terms of any field
e-journals	- journals which can be accessed and read online
e-learning	- learning through the use of internet-based resources such as web-based course tools, online courses…etc.
e-library	- Internet-based library that provides full-text, online access to eBooks of reference (e.g. netLibrary)
e-mail	- Electronic mail is the Internet service most widely used. By sending an email, a file is created that will be transmitted and delivered to the electronic mailbox of the person you address. Can also be used to transfer files containing other information such as documents, programs and multi-media data. (http://www.hcgrp.com/eGlossary/eglossary.html)
e-quiz	- on-line exercises to test students' comprehension of a lesson or course.
e-text	- a text which is published online
e-University	- a tertiary institution which is distinctive from normal university that all the teaching and learning are internet-based.

Acronymy

Acronymy involves the shortening and abbreviation of words and phrases that are commonly used in real-life communications. Based on sample messages collected from CMC users in Hong Kong (Lee 2002a, Lee 2002b), in which each of 72 subjects was asked to provide five shortened expressions which they used most often in computer-mediated messages, the following set of common CMC shortenings have been identified (Table 3).

It is found that BTW ('by the way'), U ('you'), and ASAP ('as soon as possible') are the three most commonly used CMC shortenings. Unlike traditional classification of abbreviations in language, CMC shortenings are no longer restricted to acronyms and initialisms. While many of the studies of CMC language are able to discuss acronyms and initialisms, more methods of forming shortened expressions in CMC have been identified (Table 4).

Table 3. Examples of CMC shortenings provided by the respondents

Shortening	Target Expression	N=72
BTW	by the way	19
U / u	You	19
Asap	as soon as possible	13
Bb / BB	bye bye	12
CU	see you	12
TMR	Tomorrow	9
ic	I see	8
B4 / b4	Before	6

This study has found that there is indeed no 'fixed' set of acronyms in CMC across cultures. Some of these might be more 'universal', or more widespread, such as CU for 'see you', or BB for 'bye-bye'. However, some might be foreign to Hong Kong users, such as the acronym for 'Talk to you later'. Most CMC users in Hong Kong will choose to use TTUL but in the West, it is more likely to be TTUL8R. Notice the use of 8 and R, which reflects different accents in different varieties of English. Some ESL speakers in Hong Kong rarely pronounce the diphthong in 8 as /eI/ but as /E/ instead, nor do they pronounce the final /r/. The interesting point I am trying to make is that acronymy is very much associated with the linguistic background of individuals. A native English speaker may be able to develop a more comprehensive set of acronyms while a smaller but distinctive set may be created by Hong Kong users. For instance, the expression 88 ('bye-bye') is created in terms of the resemblance between the Chinese pronunciation of 'eight' bat3 and bye-bye in English. This kind of variation shows that the notion of setting or context is very crucial to every literacy event. This is yet another way of demonstrating how the

Table 4. New approaches to shortenings

Formation	Example
Acronym of sentence	GTG ('I've Got To Go"), BRB ('I'll Be Right Back')
Letter homophone	U ('you"), R ('are")
Number homophone (or of similar pronunciation)	88 ('Bye Bye' in English resembles the pronunciation of '8' in Chinese)
Combination of letter and number homophone	b4 ('before')
Reduction of individual word	tmr ('tomorrow'), coz/cos ('because'), gd nite ('good night')
Combination of letter initial and letter homophone	TTUL/TTYL ('talk to you later'), OIC ('oh I see')

Figure 2. An Extract from a Personal E-mail Message

```
[…]OKOK .... 25/12 and 26/12 .. for now .. I am OK . he he
....also.. the performance .. I must go... he he .... but how
come so many days?? which day you want me to come??   For me....
also the point ... BUSY and BUSY.... very tired ....but just feel
happy ar.... because ..... I feel that many things is one the
right track la .... I am doing the things I want to do la ...
also .. thank you for you so support me ... really thanks !![…]
```

model of TeLCU is in line with the NLS framework, the context-sensitive approach to literacy (Lee 2002a).

Punctuation

In written language, punctuation is used to separate units of grammar such as sentences, clauses, phrases, and words from each other. It is interesting to note that new ways of using punctuation have been developed by users of online communication. These include the following:

Ellipsis

In standard forms of writing, ellipsis may be used to indicate omission in discourse, the omission itself may involve identical items in the sentence, e.g. "I do like food and John does too". Here we don't need to say: "...and John does like food too". The elliptical or omitted items may also be obvious things or actions that follow from previous actions and which are assumed to be obvious to the reader, e.g. "John and his wife put their lips together and ..." The elliptical dots represent the widely assumed action of kissing that would obviously follow from two people putting their lips together in Western cultures. In electronic communication, there tends to be a pervasive use of "ellipsis" within a single message.

In Figure 2, the writer has used ellipsis 18 times. Some of these are not used to mark omission but to indicate the end of (a part of) a sentence. It should also be noticed that ellipses here do not always appear in the standard three-dot form (...). They are not used conventionally but in a rather spontaneous and improvised manner.

Figure 3. An Example of an ICQ Message Exchange

A:	*wei. last Sun you go hiking with a boy, your BF?*
B:	*ha?! who ar???? *.**
A:	*you go hiking ma? woh, not, it is Sat.*
B:	*yes.... i go on Sat la... but who's that "boy"???? :P*
A:	*ha ha..... should not be my WRONG looking ga.....*
B:	*who ar??? i don't know I've been with a boy last Sat??? where?? form the starting point or finished pt??*
A:	*mum..... finish point.*
B:	*hahahaha!!!!1 i know la... u mean I'm standing at the entrance with a man??...*

Exclamation marks and Question marks

Another type of non-standard use of punctuation concerns exclamation marks and question marks.

In the series of ICQ messages (Figure 3), exclamation marks and question marks are used in the following ways:

Repetition of question mark: to emphasize that the message writer is curious and eager to know the truth;
Repetition of exclamation mark: to emphasize the tone of utterance;
Combination of ? and ! (?! or !?): to show surprises and doubts at the same time.

There seems to be no 'rules' for using punctuation in CMC, at least, in informal communication, as shown in the above extracts. The non-standard ways of using punctuation may serve the following functions:

For the ease of communication;
To indicate hesitations and thoughts;
To express emotion and mood of the author;
To indicate incompleteness of sentences; and
To show informality and familiarity in informal situations

Emoticonymy (Smileys)

Since emotions and facial expressions of a speaker/writer cannot easily be encoded in writing, a set of symbols which uses keyboard strokes is developed for expressing one's emotions in online communication, and now there are even images incorporated into CMC gadgets representing various moods, representing an important aspect of visual language. This is where emoticonymy comes in. By emoticonymy, I mean a subfield of CMC which involves the analysis and practices of employing smileys and related icons for conveying emotions and other linguistic and kinesic features intended by the author. The following (Table 5) is a list of the most common emoticons provided by the 72 subjects mentioned above:

The practices of using acronyms and other symbols in CMC have been noted in a number of linguistic studies of text-based computer-mediated communication (e.g. Werry 1996, Crystal 2001, and Shortis 2001). However, I would like to apply these new uses of language to the TeLCU model within the NLS framework. These new ways of using language, which are brought about as a result of the introduction of new ICTs, are very much dependent on their contexts. In this case, new ICTs require these new forms of language that involve users' creativity and self-identification in electronic communicative events.

Mobile Phone Language

The mobile phone is one of the most revolutionary tools or media of communication in recent times. The pervasive use of mobile phones in the society can in turn be the basis of new ways of using language in phone conversations. This part of the chapter focuses on the use of 'Mobile phone Cantonese' in Hong Kong, which refers to a set of distinctive forms of Cantonese expressions that are associated with communication through the mobile phone. It shares features of general telephone language but some are considered to be distinctive or have different degrees of occurrence.

A small-scale study was carried out in order to observe linguistic features in mobile phone communication. Some recurring expressions and their associated features are observed:

Request for information: Participants in a mobile phone conversation often ask for information on location, as shown in examples (1) and (2) in Figure 4, and duration as in example (3) in Figure 4. Another type of requests which mobile phone users may make from time to time is due to technical problems, presumably problems regarding the reception of the mobile phone, such as example (4) in Figure 4.

Action in progress: One of the obvious features within the contents of mobile phone Cantonese is that most of these expressions denote actions in progress, particularly about the 'mobility' of the speaker. These expressions are often marked by action verbs in progressive forms like daap3-gan2 ce1 ('traveling on a vehicle'), zau2-gan2 ('leaving'), heoi3-gan2 ('going'), as in examples (5), (6), and (7) in Figure 5.

Temporal - present progressive: There is a high tendency of using the progressive marker gan2 and the adverb ji4-gaa1 'now' to indicate the immediacy of the

Table 5: Examples of Emoticons Provided by the Respondents

EMOTICON	target expression	N
:)	happy / smiling	17
: (	Sad	10
^_^	Happy	8
-_-	not in good mood / unhappy	8
>.<	angry / disappointed / unhappy / embarrassed	7
= p	happy / joking / mischief / being naughty	7
^o^	happy / in good mood / OK / understood	6
: p	Happy	6
=)	Happy	6
>_<	annoying / angry / dissatisfied	5
:-)	Happy	5
= (	Unhappy	5
^v^	smiling / happy / that's ok	4
^.^	user is in good mood	3
@.@	idle / sleepy / feeling bored	3
>o<	angry	2
> <	annoying / angry / dissatisfied	2
^3^	Kiss	2
^ ^	Smiling	2
:✳	Dead	2
:-(	Sad	2
: D	Laughing	2
.	Confused	2
@_@	Bored	2
z_z	Sleepy	1
Y_Y	Crying	1
o_o	Bored	1
N_N	not in mood to talk to the person	1
> < '"	Confused	1
>_<:::	Crying	1
>(^v^)<	Happy	1
~_~"	Embarrassed	1
^.~	Blink	1
^^"	Reluctant	1
?_?	Confused / puzzled	1
: {}	Kiss	1

Figure 4. Examples (1)-(4)

(1)	你	喺	邊	呀					
	Nei5	*hai2*	*bin1*	aa3					
	2.SG	at	where	PART					
	'Where are you?'								
(2)	你	宜家	去	到	邊度	啊			
	nei5	*ji4gaa1*	*heoi3*	*dou3*	*bin1dou6*	*aa3*			
	2.SG	now	go	arrive	where	PART			
	'Where have you reached?'								
(3)	你	仲	要	幾	耐	至	嚟	到	啊
	Nei5	*zung6*	*jiu3*	*gei2*	*noi6*	*zi3*	*lai4*	*dou3*	*aa3*
	2.SG	still	need	how	long	to	come	arrive	PART
	'How much longer will it take for you to come?								
(4)	吓 ?	咩話?	喂?	講	大	聲	啲	啦	
	haa2	*me1waa2*	*wai2*	*gong2*	*daai6*	*seng1*	*di1*	*laa1*	
	what	what say	hello	speak	big	voice	more	PART	
	'What? What are you saying? Hello? Can you speak louder?'								

actions that are inferred in the conversation. This can be exemplified by examples (8), (9), (10), and (11) in Figure 6.

Locatives: There are high occurrences of words that denote locations, usually indicated by adverbials e.g. hai2 zim1-saa1-zeoi2 ('in Tsim Sha Tsui'), hai2 gong2 daai6 ('in HKU'). Speakers often give information such as their location in a particular street, or probably the location of the means of transportation they are using, as illustrated in (12), (13), and (14) in Figure 7.

Arrival Prediction: Very Often, a speaker will try to predict and indicate when or how long it will take for them to arrive at a certain destination from their current

Figure 5. Examples (5)-(7)

(5)	搭	𡁵	車	呀		
	daap5	*-gan2*	*ce1*	*aa3*		
	take	-PRG	car	PART		
	'(I'm) traveling in/on a [vehicle].'					
(6)	我	而 家	塞	𡁵	車	呀
	Ngo5	*ji4-gaa1*	*sak1*	*-gan2*	*ce1*	*aa3*
	1.SG	now	block	-PRG	car	PART
	'I'm in a traffic jam now.'					
(7)	我	而家	喺	港大	走𡁵	
	Ngo5	*ji4-gaa1*	*hai2*	*gong2-daai6*	*zau2-gan2*	
	1.SG	now	at	HKU	leave-PRG	
	'I'm now in HKU and I'm leaving.'					

Figure 6. Examples (8)-(11)

(8)	我	而家	去 緊	銅鑼灣	喇
	Ngo5	*ji4-gaa1*	*heoi3-gan2*	*tung4-lo4-waan1*	*laa3*
	1.SG	now	go-PRG	Causeway Bay	PART
	'I'm on my way to Causeway Bay now.'				

(9)	返 緊	嚟	喇
	Faan1-gan2	lei4	laa4
	return-PRG	come	PART
	'(I'm) on my way back (home or work)'.		

(10)	我	而家	嚟 緊	喇
	Ngo5	*ji4-gaa1*	*lei -gan3*	*laa3*
	1.SG	now	come-PRG	PART
	'I'm on my way'.			

(11)	而 家	返 緊	屋企	喇
	ji4-gaa1	*faan1-gan2*	*ngok1-kei2*	*laa3*
	now	go-PRG	home	PART
	'(I'm) on my way home now.'			

location or point in time. This type of mobile phone Cantonese is often marked by expressions like zung6-jiu3 ('still-need') or zung6-jau5 ('still-have'), plus an indication of time and space such as sap6 fan1 zung1 ('ten minutes') and loeng5 go3 zaam6 ('two stations'). In addition, zau6 ('almost') is often used as a marker of futurity in the context of this type of mobile phone Cantonese. These are shown in examples (15), (16), and (17) in Figure 8.

Figure 7. Examples (12)-(14)

(12)	我	喺	街	呀
	Ngo5	*hai2*	*gaai1*	*aa3*
	1.SG	at	street	PART
	'I'm in the street'. (I'm out.)[1]			

(13)	同	朋友	行緊	街…	喺	尖沙咀
	Tong4	*pang4-jau5*	*heng4-gan2*	*gaai1*	*hai2*	*zim1-saa1-zeoi2*
	With	friend	walk-PRG	street	in/at	Tsim Sha Tsui
	'I'm shopping with my friend in Tsim Sha Tsui.'					

(14)	我	而家	去	到	太子	㗎喇
	Ngo5	*ji4-gaa1*	*heoi3*	*dou3*	*taai3- zi2*	*gaa3 laa3*
	1.SG	now	go	arrive	Prince Edward	PART PART
	'I am now at Prince Edward MTR station.' (speaker still traveling on the train)					

Figure 8. Examples (15)-(17)

(15)	我	諗	我	仲	要	十	分鐘	先	嚟	到
	Ngo5	*nam2*	*ngo5*	*zung6*	*jiu3*	*sap6*	*fan1-zung1*	*sin1*	*lei4*	*dou*
	1.SG	think	1.SG	still	need	ten	minute	in order to	come	arri'

'I think it'll take ten more minutes to come.'

(16)	仲	有	兩個	站	就	到	㗎	喇
	zung5	*jau5*	*loeng5 go3*	*zaam6*	*zau6*	*dou3*	*gaa3*	*laa3*
	still	have	two CL	station	almost	arrive	PART	PART

'Only two stations ahead.'

(17)	我	就	到	㗎	喇
	Ngo5	*zau6*	*dou3*	*gaa3*	*laa3*
	1.SG	almost	arrive	PART	PART

'I'll be arriving soon'

Some of the above expressions may also be found in normal telephone talks. These expressions, however, tend to be more recurrent in mobile phone talks than in normal stationary phone calls. Therefore, when talking of new forms of language, we do not only deal with new inventions but also consider new ways of using existing forms.

From our observation, it is also apparent that there is usually a lack of speaker identification since both the addresser and recipient are understood in the context of a mobile phone talk. Since speakers are more identifiable in mobile phone communications, especially with the pervasive use of caller-display functions, opening expressions like examples (18) and (19) in Figure 9 are rare. In other words, the caller in a mobile phone conversation seldom verifies the intended recipient.

As mentioned earlier, most of the findings are based on observations in a limited period of time under certain fixed contexts, and may not be adequate to account for mobile phone language as a whole. However, this study can all the same contribute

Figure 9. Examples (18) and (19)

(18)	喂	唔該	[...]	呀
	wai2,	*m4-goil*	[...]	*aal*
	Hello	please	[recipient]	PART

'Hello, can I speak to (recipient), please?'

(19)	喂	[...]	喺唔喺	度	呀
	wai2	[...]	*hai2-m4-hai2*	*dou6*	*aa3*
	Hello	[recipient]	at-not-at	there	PART

'Hello, is (recipient) there?'

to the idea that new media of communication can actually change our ways of using language to a considerable extent.

NEW LITERACIES: DIGITAL LITERACY AND MULTILITERACIES

As previously mentioned, the meaning of literacy can be extended to various types of literacy such as vernacular literacy and information literacy. In this section, I argue that new ICTs do not only generate new forms and uses of language, but also new forms of literacy, or what is called new literacies. The concept of new literacies is different from that of New Literacy Studies. NLS is a new approach to the study of literacy, or what is called a social approach to literacy, whilst new literacies are the various types of literacy which are associated with the introduction and uses of new ICTs.

Digital literacy is one of the various forms of new literacies that have emerged under the impact of new communication tools. The concept is defined as 'the ability to understand and use information in multiple formats from a wide range of sources when it is presented via computers' (Gilster 1997:1). The concept is then extended to the integration of ICTs into pedagogy. It focuses on the changing relationship between literacy and schooling within the context of new communication technologies, and places literacy within the social and historical contexts that expand its potential to enrich teaching and learning in an information age (Tyner 1998). Together with digital literacy, other concepts like media literacy, visual literacy, electronic literacy (Warschauer 1998, 2006)...etc. are also new literacies in the information age.

Within literacy studies, the idea of multiliteracies has become more and more prominent with the rapid development of new ICT tools. Basically, like the various new literacies mentioned above, 'multiliteracies' is also a product of the technologization of our society. The concept aims to capture the multiplicity of communication skills involved in information processing. Technological change has created new forms of communication which often involve many forms of literacy that go beyond solely reading and writing, such as information literacy, computer literacy, visual literacy, and media literacy. As Tyner (1998:64) notes, [m]ultiliteracies related to communication and information, notably media literacy, computer literacy, visual literacy, information literacy, network literacy, and technological literacy, have stepped forward to define the changing, amorphous shape of communication needs for a society awash in electronic sounds, images, icons, and texts.

The new literacies presented above, namely digital literacy and multiliteracies, have become prevalent concepts in present-day society, especially in the domain of education, due to the increasing use of ICTs in pedagogy. The ever-changing technology, or rapid dynamism, involves both new inventions and new ways of using the existing invention. Their existence is constantly changing. A consequence of this, then, is the necessity of more training and re-training in order to keep pace with rapid developments in the ICT field.

CONCLUDING REMARKS

In this chapter, to provide a model for demonstrating our major theme of CMC technology engendering changes in language and literacy forms and uses, I have introduced the framework of TeLCU as a theory of the relationship between these new communication technologies and new forms of language and literacy. This causal relationship illustrates that new tools and media of communication generate changes in language as well as new ways of using language. I have then identified several aspects of new forms of language such as technobabble and mobile phone Cantonese. I have also discussed the idea of new literacies in which digital literacy has emerged under the influence of the pervasive use of new ICTs.

Are there any substantial implications of the TeLCU model of the relationship between technology and language for the theory of literacy studies? Within the field of literacy studies, there is a widespread shift away from autonomous approaches to a more context-sensitive approach to the analysis of literacy. I believe that the proposed model adds a new dimension to context-sensitive approaches to literacy, such as NLS. Nevertheless, context-sensitivity is in itself rather broad, and if we must really understand the import of the approach, we need to go beyond this broad specification and provide a detailed specification of what exactly constitutes context-sensitivity. TeLCU is obviously a technology-sensitive approach to the dynamics of language and literacy studies. I consider this technology-sensitive approach as a particular and more specific instantiation of context-sensitivity. The obvious implication here is that for an adequate study of the dynamics of language and literacy forms and uses in the 21st century, linguists and literacy researchers and practitioners need to pay a careful attention to the role of Information Communications Technology. A Technology-conditioned approach to Language Change and Use (TeLCU), as proposed in this chapter, provides the bearings towards such an orientation.

The implications I have discussed are mainly theoretical but there may in fact be some practical implications for the design of new technology or upgrades of existing technology, for the design of new forms of language, and for how one may prepare and train technology-literate users. To give brief illustrations, in designing keyboard layouts and shortcuts to typing particular chunks of words in each language, for instance, a computer or mobile phone keyboard designer may need to know the most frequent shortenings that computer and mobile phone users employ in the particular language. For example, in a study I did on mobile phone language that has been reported in this chapter and at various other places in the book some of the most frequent words that SMS users employ include U, Btw, CU, etc. The most efficient SMS system would be one that can provide shortcut keys and predictive texts involving the most frequently used combinations of letters or characters in the language. Some knowledge of the intricate relationship between new technology and language structure is therefore essential. There are many computer systems designers that follow what linguistic innovations take place among communities of users. For instance, new emoticons and other forms of visual language are constantly being invented in such communities. Systems upgrades involving existing computer and mobile phone systems must include ways to incorporate these forms of language and communication. Finally, my model can be used to teach and train technology-literate users about the important relationship between new technologies and new forms of language and communication. There are, for example, many dictionaries and wordlists both online and in print explaining the meaning of new emoticons. New users of chat systems, social network systems, and mobile phone texters can all make use of these resources to learn how new forms of language and communication are closely tied to the invention, creation, and upgrade of new computer-mediated communication systems.

Chapters 1 to 4 may be regarded as foundational chapters of this book. In the first three chapters I presented discussions about our major theme of the relationship between CMC technologies and how they impact natural language and their implications for education. In this chapter I built on the discussion by providing a model and a theoretical outline of the ICT – natural language relationship. TeLCU provides a technology-sensitive ecology for discussing evolving forms and uses of natural language. In chapter 5 and subsequent ones, I shall now illustrate this relationship by focusing on one technological gadget or medium at a time.

REFERENCES

Adams, A. (1996). Language awareness and information technology. *Curriculum and Teaching, 11*(2), 69–76.

Baron, N. S. (1984). Computer mediated communication as a force in language change. *Visible Language, XVIII*(2), 118–141.

Barry, J. A. (1991). *Technobabble.* Cambridge, Mass.: MIT Press.

Barton, D. (1994). *Literacy: An Introduction to the Ecology of Written Language.* Oxford UK and Cambridge USA: Blackwell.

Barton, D., & Hamilton, M. (2000). Literacy practices. In D. Barton, M. Hamilton, and R. Ivanic (Eds) *Situated Literacies: Reading and Writing in Context* (pp. 7-15). London and New York: Routledge.

Barton, D, M. Hamilton and R. Ivanic< (Eds). (2000). *Situated Literacies: Reading and writing in context.* London and New York: Routledge.

Baynham, M. (1993). Code switching and mode switching: community interpreters and mediators of literacy. In B. Street (Ed.), *Cross Cultural Approaches to Literacy* (pp. 294-314.). Cambridge, UK: CUP.

Blurton, C. (1999). *New directions of ICT-use in education.* UNESCO's World Communication and Information Report 1999. on-line available at http://www.unesco.org/education/educprog/lwf/d1/edict.pdf

Bodomo, A., & Lee, C. (2002). Changing forms of language and literacy: technobabble and mobile phone communication. *Literacy and Numeracy Studies: An International Journal in Education and Training of Adults, 12*(1), 23–44.

Bodomo, A. B. (2000) Lecture notes on Language and Literacy. *Computer literacy and language educational technology*. The University of Hong Kong. URL: http://ecourse.hku.hk:8900/public/LING2011

Bodomo, A. B., & Lee, C. (2001) Changing forms of language and literacy in the information age. Paper presented at the 8th International Literacy and Education Research Network Conference on Learning. Spetses, Greece.

Crystal, D. (1997). *The Cambridge Encyclopedia of Language* (2nd Ed.). Cambridge, UK: CUP.

Crystal, D. (2001). *Language and the Internet*. Cambridge, UK: CUP.

de Pourbaix, R. (2000). Emergent literacy practices in an electronic community. In D. Barton, M. Hamilton, and R. Ivanic (Eds.), *Situated Literacies: Reading and Writing in Context* (pp. 125-148). London and New York: Routledge.

Gee, J. P. (1996) [1990] *Sociolinguistics and Literacies* (2nd ed.). London: Taylor and Francis.

Gilster, P. (1997). *Digital literacy*. New York: John Wiley & Sons.

Goody, J., & Watt, I. (1968). The consequences of literacy. In J. Goody. Cambridge (Ed.), *Literacy in Traditional Societies* (pp. 27-68). CUP.

Halliday, M. A. K. (1985). *Spoken and Written Language*. Deakin University.

Heath, S. B. (1983). *Ways with words.* Cambridge: Cambridge University Press.

Kress, G. (1998). Visual and verbal mode of representation in electronically mediated communication: the potentials of new forms of text. In I. Synder (Ed.), *Page to Screen: Taking Literacy into the Electronic Era* (pp. 53-79). London and New York: Routledge.

Lakoff, R. T. (1982). Some of my favorite writers are literate: the mingling of oral and literate strategies in written communication. In D. Tannen. Norwood (Ed.), *Spoken and Written Language* (pp. 239-260). New Jersey: Ablex.

Lee, C. K. M. (2002a). *Chinese and English Computer-mediated Communication in the Context of New Literacy Studies.* Unpublished M.Phil. dissertation. Hong Kong: The University of Hong Kong.

Lee, C. K. M. (2002b). Literacy practices in computer-mediated communication in Hong Kong. *The Reading Matrix, Special Issue: Literacy and the Web.* Reading Matrix.com. Available online at: http://www.readingmatrix.com/articles/lee/article.pdf

Luke, A. (1997). Critical approaches to literacy. In V. Edwards and D. Corson (Eds.), *Encyclopedia of Language and Education, Volume 2: Literacy* (pp. 143-151) The Netherlands: Kluwer Academic Publishers.

Luke, C. (2000). Cyber-schooling and technological change: multiliteracies for new times. In B. Cope and M. Kalantzis (Eds.), *Multiliteracies: Literacy Learning and the Design of Social Futures* (pp. 69-91). Australia: Macmillan.

Ong, W. J. (1982). *Orality and Literacy: the Technologizing of the Word.* Routledge.

Scribner, S., & Cole, M. (1981). *The Psychology of Literacy.* Cambridge, MA: Harvard University Press.

Shortis, T. (2001). *The Language of ICT: Information and Communication Technology.* London and New York: Rouledge.

Street, B. (1984). *Literacy in Theory and Practice.* Cambridge: Cambridge University Press.

Tusting, K., Ivanicà, R., & Wilson, A. (2000). New Literacy Studies at the interchange. In D. Barton, M. Hamilton, and R. Ivanic (Eds.), *Situated Literacies: Reading and Writing in Context* (pp. 210-218). London and New York: Routledge.

Tyner, K. (1998). *Literacy in a Digital World: Teaching and Learning in the Age of Information*. London: Lawrence Erlbaum Associates.

Warschauer, M. (1998). Electronic literacies: Language, culture and power in online education. Mahwah, NJ: Lawrence Erlbaum Associates.

Warschauer, M. (2006). Laptops and literacy: learning in the wireless classroom: Teacher's College, Columbia University.

Werry, C. (1996). Linguistic and interactional features of Internet relay chat. In *Computer-Mediated Communication: Linguistic, Social and Cross-Cultural Perspectives* (pp. 47-64), Amsterdam: John Benjamin's. [1] The data and analyses contained in this chapter first appeared in the journal *Literacy and Numeracy Studies* as Bodomo and Lee (2002). I thank Carmen Lee, my former student and collaborator, for almost a decade of collaborative research on many aspects of Language, Literacy, and IT since 2000.

ENDNOTE

[1] The data and analyses contained in this chapter first appeared in the journal *Literacy and Numeracy Studies* as Bodomo and Lee (2002). I thank Carmen Lee, my former student and collaborator, for almost a decade of collaborative research on many aspects of Language, Literacy, and IT since 2000.

Chapter 5
Insights from an MSN Corpus

INTRODUCTION

In this chapter I look closely at one type of Computer-Mediated Communication (CMC), Microsoft Network (MSN) instant messaging. MSN instant communication is quite popular among the youth in most parts of the world. In Hong Kong, it has fast replaced I-seek-you (ICQ) and QQ (in mainland China) as one of the primary instant communication tools among the youth.

In this chapter I will be looking particularly at the linguistic features of this kind of communication. More specifically, Hong Kong being a trilingual and biliterate society, one would want to see how participants juggle these languages within the medium of the MSN instant messaging and what linguistic features are thus produced. As is usual with most chapters our observations are based on a case study. So first I

DOI: 10.4018/978-1-60566-868-0.ch005

present this study in the next section. A distinctive feature of this chapter from some others is that I present and describe a comprehensive corpus as an example of an actual CMC exchanges. This will give us insights into the actual communication sessions of young people who use MSN in Hong Kong.

CASE STUDY: LINGUISTIC FEATURES OF MSN COMMUNICATION[1]

Hong Kong is one of the most densely populated areas in the world, and it is also one of the most wired cities in the world. There is thus the potential for a very active environment of CMC activities. According to a survey done by Hong Kong Census and Statistics Department (2007), 90.4% of students in Hong Kong, including 81.1% of secondary students and 95.4% of tertiary students, use the internet service to communicate with other people. This case study (mainly based on Yee, 2008) will particularly look at the linguistic features produced by tertiary-level students in Hong Kong in the course of these text-based computer-mediated communication practices.

The case study is organized as follows. First I introduce this scenario of prevalent use of MSN and the nature of MSN itself. Next I look at what studies have been done in the area, especially in relation to Hong Kong. Following this I describe the methodology used to collect and analyze the data before presenting the linguistic features identified from the data with focus on the linguistic features specific to Hong Kong contexts.

What is MSN?

MSN (The Microsoft Network) is a free instant messaging service allowing Internet users to communicate via personalized text messages with friends, family members, and business colleagues in real time. The first version of MSN was released by Microsoft in 1995 for mainly home users whose computers were compatible with the Microsoft Windows 95 operating system. It was renamed Windows Live Messenger (WLM) in February 2006. In Hong Kong, the use of MSN has significantly decreased the use of ICQ because the functions and varieties of MSN outweigh those of ICQ. In the latest version of MSN Messenger 8.5, various functions like video conferences, audio conversations, online games and exchange of video clips, music files and photos are allowed through MSN. There are also over 10,000 emotion icons such as pictures, symbols, and cartoons for users to choose from. It seems that we gradually thus moved from a purely textbased CMC to include video-based

CMC even though the text-based aspects are still prevalent. Nevertheless, MSN is a largely text-based CMC as compared to YouTube and Skype, for instance.

MSN Messenger is a private synchronous and one-to-one communication system. A registered user can choose to chat with certain partners by selecting the person whose status (e.g. online, offline, away) is shown in the contact list. The contact list is automatically opened as a pop-up window when the users log into the MSN account, and then the dialogue box will be opened up for users to type and send the texts. MSN users can open two or more dialogue boxes and communicate with more than one person simultaneously. They can also see whether one is in the process of typing within the communication situation. On the other hand, MSN can also be seen as an asynchronous CMC as well because users may leave the messages to their conversation partners even though they may be offline and the messages will be sent to them once the targeted receivers log into their MSN accounts. Again, this underlies our earlier mention in chapter 1 of the fact that the distinction between synchronous and asynchronous CMC is getting blurred.

Literature Review

Much research is devoted to the explorations and comparisons of linguistic features of ICT tools such as ICQ, MSN, Email, blogs, online chat rooms and mobile phones (Baron, 2001, Wu, 2000 & Lee, 2001). However, much of the research on linguistic features of Instant messaging texts focus on the features of English written by second language learners (e.g. Cherny, 1999; Crystal, 2006, Herring, 2001) rather than on exploring linguistic features of non-Western languages (e.g. Sugimoto & Levin, 2000). Though there are studies on the linguistic features of Cantonese in different ICT tools, these are not many. In Lee (2002), linguistic features of MSN specific to Hong Kong were investigated in great detail in regards to email, which provide evidences of how language changes are influenced by the specific social and cultural contexts in Hong Kong based on the TeLCU model as developed in chapter 4 of this book. For example, Lee found out that romanization of Cantonese and code-mixing are very common in MSN texts and these two linguistic features are well-known phenomena in Hong Kong. Lee (2007a, 2007b) also discusses the language-related issues of MSN such as how people express themselves in spoken Cantonese with regards to the sociolinguistic contexts in Hong Kong. This case study focuses more on how Cantonese, the local variety of Chinese, is handled in this CMC context, though it also addresses issues such as the frequency of use, the reasons why students use MSN, and how students invent their own Romanization schemas to express themselves.

Table 1. Age

21	17
21-24	4

Table 2. Gender

Female	16
Male	5

Table 3. Native language

Cantonese	21
English	0

Methodology

Questionnaires were administered to 21 undergraduate students studying at the University of Hong Kong in the presence of the investigator. Their demographic data and background information are shown in Tables 1, 2, 3, 4.

Questionnaire

The purpose of administering a questionnaire was to examine the linguistic features of MSN among university students in Hong Kong. The questionnaire consisted of 13 questions and it was written in English only. The questionnaire was completed in the presence of the investigator. The first part of the questionnaire collects the

Table 4. Second language

Cantonese	0
English	21

Figure 1. Frequency of using MSN among university students

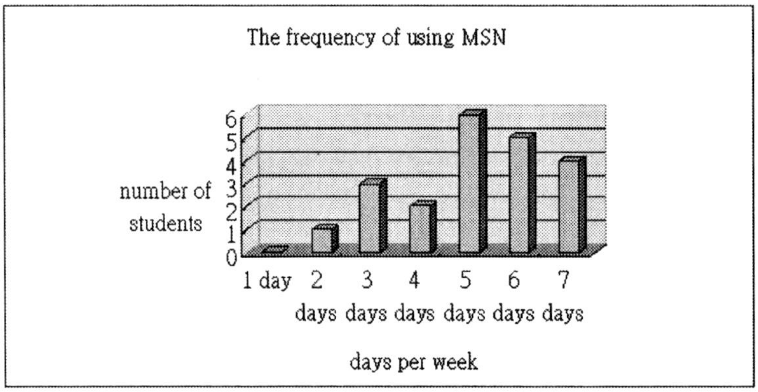

demographic data of subjects. The second part collects information about the daily practices and uses of MSN and in the third part, all subjects were asked to voluntarily provide the textual data saved in the MSN history. All data collected were processed in MS Word and data obtained were summarized and organized by charts and tables.

The Uses of MSN among Hong Kong University Students

Students were asked how frequently they use MSN per week. Only one student uses it less than two days per week. Five students use it three to four days a week and totally 15 students use it five days or above per week. These statistics show that about 71% Hong Kong Tertiary students use MSN frequently, at least five or more days per week (Figure 1).

Participants do a lot of activities such as sharing documents, music, and photos; playing online games; and chatting through MSN. Subjects in this study were asked about their main purposes of using MSN (Figure 2).

The results show that all students use MSN to have free chat with familiar people like friends. 15 out of 21 use MSN to talk about schoolwork like presentations and group projects. 10 students use MSN to exchange videos, music, pictures and photos. Only eight of them reported they play games through MSN. One subject reported that he uses MSN to find people if he lost their phone numbers and one subject reported that he uses MSN to convey messages when he does not want to talk or when he thinks there is something which is hard to say in face-to-face situations.

The results show that they use MSN to chat freely with friends and this may help explain why language forms and uses of MSN texts are always informal. It is because communications with friends are always conducted in a casual manner. Dur-

Figure 2. Different uses of MSN among university students

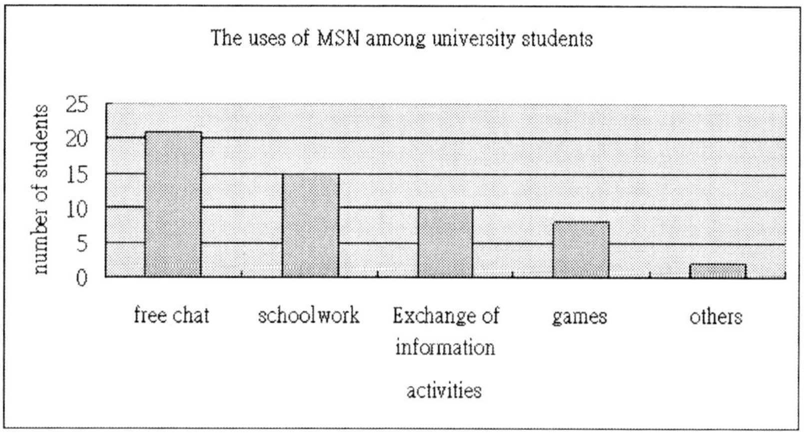

ing the instant messaging exchanges, people may be concerned more with whether the words can convey the meanings and the speed of message transmission rather than the accuracies of lexical choice or grammar.

Linguistic Features and Discussions of MSN Texts of Hong Kong University Students

Abbreviations

Shortening is a common feature which can be found in the MSN texts. Some of the shortenings such as ASAP used in MSN texts can also be found before the uses of MSN in various contexts like in letters and invitation cards. However, many of them are created and used pervasively only after the invention of the instant messaging tools like MSN. Examples of shortenings occurring in our MSN corpus are provided by respondents in Table 5.

U/u (you), tmr/tml (tomorrow), thx (thanks), 88 (Bye-bye), and bc/coz (because) were the top five commonly used shortenings in MSN. Though shortenings are always done by acronyms, a word formation process, which forms words by combining the initial letters in a phrase or sentences, shortenings in MSN are not limited to this formation method. As the popularity of MSN is growing, language forms of MSN texts are becoming richer through the development of new methods to form shortenings by numerous MSN users. All shortenings in MSN texts collected in this study are classified in accordance to different approaches of shortenings formation as shown in Table 6.

Table 5. Examples of MSN abbreviations and shortenings provided by respondents

U/u	you	17
tmr/tml	tomorrow	12
Thx, thx	thanks	11
88	bye bye	11
bc/coz	because	11
ic	I see	10
2	to/too	10
B4/b4	before	9
c u/cya	see you	9
LoL/lol	laughing out loud	8
of coz	of course	7
tonite/29	tonight	7
Ttul/ttul	talk to you later	7
Hv/hv	have	6
R/r	are	6
wt/wht	what	6
GTG	got to go	6
u're	you're	5
gd	good	5
wanna	want to	5
ppl	people	5
ur	your	5
wt's up	what's up	4
pls	please	4
da	the	4
gd nite	good night	2
reli/rili	really	2
y	why	2
slp	sleep	1
asap	as soon as possible	1
Dept	department	1
scl	school	1
alrite	alright	1
juz	just	1
5366	I want to have a talk ()	1
GG	elder brother ()	1

Table 5. (continued)

JJ	elder sister ()	1
MM	younger sister ()	1
PMP	flatter ()	1
PL	pretty ()	1
BT	abnormal ()	1
520	I love you ()	1

Table 6. Approaches to shortenings

Approaches	*Examples*
Letter homophone	U (you), c u/ cya (see you), R/r (are), da (the), ur (your), u're (you are)
Number homophone (or similar to pronunciation)	520 (我爱你), 2 (to/too), 88 (Bye bye)
Combination of letter and number homophone	2nit (tonight), b4 (before)
Combination of letter initial	Lol (laughing out loud), GTG (got to go), asap (as soon as possible)
Combination of letter initial and letter homophone	Ic (I see), Ttul (talk to you later)
Reduction of individual word	Bc (because), thx (thanks), hv (have), of coz (of course), juz (just), wanna (want to), wt/wht (what), ppl (people), gd (good), pls (please), slp (sleep), reli (really), dept (department), psy (psychology), prof (professor), scl (school), ref (reference)
Reduction of phrase	港大 (香港大学)

The approaches of shortening formations can mainly be divided into two types. The first one is the shortenings produced at the morphological level and the second one is produced at the phonological level as suggested by Crystal (2006). At the morphological level, an individual word is produced as short as possible with retention of its meanings. For example, the reduction of individual words like bc (because), thx (thanks), hv (have), of coz (of course), wt/wht (what), ppl (people), gd (good), and pls (please) are words with a few letters deleted but we still can recognize and infer the meanings from the remaining letters. At the phonological level, half of the approaches, including letter homophone, number homophone, and combination of letter and number homophone, form shortenings by replacing longer letters or words by shorter letters and words which are either the same as or near to the pronunciations of the longer words. It is interesting to notice that one student used shortenings originated from Mandarin Pinyin to make number homophones. 520 (wu3 er4 ling2) resembles the pronunciation of the expression 我爱你 (wo3 ai4 ni3) in terms of the initial consonant /w/ of 5 and 我; and the tone of 2 and 爱. There are

Figure 3. Proportion of students who have ever used romanization of Cantonese in MSN texts

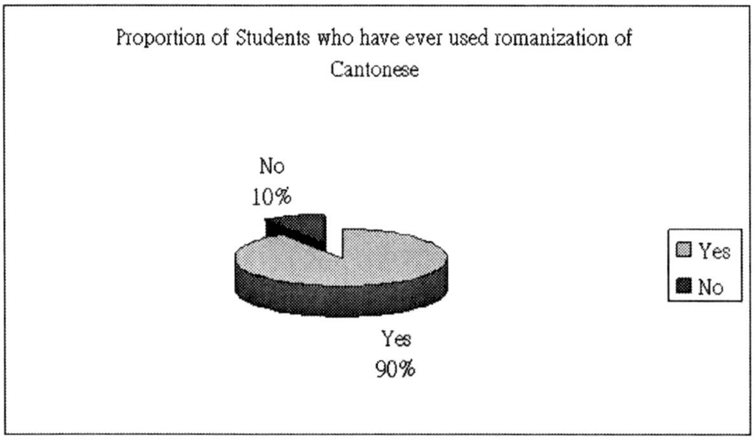

other examples of formation of shortenings by initialization of Pinyin in Mandarin like GG (ge1 ge1), JJ (jie3jie) and MM (mei4mei) provided by the subject.

Romanization of Cantonese Characters: Linguistic Features Specific to Cantonese

Romanization is a system to represent a word or sentence with the Roman alphabet. Over 90% of subjects reported that they have ever used romanization of Cantonese

Figure 4. Proportion of students who have invented romanization of Cantonese

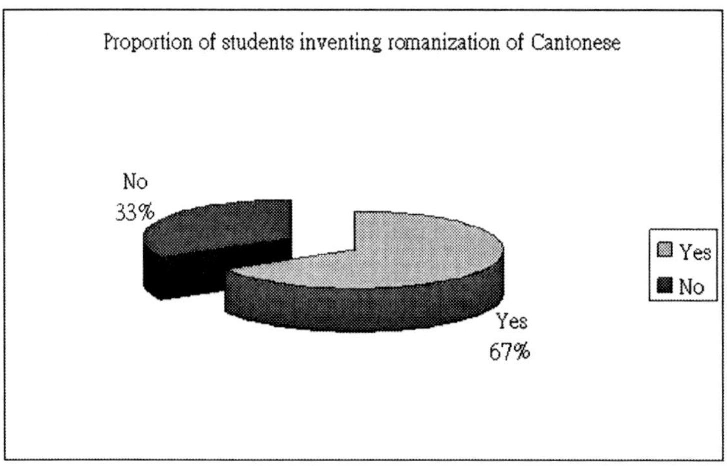

Figure 5. Reasons for using romanization of Cantonese to represent Cantonese character

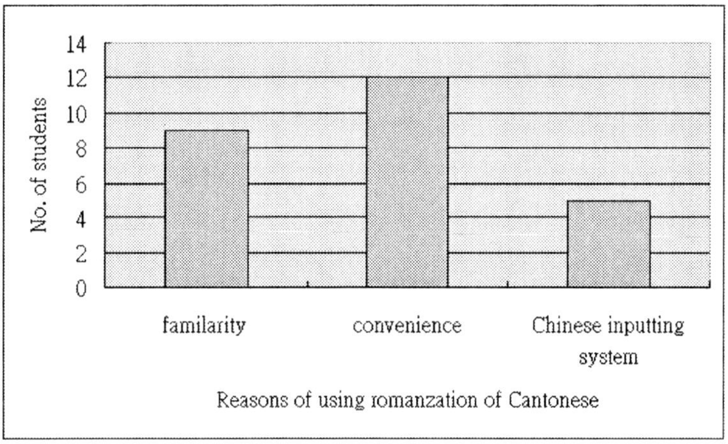

in expressions in this study (Figure 3). 67% of them reported that they have invented the romanization by themselves (Figures 3 and 4). They also mentioned three main reasons why they invent and represent Cantonese characters in romanization by themselves (Figure 5).

As shown by the statistics, 12 subjects find that it is more convenient and quicker to romanize and type the Cantonese words in English instead of alternating the inputting system and typing them by using the Chinese inputting system. Nine subjects think that the uses of romanized Cantonese can increase a sense of familiarity between the senders as they mainly use Cantonese in daily conversations. There are also five subjects reporting that either they have not installed the inputting system for typing spoken Cantonese or they have difficulty in using the Chinese inputting system, which may indicate that the limitations in the computer literacy of tertiary students affects how they choose and use language in MSN texts.

Romanization of Sentence Final Particle

There are three major kinds of romanization in Cantonese observed from the data. The first one is romanized sentence final particle. These sentence final particles do not have specific meanings in Cantonese. They are always used in the sentence final position to express emotions or emphasize the tone of expression. Here are examples from the data collected (Table 7).

Table 7. Romanized sentence final particle in Cantonese

Romanization	*Target*	*Examples*
la	啦(laa13)	hope they won't take gpa too seriously ***la***...^^
ga la	架啦 (gaa-3laa3)	Don't think too much la. slp early. I am sure u will be ok tml ***ga la***, if she wrote you a very good ref.... that helps a lot ***ga la***
ma	嘛 (maa3)	Two ref ***ma***, need to see another prof's ref, coz i think that prof also has good impression on me
ar	呀 (aa1)	Hi. Miss Chan ***ar***, hou ***ar***
wor	喔 (o1)	No ***wor***, mm g ***wor***...
leh	呢 (ne1)	ging ***leh***!

Romanization of Cantonese Interjections

The second one is romanization of Cantonese interjections, which are always used to express strong and sudden emotions such as "Hey" and "Wow" in English. Interjections always occur at the beginning of the sentence (Table 8).

Romanization of Other Cantonese Expressions

The third one is romanization of other Cantonese expressions (Table 9). MSN users produce common Cantonese words and phrases freely. However, there may be a problem of decoding the romanized Cantonese when people try to be creative and end up producing an inaccurate romanization. Students were asked in the questionnaire whether they have had difficulties in recognizing the romanized Cantonese in the MSN messages received from others and the reasons behind the difficulties. The results show that 48% students reported to have this problem (Figure 6) for the following reasons. First, the receivers think that the senders romanize Cantonese words with inappropriate letters which make it difficult to be read; second, it is hard to read a long sentence with every word typed in romanization; and third, it is difficult to understand the romanized words when they are not relevant to the topics which are being discussed.

Table 8. Romanized interjections in Cantonese

Romanization	*Target*	*Examples*
ai	哎 (aa1)	***ai ai***...yes ar.
o	哦 (ngo3)	***ooooo***
wei/wai	喂 (wai3)	***wei.*** Did u download the film yesterday? ***wai***, yesterday my prof told me secretly that in the mphil application,

Table 9. Romanized expressions in Cantonese

Romanization	***Target***	***Examples***
ngo	我(ngo3) I 1st Sig.	download? then ***ngo*** am not too interested in korean film
ho chi	好似 (hou2 ci3) seems	that was a good movie:P, fighting ***ho chi*** pretty cool ah, but the story is ... boring
sin **gong**	先(sin1) first 講(gong2) speak	I c u ***sin gong***.
hou	好(hou2) good	***hou*** ar
ging	勁(ging3) great/powerful	***ging*** leh!
mm **g** **hai**	唔(m4) don't 知(zi1) know 系(hai6) tie up	***mm g*** wor, ***mm hai*** very difficult terms XD. i just need to put the words together XD.
chi **jor** **sin**	痴(ci1) foolish 左(zo2) completed action marker 線(sin3) clue	u ***chi jor sin***......

Reduplication of English Words —The Impact of Cantonese on MSN Texts

Reduplication of words is a common linguistic feature of Cantonese such as 睇睇(tai2 tai2), 跳跳紮(tiu3 tiu3 zaat3) and 好好味 (hou2 hou2 mei6). Under the influences of reduplication of Cantonese words, Cantonese words are translated into English

Figure 6. Proportion of students who have ever had difficulties in recognizing romanized Cantonese

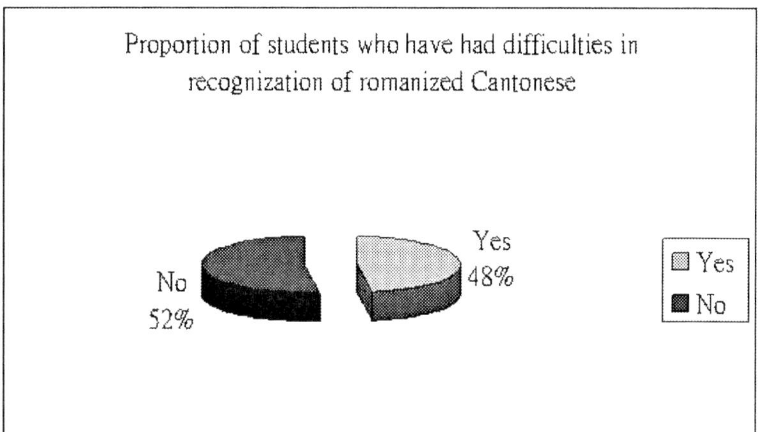

Table 10. Reduplication of English words

Reduplicated words	*Examples*	*Corresponding Cantonese reduplications*
ask	Hey, may you **ask ask** ur family members what is the meaning of the quality of life for them. I am doing health psy admission ticket > <	問問
try	i think i would **try try** to seek SKH in next week, & if rili cant find one, i would do wht u suggest	試試

words directly and they keep this Cantonese-specific linguistic feature in English. Table 10 contains a list of examples extracted from the data.

Morpheme to Morpheme Translation (Transliteration) — The Impact of Cantonese on MSN Texts

Other than romanization, it is found that people may use morpheme to morpheme translation, which refers to the expressions being translated word by word from Cantonese to English, in MSN texts (Table 11).

Character Representation of Spoken Cantonese — Cantonese-Specific Feature

People may use other characters to represent spoken Cantonese words which are homonyms due to the lack of supplementary character inputting system of spoken Cantonese. There are examples extracted from the data collected in Table 12.

GRAMMATICAL ERRORS

It is found that MSN texts are full of common grammatical errors such as omission of subjects, inappropriate tense forms, sentence structure errors and spelling mistakes. 17 out of 21 subjects reported that they do not pay attention to the grammars in MSN texts. Only four of them pay attention to their grammars (Figure 7).

Table 11. Morpheme-to-morpheme translation for expressions in Cantonese

Morpheme-to-morpheme translation	*Target*	*Examples*
Add oil 加 油 (gaa1 jau2)	Work hard	> ...Yes. **Add oil** together! > ok. **Add oil!**

Table 12. Character representation of spoken Cantonese

Character representation	*Target*	*Examples*
個	個	<A>weiwei~~besides 個news 之外, do u have the essay copy and pls send to me 呀 個: (go3) classifer piece
下	吓	…… <A> i finally finish all essay. <B>下!!!!!!!!!!???? Really? (What? Really?) 下: (haa2) below <A>Yup. <B>I haven't yet finished it. Wait, I ttul, ok?

Omissions of Subjects

The subjects of a complete sentence are always omitted by MSN users.

The omission of subjects may be due to three main reasons. First, people may be affected by the syntax of Cantonese which is a subject pro-drop language, meaning that the overt subject is not a necessity in forming a grammatical sentence. Second, the subject identity is always acknowledged before the conversations when one initiates the conversations by choosing the partner from the contact list and thus there is no need to repeat the singular pronoun like 'I' and 'you' repetitively in subsequent messages. Third, as shown by participants, MSN communication serves mainly as communication between friends and other people with whom participants are familiar, and thus no formal address of name is needed. Here are examples of subject omission extracted from the data (Excerpt 1). In the excerpt, subjects in the

Figure 7. Grammar awareness in using MSN

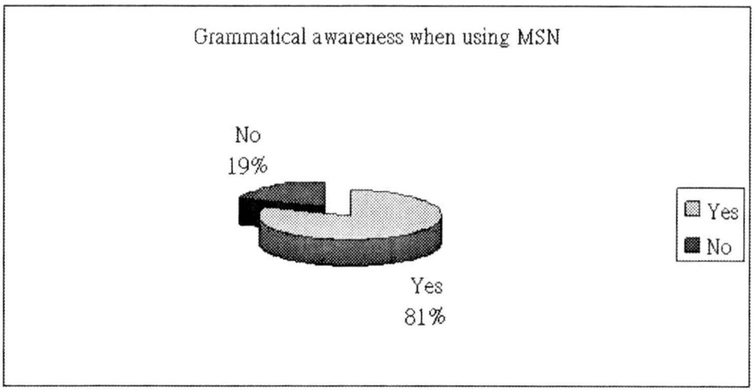

bracket are originally left out in the data provided by subjects.

Excerpt 1

*<A> i think i would try to seek SKH in next week, & if***[I]***rili cant find one, i would do wht u suggest*

<B> hmmm ic, **[you]***ask jacky's help la, he's very experienced in doing these*

Inappropriate Tense Uses of Verb

It is found that there is inappropriate use of tense forms in MSN texts (Excerpt 2). As students mostly use MSN to chat with friends, communication through MSN is always conducted in casual manner. People may be more concerned with whether the words can convey the meanings and the speed of message transmissions rather than the accuracies of grammars.

Excerpt 2 Expressions Target > i finally ***finish*** all essays. I ***have*** finally ***finished*** all essays.

> **do** *u see the webstie in my tag?* **Did** *you see the website in my tag?*

Spelling Errors

The reasons for the occurrences of spelling errors may be attributed to the low awareness of grammar of the senders who always communicate with friends through MSN and the high consciousness of speed during message transmissions. No embarrassment results from committing typing errors because spelling accuracy is not expected to occur in MSN texts by people familiar with another like friends during free chatting. For example, capitalization of the first letter of country names, capitalization of the first singular person pronoun and capitalization of the first letter at the beginning of a sentence in Standard English are not strictly followed when typing MSN texts. Examples of spelling errors in MSN texts extracted from the data are shown in Excerpt 4.

Excerpt 4 Expressions Target

> *do u see the* ***webstie*** *in my tag? website*

Table 13. Code-mixings in MSN texts

Code-mixings	*Examples*
Within sentence	>u can still try try to phone the 校務處 校 務 處 JP: haau6 mou6 cyu3 MPH: school affairs locale >Hey, Kitty, do u know lei, last time i met that prof in the lift at 港大 la, and she said hi to me, but then i didnt recognize her immediately -_- 港 大 JP: gong2 dai6 MPH: harbor big
Between sentences	> <A> I don't wanna talk to u. <B> What do you mean by 'never meet again' this morning? <A> 你知道我講咩。 你 知 道 我 講 咩 JP: nei5 zi1 dou3 ngo3 gong3 me1. MPH: You (Sig) know road I (Sig) speak what

> *<A> lol **i've** waited for a longggg time for a **kroean** film. I've, Korean*

> *<B> it is a **korean** film with "food" as the theme Korean*

> ***all* i** *remember is that **zhang ziyi** was kept being raped. lol All, I, Zhang Ziyi*

> ***i** want to ask whether the data has all been entered or not? I*

Code-Mixing — A Popular Linguistic Phenomenon in Hong Kong

Code-mixing obtains when one uses more than one language either within one sentence or between the sentences of a chat sessions. Code-mixing is a well-known linguistic phenomenon in Hong Kong (Li, 2002, Wu, 2000). Here are some examples extracted from the data (Table 13).

Mode-Mixing

Mode-mixing exists when codes and symbolic systems are combined in the messages such as the combination of letters and numbers (Bodomo & Lee, 2002). An example from the data is shown in Table 14.

Table 14. Mode-mixing in MSN texts

Mode-mixing	*Example*
Words/letters + numbers	> I go to have lesson la, c u 2nit. (I go to have lesson now. See you tonight.)

Emoticonymy

As people cannot always communicate face-to-face, no facial expressions and emotions can be seen. MSN users then invent some ways to express their facial expressions through combinations of letters and strokes in the keyboard. Rather than a specific linguistic feature of Hong Kong, emoticons are used worldwide and are developed into a quite standardized symbol system to express some common face expressions or emotions, for example, :) represents happy and :(represents unhappy. Table 15 lists emoticons found in the data collected.

PRESENTATION AND DESCRIPTION OF A COMPLETE MSN CORPUS

In this part of book I present a real example of a full CMC corpus to enable us get insights into the world of CMC. I first present background information showing when the data was collected and by who. I also explain certain abbreviations used and try to illustrate the glossing scheme used. Then I present 21 sessions of CMC conversations, mostly unedited, but making sure that identities of the participants are not revealed. There are very rare cases of actual CMC texts published and the one presented here would be a welcome addition to the CMC community of researchers because it goes a long way to provide insights into the structure of instant CMC conversations.

Table 15. Emoticons in MSN texts

Emoticons	*Target expressions*
:P	happy
^^	Smiling
-_-	not in good mood
:)	happy
><	annoying

Table 16. All MSN abbreviations and shortcuts found in the Corpus

abbreviations	Target	abbreviations	target
slp	sleep	u	you
juz	just	ur	your
reli/rili	really	u're	you're
gd	good	wt/wht	what
wanna	want to	bc/ coz	because
88	bye bye	b4	before
thx	thanks	2	to/too
wt's up	what's up	tonite/ 29	tonight
c u/cya	see you	hv	have
lol	laughing out loud	r	are
ttul	talk to you later	pls	please
da	the	tmr/ tml	tomorrow
asap	as soon as possible	ic	I see
psy	psychology	of coz	of course
prof	professor	dept	department
ppl	people	scl	school
gd nite	good night	GTG	got to go
ref	reference	5366	I want to have a talk
港大	香港大学	sori	sorry
y	Why		

Background Information of the Corpus

Type of text: Text messages of MSN (Microsoft Network)

Period of data collection: April 4, 2008-April 11, 2008
Data collected by: Yee Wai Man
Format specifications:
Font size: 12
Margin: 2.54 cm on top and down sides and 3.17 cm on left and right sides of an A4 sheet
Line Spacing: 15 pt (a blank line is left between two MSN messages)
Software: Microsoft Word 2000 / XP
No. of messages: 21 (each message is labeled with the format of <MSN 1> to <MSN 21>

Glossing Scheme

Data in this corpus are glossed as follows (see Table 16):

1. FT: Free Translation
2. NB: Additional notes/explanation on the message / expression
3. Complete or self-explanatory messages in English will not be glossed and are marked by *.
4. MPH: Morpheme-for-Morpheme glossing
5. JP: JyutPing Cantonese Romanization
6. All names in the data are replaced by letters <A> and <B> which represent the two persons who engage in the conversations.

All MSN Abbreviations and Shortcuts Found in the Corpus

Remarks

CL= Classifier
1st Sig = 1st person singular pronoun
2nd Sig = 2nd person singular pronoun

The Corpus

<MSN 1>

<A>Hey, Suki asked me to give u a call.
FT: Hey, PN asked me to give you a call.
NB: u = you
<B>She said it's troublesome if it's not old boys collect the data
FT: She said it would be a problem if it's not old boys who are going to collect the data.
<A>right? my fd asked the teacher who's said that la, he only said it's more complicated
FT: Right? My friend asked the teacher who said that. He only said it's more complicated.
NB: fd = friend, la is a sentence final particle and it is romanized by the author here.
<B>then...wht should i do??
FT: So, what should I do then?
NB: wht = what

<A>u can still try try to phone the 校務處

NB: u = you

JP: 校(haau6) 務(mou6) 處(cyu3)

MPH: school affairs locale

FT: You can still try to phone the School Office.

<B>OKOK~~thx!!

FT: Ok, thanks!

NB: thx = thanks!

<A>good luck la:)

FT: Good luck!

NB: la is a sentence final particle and it is romanized by the author here,:) = happy

<B>u2

FT: You too.

NB: u = you, 2 = too

<MSN 2>

<A>Johnson!!! i called you b4

FT: Johnson, I called you before!

NB: b4 = before

<B>i missed the call. Sorry. Wt's up?

FT: I missed your call. Sorry. What'up?

NB: wt's = what's up

<A>i want to ask whether the data has all been entered or not?

FT: I want to ask whether or not all the data have been entered?

<B>i remember there're still a few sets, but almost done la

FT: I remember there are still a few sets [of data], but they are almost done.

NB: la is a sentence final particle and it is romanized by the author here.

<A>coz i will come back tmr morning... but not free on afternoon... so see what i can help

FT: I will see what I can do to help because I will come back tomorrow morning but I will not be free in the afternoon.

NB: coz = because, tmr = tomorrow

<B>then probably u may come back later la coz no one in the lab in the morning, maybe i ask josephine sin, leave u a SMS later tonite

FT: Then it's probably better for you to come back later because no one is in the laboratory in the morning. Maybe I'll ask PN first and leave you an SMS message later tonight.

NB: u = you, coz = because, lab = laboratory, tonite, tonight, sin = first/before in Cantonese.

<A>ok.

<MSN 3>

<A>hi*

<B>Sori for not joining ur party yesterday. Did u have a gd time in the farewell party?

FT: Sorry for not joining your party yesterday. Did you have a good time in the farewell party?

NB: Sori = sorry, ur = your, u = you, gd = good

<A>Yes. How do u feel today?

FT: Yes. How do you feel today?

NB: u = you

<B>I feel better now. Thx!

FT: I feel better now. Thanks!

NB: Thx = thanks

<B>I go to bed sin, ttul.

FT: I am going to bed now. Talk to you later.

NB: sin, as mentioned earlier is a romanized Cantonese word '先' (sin1)

MPH: sin1 = first

<A>ok. don't disturb u la. gd nit.

FT: ok. [I] will not disturb you. Good night.

NB: gd = good, nit = night.

<MSN 4>

<A>hold on for a sec. I am sending a picture to you.

FT: Hold on for a second. I am sending a picture to you.

NB: sec = second

<B>Have u received it?

FT: Have you received it?

NB: u = you

<A>Yup.

FT: Yes

<B>Thx la!

FT: Thanks!

<A>u're welcome!

FT: You are welcome!

NB: u're = you' re

<MSN 5>

<A>weiwei

FT: hihi

NB: wei is used for calling someone in Cantonese and it is romanized by the author here.

<B>do u see the webstie in my tag?

FT: Did you see the website in my tag?

NB: u = you

<A>no ah lol

FT: I have not seen it.

NB: ah is a sentence final particle and it is romanized by the author here.

<B>too lazy XD

FT: You are too lazy.

NB: XD = laugh crazily

<A>it is a korean film with "food" as the theme

FT: It is a Korean film with "food" as the theme.

<B>lol i've waited for a longggg time for a kroean film. lol

FT: I have waited for a Korean film for a long time.

NB: lol = laughing out loud, longggg is used to emphasize the long time the author waited.

<A>download it and watch together?

FT: [I] will download the film, do you want to watch it together with me?

<B>download? then ngo am not too interested in korean film ^^

FT: [You] will download the film? Well, I am not too interested in Korean films.

NB: ngo is a romanized Cantonese word '我' (ngo3), ^^ = smiling

MPH: ngo3 = I (1st Sig)

<A>GTG□ou don't have lecture today?

FT: [I]'ve got to go. You don't have a lecture today?

<B>no. cya

FT: no. See you.

NB: c = see, ya = you

<MSN 6>

<A>wei. Did u download the film yesterday?

FT: Hey, did you download the film yesterday?

NB: wei is used for calling someone in Cantonese and it is romanized by the author here. u = you

<B>No. btw, what movie you watched in cinema recently?

FT: No. By the way, what movie did you watch at the cinema recently?

NB: btw = by the way

<A>十面埋伏

FT: A Chinese film!

<B>that was a good movie:P, fighting ho chi pretty cool ah, but the story is ... boring -_-

FT: That was a good movie. The fighting scene seems to be pretty good. But the story is boring.

NB::P = happy, ho chi (hou2 ci5) = seems to be, ah is a sentence final particle and it is romanized by the author here. -_- = boring

<A>all i remember is that zhang ziyi was kept being raped. lol

FT: All I remember is that Zhang Ziyi was being raped continuously.

NB: lol = laughing out loud

<B>HAHAHAHAHA. XDXDXD.

FT: Hahahahaha.

NB: XD = laugh crazily.

<A>XD. I go to have meeting la, c u 2nit.

FT: I am going to attend a meeting now. See you tonight.

NB: XD = laugh crazily, c = see, u = you, 2nit. = tonight

<A>88

FT: Bye bye.

NB: 88 = bye bye

<MSN 7>

<A>Baby. Where have you been?

<B>我剛睡醒

JP: ngo3 gong1 seoi6 sing2

MPH: I (1st Sig) just sleep wake up

FT: I have just woken up.

<A>我現在回學校,遲些打電話給你。520.

JP: ngo3 jin6 zoi6 wui4 hok6 haau6,

MPH: I (1st Sig) appears at return learn school.

JP: ci4 se1 daa2 din6 waa6 kap1 nei2

MPH: late little beat electricity speech give you(2nd Sig)

NB: 520 = I love you

FT: I am going to school now. I will call you later. I love you.

<B>ok. 520.

FT: ok. I love you.

NB: 520 = I love you

<MSN 8>

<A>wai, yesterday my prof told me secretly that in the mphil application, i am shortlisted XDD by the dept, and now the list sent back to the graduate scl to further consider... depends on the ref letters and my gpa and my prof said she has written me a very good ref.

FT: Hi, my professor told me in confidence yesterday that I am in the shortlist for MPhil application by the department and the list has been sent back to the graduate school for further consideration. [The admission result] depends on the reference letters and my GPA. My professor said she has written me a very good reference letter.

NB: wai = hey, prof = professor, dept = department, scl = school, ref = reference

<B>Congratulations!*

<A>hope it won't take gpa too seriously la...^^

FT: I hope [the graduate school] would not take my GPA into account for admission selection in a serious way.

^^ = smiling

<B>if she wrote you a very good ref.... that helps a lot ga la

FT: It helps you a lot to get admitted to the MPhil programme if your professor gave you a very good reference letter.

NB: ga and la are sentence final particles and they are romanized by the author here. ref = reference

<A>two ref ma, need to see another prof's ref, coz i think that prof also has good impression on me

FT: There are two reference letters. I need to see the other professor's reference letter but I think that professor also has a good impression about me.

NB: ma is a sentence final particle and it is romanized by the author here. ref = reference, prof = professor, prof's = professsor's, coz = because

<B>reli?

FT: Really?

NB: reli = really

<B>Are u in hall now?

FT: Are you in hall now?

NB: u = you

<A>Yes.*

<B>I now find u and c u sin gong.

FT: I will [go to] find you now. Let's talk when I see you.

NB: c = see, u = you, sin is a romanized Cantonese word '先' (sin1), gong is a romanized Cantonese word '講' (gong2)

MPH: sin1 = first, gong2 = speak

<A>hou ar

FT: Alright.

NB: hou is a romanized Cantonese word '好' (hou2), ar is a sentence final particle and it is romanized by the author here.

MPH: hou2 = good

<B> 8.

FT: Bye.

NB: 8 = bye

<MSN 9>

<A>I want to have a talk.*

<B>I don't wanna talk to u.

FT: I do not want to talk to you.

NB: u = you

<A>What do you mean by 'never meet again' this morning?*

<B>你知道我講咩。

JP: nei5 zi1 dou3 ngo3 gong3 me1.

MPH: You (2nd Sig) know arrive I (1st Sig) speck Interjection

FT: You know what I said.

<B>There is no need to talk again.

<MSN 10>

<A>Hey, Kitty, do u know lei, last time i met that prof in the lift at 港大 la, and she said hi to me, but then i didnt recognize her immediately -_-

FT: Hey, Kitty. Do you know something? Last time, I met that professor in the lift at the University of Hong Kong and she said hi to me. But I did not recognize her immediately.

NB: u = you, prof = professor, 港大(gong2 dai6) is the shortening of 香港大学 in Cantonese = The University of Hong Kong, la is a sentence final particle and it is romanized by the author here. -_- = not in good mood

<A>i only said.... "... hi"

FT: I only said "... hi".

<B>lol!!

NB: lol = laughing out loud

<B>poor face recognition XDDD

FT: [You have] poor face recognition ability.

NB: XDDD = laughing out loud

<A>i dont know y i cant recognize her immediately

FT: I do not know why I could not recognize her immediately.

NB: y = why

<B>"degenerate cognitive recognition ability".

FT: [You have] 'degenerating cognitive recognition ability'.

<A>How come so many terms? Where u learn them?

FT: How come you mention so many terms? Where did you learn them?

NB: u = you

<B>ging leh!

FT: [I am] great, right?

NB: ging is a romanized Cantonese word'勁' (ging3)

MPH: ging3 = great

<B>mm g wor, mm hai very difficult terms XD. i just need to put the words together XD.

NB: mm is a romanized Cantonese word '唔' (m4), g is'知' (zi1), hai is'係' (hai6)

MPH: m4 = don't, zi1 = know, hai6 = tie up

FT: I do not know. They are not very difficult terms. I just put the words together.

<A>u chi jor sin…

NB: chi is a romanized Cantonese word '痴' (ci1), jor is '左' (zo2), sin is'線' (sin3)

MPH: ci1 = foolish, zo2 = left, sin3 = clue

FT: You go crazy.

NB: u = you

<B>XDDD, wait wait…

FT: Wait.

NB: XDDD = laughing out loud

<MSN 11>

<A>Hihi.*

<B>Yup.*

<A>i finally finish all essay.

FT: I have finally finished all essays.

<B>下!!!!!!!!!????

JP: haa2:

MPH: below

FT: What?.

<MSN 12>

<A>hi, do u know 29 hand in

FT: Hi, do you know [we should hand in the essay by] 29.

NB: u = you

<A>I have done 500 out of 2500 words

<B>ooooo

FT: oh

NB: ooooo is a romanized Cantonese interjection '哦' (ngo3)

<A>But the introduction part is not well-written

FT: But the introduction part [of my essay] is not well-written.

<A>afraid

FT: [I am] afraid [of low marks].

<B>ai□es ar

FT: Yes.

NB: ai is a particle and is romanized by the author here. ar is a sentence final particle and it is romanized by the author here.

<A>how about your essay?

FT: How about your essay?

NB: u = you

<B>ok la, doing now, tmr is deadline, talk later la

FT: It's ok. [I am] doing it now. The deadline is tomorrow. Let's talk later.

tmr = tomorrow

<A>ok. Add oil!

FT: ok. Work hard!

NB: Add oil is a morpheme to morpheme translation. Add = 加 (gaa1), oil = 油 (jau4).

<B>u2.

FT: You too.

NB: u = you, 2 = too

<MSN 13>

<A>weiwei~~besides 個news 之外, do u have the essay copy and pls send to me 呀

NB: 個(CL) = piece. 呀(aa1) is a sentence final particle of Cantonese.

JP: 之(zi1) 外 (ngoi6)

MPH: 's outside

FT: Hey. Besides the news, do you have a copy of the essay? Please send it to me.

<B>no wor

FT: No.

NB: wor is a sentence final particle and it is romanized by the author here.

<A>u don't have another copy 咩?

FT: You do not have another copy?

NB: u = you, 咩 (me1) is a sentence final particle of Cantonese.

<B>No.*

<A>Ok. Thx!

<MSN 14>

<A>thx a lot.

FT: Thanks a lot.

NB: thx = thanks

<B>welcome.

FT: You are welcome!

<MSN 15>

<A>Hey, Tracy, I reli wanna die bc of too many essays!

FT: Hey, Tracy. I really want to die because of too many essays.

NB: reli = really, bc = because, wanna = want to

<B>me 2

FT: me too.

NB: 2 = too

<A>when u grad. what do u wanna do?

FT: What do you want to do when you graduate?

NB: u = you, grad. = graduate, wanna = want to

<B>whatever as long as I can find a job. I am reli worried about the future.

FT: Whatever. as long as I can find a job. I am really worried about the future.

NB: reli = really

<A>Don't think too much la. slp early. I am sure u will be ok tml ga la.

FT: Don't think too much. [If you] sleep early, I am sure you will be ok tomorrow.

NB: slp = sleep, u = you, tml = tomorrow, la and ga are sentence final particles and they are romanized by the author here.

<B>can't slp

FT: [I] cannot sleep.

NB: slp = sleep
<A>then drink some milk ar
FT: Then [you] may drink some milk
NB: ar is a sentence final particle and is romanized by the author here
<B>yup. Thx. don't disturb u la, 88
FT: Yup. Thanks. [I] don't disturb you. Bye bye.
NB: u = you, thx = thanks, 88 = Bye bye.
<A>88
FT: Bye bye.
NB: 88 = Bye bye.

<MSN 16>

<A>Hey, may you ask ask ur family members what is the meaning of the quality of life for them. I am doing health psy admission ticket > <

FT: Hey, may you ask your family members what the term "quality of life" means to them?. I am doing a health psychology admission questionnaire.

NB: ur = your, psy = psychology, > < = annoying
<B>my mum? ok.*
<A>thx!
FT: thanks!
NB: thx = thanks
<A>and also plz let me know her age, income and occupation
FT: And please also let me know her age, income and occupation
NB: plz = please

<B>my mom said it can be divided into material and psychological parts. for materials, living environment is importance (e.g. air quality, space, hygiene). These material things help emotional stability and psychological health. age= 59, income = $0 as she is a housewife.

FT: My mom said [the term "quality of life"] can be divided into material and psychological parts. For the material part, living environment is important (e.g. air quality, space, and hygiene). These material things may affect emotional stability and psychological health. [Her] age= 59, income = $0 as she is a housewife.

<A>ok thx
FT: Ok. Thanks.
NB: thx = thanks
<B>:)
NB::) = happy
<MSN 17>
<A>Kelly, have u sent me the file?

FT: Kelly, have you sent me the file?

NB: u = you

<B>I haven't.*

<A>Pls send it 2 me b4 11:00p.m. 29 or I will not give u a check.

FT: Please, send it to me before 11:00 p.m. to night or I will not send out a check.

NB: pls = please, 2 = to, b4 = before, 29 = tonight, u = you

<B>Sure. I will send u asap.

FT: Sure. I will send it you as soon as possible.

NB: u = you, asap = as soon as possible

<A>ok. ttul

FT: Ok. Talk to you later

NB: ttul = talk to you later

<MSN 18>

<A>Hi. Miss Chan ar.

FT: Hi. I am Miss Chan.

NB: ar is a sentence final particle and it is romanized by the author here.

<A>We have a private tutorial tml. 12:30p.m., ok?

FT: We have a private tutorial tomorrow. 12:30p.m., ok?

NB: tml = tomorrow

<B>ok.*

<A>ok.*

<MSN 19>

<A>wei. i m still thinking about my personal statement now

FT: I am still thinking about my personal statement now.

NB: m = am, wei is sentence final particle and is romanized by the author here.

<B>i should write it but i dun hv da mood, thinking of seeking help

FT: I should write it but I am not in the mood. [I am] thinking of seeking help.

NB: dun = don't, hv = have, da = the

<A>i know that u might not want to stand in front of the scl and ask ppl to fill in the forms for u, but wt if in the end u really cannot find a scl??

FT: I know that you might not want to stand in front of the School and ask people to fill in the forms for you. But what if in the end you really cannot find a School?

NB: u = you, scl = school, ppl = people, wt = what

<B>i think i would try to seek SKH in next week, & if rili cant find one, i would do wht u suggest

FT: I think I would try to contact SKU next week. If [I] really can't find one, I would do what you suggest.

NB: rili = really, wht = what, u = you

<A>hmmm ic, ask jacky's help la, he's very experienced in doing these

FT: I see. [You may] ask Jacky for help. He is very experienced in these things.

NB: ic = I see, la is a sentence final particle and it is romanized by the author here.

<B>okoki will ga la~~

FT: Ok, ok. I will.

NB: ga and la are sentence final particles and they are romanized by the author here.

<B>Raymond??

FT: PN?

<A>Yes?

<B>If u need, I will offer help.

FT: If you need, I will offer help.

NB: u = you

<A>thx. I will if u need as well.

FT: Thanks. I will [offer help] if you need it.

NB: thx = Thanks, u = you

<A>juz try ur best and do it within these few days

FT: Just try your best and do it within these few days.

NB: juz = just, ur = your

<B>Yes. Add oil together!

FT: Yes. Let us work hard together!

NB: Add oil is a morpheme to morpheme translation. Add = 加 (gaa1), oil = 油 (jau4).

<MSN 20>

<A>hello*

<B>Tmr, will you come? They will farewell me at campus.

FT: Will you come tomorrow? They will hold a farewell party for me on campus.

NB: Tmr = tomorrow

<A>well.I will try my best to come coz I'm not well today.

FT: Well, I will try my best to come because I'm not feeling well today.

NB: coz = because
<B>Ok. Hope to c u tmr. 88
FT: Ok. I hope to see you tomorrow. Bye bye!
NB: c = see, u = you, tmr = tomorrow, 88 = bye bye

<MSN 21>

<A>Hi, Victor, long time no c. How r u these days?
FT: Hi, Victor. Long time no see. How have you been of late?
NB: c = see, r = are, u = you
<B>ok la, u lei?
FT: Ok. [How about] you?
NB: u = you, la and lei are sentence final particles and are romanized by the CMC user here.
<A>ok la.
FT: [I am] ok.
<B>What are you busy with these days?*
<A>essays, quizzes, exams, assignment, etc…
FT: [I am busy with] essays, quizzes, exams, assignments, etc.
<B>wow.so many works to do. I juz wondered whether to call u out lei.
FT: Wow! [You have] so much work to do. I have been wondering whether to call you or not.
NB: juz = just, u = you
<A>of coz, let's hang out together after all these works
FT: Of course! Let's hang out when all this is over.
NB: of coz = of course
<B>ok, u call me after you finish all works la. My phone number is not changed ar.
FT: Ok. You may call me after all this. My phone number is not changed.
NB: u = you, la and ar are sentence final particles and are romanized by the author here.
<A> ok. I will.*

As can be seen from the above, these 21 pieces of MSN conversations serve as a window to the world of online chat CMC among young people in Hong Kong and elsewhere. They provide us with interesting and innovative ways in which these CMC users make use of language. The style is quite distinctive. It is informal, it is free-flowing, and it is devoid of any stringent comformity to writing rules.

DISCUSSIONS AND CONCLUSION

In this chapter, I have looked closely at the linguistic features of one type of text-based CMC, the MicroSoft Network (MSN) chat system. MSN is currently very popular among the youth the world over. But I have concentrated on the way it is used in Hong Kong through a case study of a group of Hong Kong University students. Evidently, we need to do more of such studies in other parts of the world. In this way we can cross-compare these different situations.

The most innovative linguistic phenomena involve the creativity displayed by participants in trying to invent romanized alphabets to express Cantonese in the contexts of their mainly English-based alphabetic MSN system (though some MSN versions now allow direct Chinese character inputting).

One interesting finding from this study is that Hong Kong MSN users are beginning to adopt homophones and initializations of Mandarin Pinyin to represent Cantonese characters in MSN texts though this language practice is still not common among Hong Kong MSN users. However, this finding may be indicative of the extent of the influences of how exchange of information through media and closer communications between Mainland China and Hong Kong people in these 12 or more years affect language use in Hong Kong. The most typical example is the noun phrase '粉絲' (fen3 si1). This is a homonym of the English word 'fans', originally from Mainland China. It is now frequently used in Hong Kong newspapers[2] and it is getting more and more noticeable in the spoken language of people in Hong Kong. After 1997, Hongkongers began to have closer relationships with Mainlanders for economic and political reasons and Mandarin proficiency among the young generation is increasing through its promotion by the Hong Kong government, which has included Mandarin in the regular curriculum in primary and secondary schools.

As a summary, MSN text practices in Hong Kong include shortenings, romanization of Cantonese words and phrases, morpheme to morpheme translation from English to Cantonese, reduplication of English words, character representation of Cantonese words, grammatical errors (omissions of subject, verb form errors, sentence structure errors and spelling errors), code-mixing, mode-mixing and emoticonymy. Through the investigation of how MSN users try to romanize Cantonese, and the problems and issues associated with doing this, I am touching on one of the fundamental themes of the book – how the introduction of new ICT tools affects or even changes, not only the forms and structures of our languages but also how we use these languages. In the next chapter, I continue investigating this theme with another kind of ICT tool, the mobile phone, which is widely discussed in the book, both in terms of mobile phone voice communication and mobile phone texting.

REFERENCES

Baron, N. S. (2001). *Alphabet to Email: How Written Language Evolved and Where It's Heading.* London and New York: Routledge.

Bodomo, A., & Lee, C. (2002). Changing forms of language and literacy: technobabble and mobile phone communication. *Literacy and Numeracy Studies: An International Journal in Education and Training of Adults, 12*(1), 23–44.

Cherny, L. (1999). *Conversation and community: Chat in a virtual world.* California: CSLI Publications.

Crystal, D. (2006). *Language and the Internet*. Cambridge: CUP.

Hamilton (Eds.), *The handbook of discourse analysis* (pp. 612-634). Oxford: Blackwell Publishers.

Herring, S. (2001). Computer-Mediated Discourse. In D. Schiffrin, D. Tannen, & H.

Hong Kong Census and Statistics Department. (2007). *Thematic Household Survey: Information Technology Usage and Penetration.* Retrieved April 20, 2008, from www.censtatd.gov.hk/products_and_services/products/publications/index_tc.jsp

Lee, K. M. (2001, December). *On context-sensitivity and the language.* Paper presented at Annual Research Forum, Hong Kong Baptist University, Hong Kong.

Lee, K. M. (2002). Literacy practices in computer-mediated communication in Hong Kong. Reading Matrix. 2(2). Retrieved April 20, 2008, from site: http://www.readingmatrix.com/articles/lee/article.pdf.

Lee, K. M. (2007a). Linguistic features of email and ICQ instant messaging in Hong Kong. In B. Danet & S. C. Herring (Eds.), *The multilingual Internet: Language, culture and communication online* (pp. 184–208). New York: Oxford University Press.

Lee, K. M. (2007b). Text-making practices beyond the classroom context: private instant messaging in Hong Kong. *Computers and Composition, 24*(3), 285–301. doi:doi:10.1016/j.compcom.2007.05.004

Li, C. S. (2002). Cantonese-English code-switching research in Hong Kong: A survey of recent research. In K. Bolton (Ed.), *Hong Kong English: Autonomy and creativity* (pp. 79–100). Hong Kong: Hong Kong University Press.

Sugimoto, T., & Levin, J. A. (2000). Multiple literacies and multimedia: A comparison of Japanese and American uses of the Internet. In E. H. Gai & L. S. Cynthia (Eds.), *Global literacies and the world wide web* (pp. 133–153). London: Routledge.

Wu, D. Y. (2000). Orality in Hong Kong print media. In D. Li, A. Lin, & T. Tsang (Eds.), *Language and Education in Postcolonial Hong Kong*. Hong Kong: Linguistic Society of Hong Kong.

Yee Wai Man Jeff. (2008). Linguistic Features of MSN in specific Hong Kong Contexts. Term Paper for the course Language and Information Technology. HKU

ENDNOTES

1 I thank Jeff Yee Wai Man for collecting the data for this section of the book as part of a term paper (Yee 2008) for my course in Language and Information Technology.

2 from Ming Pao (28/03/2008, 11/04/2008, 16/04/2008, 18/04/2008) and from Sing Tao (22/03/2008, 23/04/2008). Retrieved April 23 from http://hk.search.yahoo.com/search/news?p=%E7%B2%89%E7%B5%B2&s=-ut&tx=d&u=B&h=N&n=15&ei=utf8

Chapter 6
Insights from Mobile Phone Voice Communication

INTRODUCTION

Mobile phone communication as an instance of computer-mediated communication (CMC) has not received as much attention as email and others. Indeed, it may be argued that it is not a typical instance of CMC as some people do not consider the mobile phone a computer and also because there is hardly any Internet Protocol (IP) address involved in sending SMS, one might claim. However, as has been mentioned by writers like Prensky (2004) the mobile phone has reached such sophistication that it is indeed more powerful than some computers in terms of processing power, and indeed internet facilities are easily incorporated into mobile phones. While many have studied email and its effects on writing and the school curriculum, among other issues (e.g. Abdullah 1998, 2003, Akers-Jordan 2002, Baron 2001, Crystal

DOI: 10.4018/978-1-60566-868-0.ch006

2001, Healey 2007, etc), little has been done on the mobile phone until now (but see Bodomo 2003, 2007, Bodomo and Lee 2002, Thurlow 2003 and Prensky 2004). This chapter and the next focus particularly on the use of mobile phones through case studies based on reports of completed research projects. There are, at least, two aspects of mobile phone communication, voice mobile phone communication and text messaging. Chapter 6 focuses mainly on voice mobile phone communication aspects while chapter 7 dwells on mobile phone texting, especially its grammatical aspects.

CASE STUDY

In this chapter, I shall mainly report on a study of mobile phone voice communication by a group of people in Hong Kong, as a way of giving us an insight into the way the mobile phone, which has become one of the most important CMC tools of our time, is being used. I shall draw in background information, data, questionnaire, analysis, and activities from a funded research project which was indeed one of the first projects in Hong Kong, if not worldwide, dedicated to understanding the way language is used through the medium of mobile phone communication.[1]

Background Information

The mobile phone is one of the most revolutionary tools or media of communication in recent times. The pervasive use of mobile phones in society may be the basis of new ways of using language in phone conversations. Mobile phone usage in Hong Kong began in the late 1980s. Since the mid-1990s, the penetration rate has grown enormously as more service providers joined the market, thus leading to lower service charges[2]. For instance, the penetration rate has increased from 20% in 1997 to 80% in 2001. In addition, a report in as early as March 2002 even shows that approximately 98% of people aged 20 to 29 in Hong Kong owned a mobile phone[3]. In Asia, Hong Kong has the highest levels of personal penetration of mobile telephones and has maintained that position over years (ACNielsen, 2001)

In this case study of mobile phone voice communication, I focus on the use of 'Mobile phone Cantonese' in Hong Kong, which refers to a set of distinctive forms of Cantonese expressions that are associated with communication through the mobile phone. It shares features of general telephone language, but some are distinctive or have different degrees of occurrence.

As part of the research project on mobile phones mentioned above, an informal study was carried out among students in a Language and Literacy class taught by

Table 1. No. of years of MP ownership

No. of years	No. of Respondents
Less than 1 yr	0
1-2 yrs	1
2-3 yrs	4
3-4 yrs	3
4-5 yrs	3
Over 5 yrs	6

the author in order to observe the structure of mobile phone Cantonese, and most common used phrases.

This preliminary survey further obtained initial findings on some linguistic features of mobile phone communication (MPC henceforth), and some general habits and practices of MPC in Hong Kong. The survey took place on February 28, 2002. In this survey, questionnaires were administered to a class of 17 university students. The following outlines the findings of the case study.

Findings

No. of questionnaire distributed: 17
No. of questionnaire collected: 17

The Subjects

Occupation and Educational Background: 16 were undergraduate students, one was postgraduate student
Gender: Four male, 13 Female
Age: All of the subjects are between 20 to 29 years old

Questions 1-7 were mainly concerned with some general habits of mobile phone usage. Question 1 asked the respondents how long they had been using a mobile phone. All of them had been using a mobile phone for at least one year. The majority had five or more years of mobile phone ownership (see Table 1 and Figure 1).

Figure 1. How Many Years You Have Been Using a Mobile Phone?

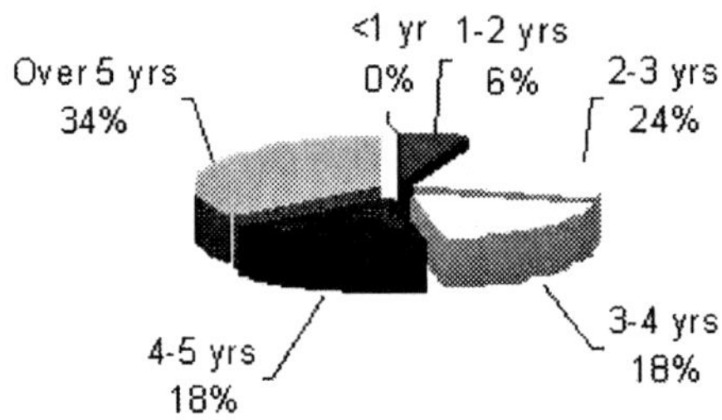

Frequency of Mobile Phone Usage

Question 2 was meant to investigate the frequency of mobile phone usage. It was found that 88% of the respondents used their mobile phone more than once a day. The results were set out in the Table 2 and Figure 2.

When and Where a Mobile Phone is Used Most Often

To help obtain representative data, we also wanted to know when and where people use their mobile phone the most. Therefore, questions 3 and 7 were meant to investigate these. Results showed that over 50% of the respondents used their mobile phone after office hours, including holidays, weekends, and evenings. On the other hand, school, public transportation, and street, were the three main places where MPCs usually took place. More detailed results could be seen in Tables 3 and 4, and Figures 3 and 4:

Table 2. Frequency of MP Usage

Frequency	No. of Respondents
Less than once a week	0
Once a week	0
2-5 times a week	1
Once a day	1
More than once a day	15

Table 3. Period of Time When a Mobile Phone is Used

Period of time	No. of Respondents
Before work / morning	0
During office hrs / afternoon	7
After office hrs / weekend/holidays/evenings	10

Figure 2. How Often Do You Use Your Mobile Phone?

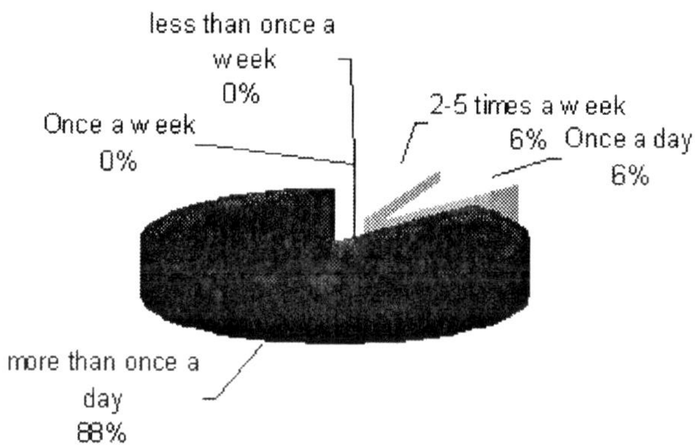

Purposes of the Mobile Phone

Questions 4a and 4b were concerned with the major purposes of the mobile phone. Respondents were first asked to choose from a list of 10 uses to which the mobile phone is put, in which they were allowed to choose more than one. It was found that respondents mainly used their mobile phone for informing people about their

Table 4. Places where MPC frequently takes place

Place	No. of Respondents
Home	2
Street	9
Public transportation	6
Restaurant	2
Office	0
School	8
Other	0

Figure 3. Which Period of Time in the Day Do You Use Mobile Phone the Most?

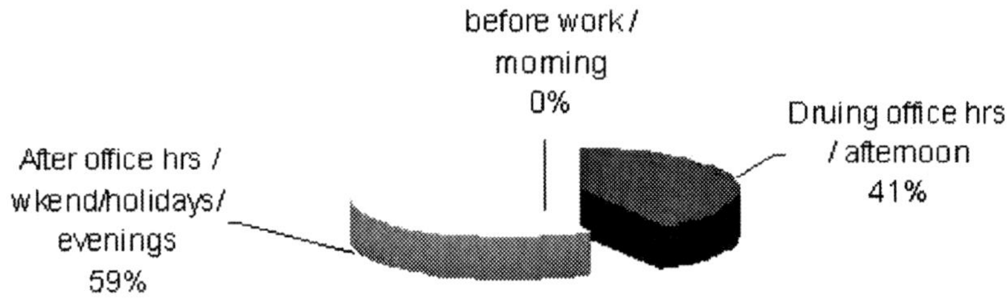

locations, making appointments (e.g. to arrange where and when to meet somebody), and chatting with people (see Table 5 and Figure 5).

In 4b, respondents were also requested to prioritize what they had chosen in 4a. Of the 16 respondents, six claimed that 'informing people about your location' was the most common purpose of their mobile phone. Another six of them said making appointments with people was the main purpose. 'Personal chats' turned out to be another important purpose, for which three respondents ranked it as the most important purpose, while five of them put it as the second most crucial purpose.

Time Spent on Phone Calls

In questions 5 and 6, respondents were asked about the amount of time they normally spend on a mobile phone call and a normal stationary phone call (SPC). The aim

Figure 4. Where Do You Use Mobile Phone the Most?

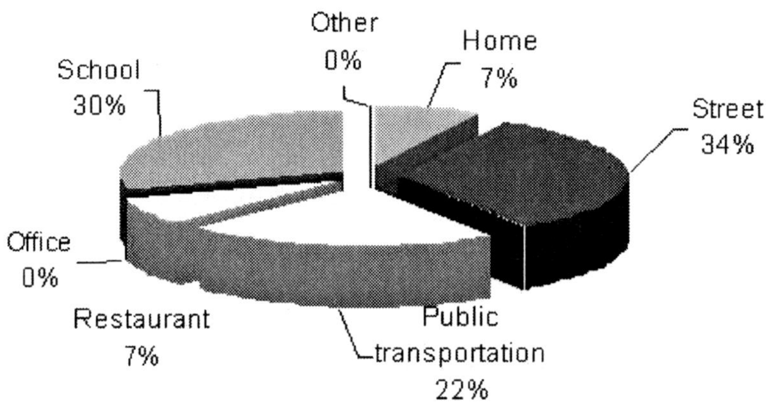

Table 5. Purposes of the Mobile Phone

Purpose	No. of Respondents
(a) Business affairs	7
(b) Informing locations	15
(c) Personal chats	14
(d) SMS	8
(e) Emergencies	9
(f) Information retrieval (news/weather)	3
(g) Entertainment	1
(h) Appointment	15
(i) Internet (WAP)	0
(j) Other*	2

* Two respondents said 'other', which represents the use of 'alarm clock' and 'checking voicemail'.

of these two questions was to compare whether the nature of these two technologies would have a certain kind of influence on how much time one would spend on each call. Not surprisingly, most respondents spent a shorter time on MPC than on SPC. In general, respondents would not spend more than five minutes on a mobile phone. Only a few spent up to 15 minutes. However, it is more common for people to spend over five minutes in an SPC. The results of are summarized in Tables 6 and 7, and Figures 6 and 7.

Figure 5. What Do You Use a Mobile Phone for?

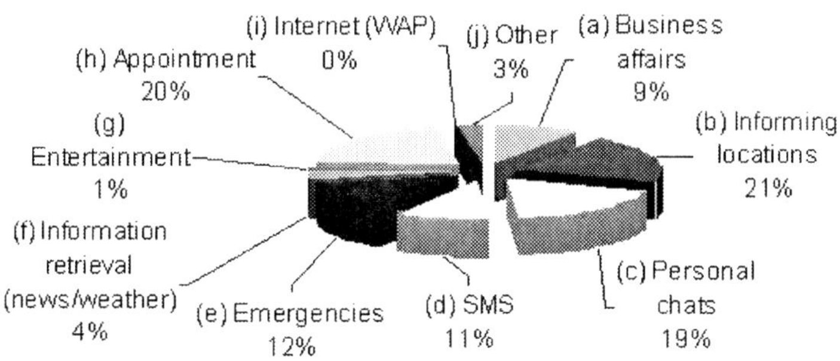

Table 6. Time spent on each mobile phone call

Time	No. of respondents
less than 5 mins	12
5-15 mins	5
15-30 mins	0
30 mins - 1 hr	0
more than 1 hr	0

Table 7. Time spent on each Stationary Phone Call

Time	No. of respondents
less than 5 mins	4
5-15 mins	8
15-30 mins	4
30 mins - 1 hr	1
more than 1 hr	0

Main Reasons for Preferring MPC to SPC

The survey was also interested in why people would rather communicate via the mobile phone than via a normal telephone. The majority of respondents thought that the 'mobility' of the mobile phone, i.e. it can be used anywhere, was the key concern. The second major reason is closely related to this, i.e. the mobile phone can be used anytime. The results are shown in Figure 8.

Figure 6. How Much Time Do You Normally Spend on Each Mobile Phone Call?

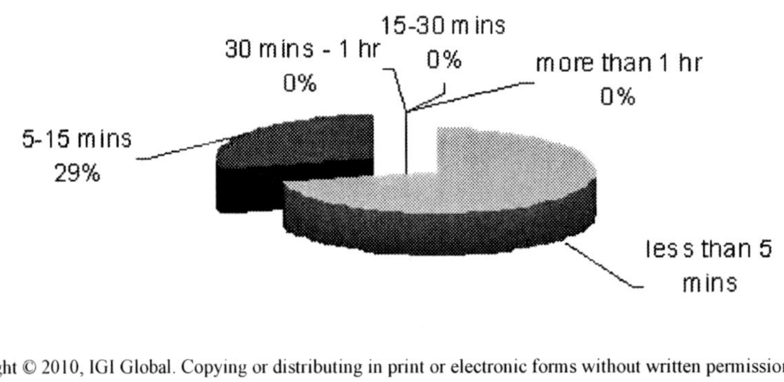

Figure 7. How Much Time Do You Normally Spend on Each Stationary Phone Call?

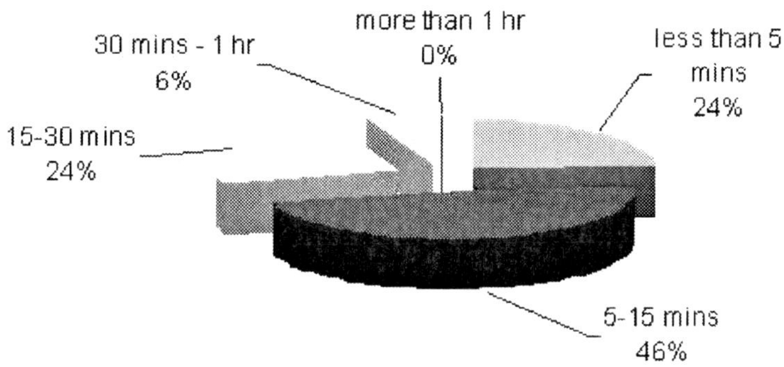

Communicative Functions Available on Mobile Phone

Respondents were asked what other communicative functions (or the so-called value-added services) they had subscribed to. The results are summarized in Figure 9.

Besides speech data, this research is also concerned with text data of SMS. Therefore, for those who claimed that SMS was available on their mobile phone, they were also asked how frequently they actually used SMS. Interestingly, even though SMS is available on most of the respondents' mobile phone, they hardly use it as a communication tool at the time of this research. As would be seen in the

Figure 8. Reasons for Using Mobile Phone to Normal Telephone

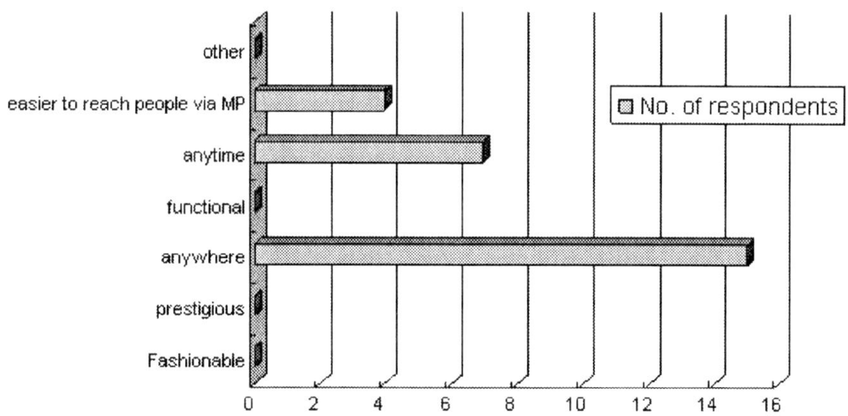

Figure 9. Communicative Functions Available on Mobile Phone

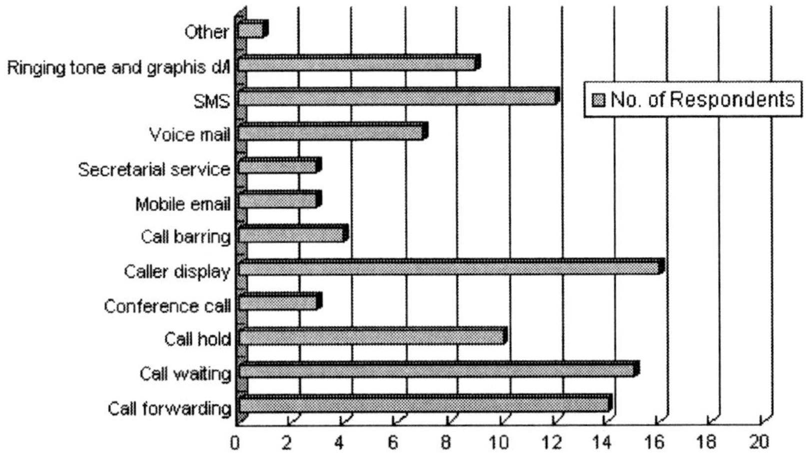

next chapter the SMS has now become a major source of CMC. The majority said they use SMS less than once a week. Only 8% of respondents claimed that they always use it (i.e. more than 10 times a week). The overall results are set out in Table 8 and Figure 10.

Judgment of Frequently Occurring Expressions in MPC and SPC

A crucial part of the survey was to find out if participants could recognize distinctive features of MPC. Respondents were given a list of 22 Cantonese sentences or expressions, some of which were identified in an earlier informal study. Their task was to identify, from the speaker's perspective, whether each of these sentences is (i) mostly said in MPC, (ii) mostly said in SPC, (iii) seldom said in MPC, (iv) seldom said in SPC, or (v) common in both MPC and SPC.

Table 8. Frequency of using SMS

Frequency	No. of Respondents
Seldom (less than once/ week)	8
Quite Often (5-10 times/week)	4
Always (more than 10 times/week)	2
Never	0

Figure 10. Frequency of Using SMS

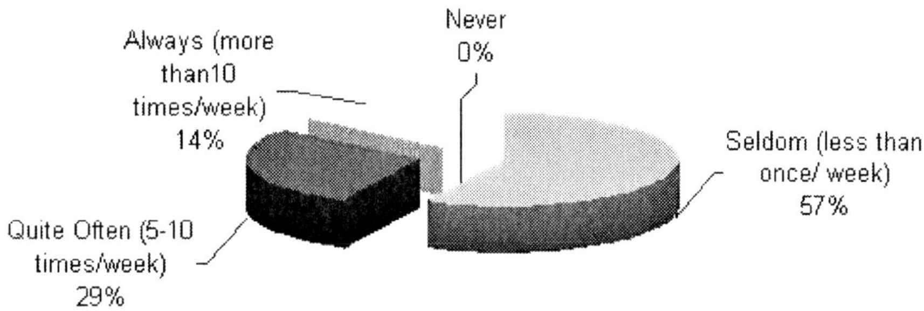

It seems that sentences (1) to (19) in Table 9 are generally treated as frequent expressions in MPC. However, results became varied for sentences (20) to (22) in Table 9, which mainly dealt with speaker's identification. Detailed counts of responses for each sentence are shown in Table 9.

Significance and Future Directions

One of the most significant findings of this survey or case study is the consistency in the result concerning the purposes of mobile phone use and those expressions which are recognized and judged by the respondents as 'most frequently said in MPC'. In other words, on the one hand, results have shown that 'informing people about locations' and 'making appointments' are the main reasons for mobile phone use; On the other hand, we have also found that those Cantonese expressions, which were categorized by the respondents as 'most frequently said in MPC', are those which are dealing with information about locations and time. All this has provided some innovative insights that could engender further research on mobile phone communication.

CONCLUSION AND IMPLICATIONS

This study of mobile phone communication, as a particular kind of computer-mediated communication, especially the linguistic and literacy aspects, is one of

Table 9. Cantonese Expressions

Cantonese Expressions	Mostly said in MPC	Mostly said in SPC	Seldom said in MPC	Seldom said in SPC	Common in BOTH
(1) 你□邊呀? (Where are you?)	16			3	3
(2) 你而家去到邊度呀 (Where have you reached now?)	16			3	1
(3) 你仲要幾耐至□到呀 (How much longer does it take for you to come?)	16			4	
(4) □? 咩話? 喂? 講大聲□啦 (What? Can you speak louder?)	11	1		4	4
(5) 我聽唔到你講咩喎 (I can't hear you)	11	1		4	4
(6) 我個電話收得唔清, 一陣再打畀你啦! (The reception is bad. I'll call you back later)	16		1	3	
(7) 搭□車呀 (I'm travelling in/on a vehicle)	17			3	
(8) 我而家塞□車呀 (I'm caught by a traffic jam)	16			3	1
(9) 我而家□港大走□ (I'm now at HKU and I'm leaving)	16	1		3	
(10) 返□□□ (I'm on my way back)	17			3	
(11) 我而家□□□ (I'm on my way)	17			3	
(12) 而家返□屋企□ (I'm on my way home)	17			3	
(13) 我而家去□銅鑼灣□ (I'm on my way to Causewaybay)	15			3	2
(14) 我□街呀 (I'm in the street)	16		1	3	
(15) 我諗我仲要十分鐘先□到呀 (I think it'll take ten more minutes to come)	16			3	
(16) 仲有兩個站就到□□ (Two more stations ahead)	16			3	1
(17) 我而家去到太子□□ (I'm now at Prince Edward MTR)	16			3	1

Cantonese Expressions	Mostly said in MPC	Mostly said in SPC	Seldom said in MPC	Seldom said in SPC	Common in BOTH
(18) 同朋友行□街…□尖沙咀	16			3	1
(I'm in hanging around TST with my friend(s))					
(19) 我就到□□	17			3	
(I'll arrive soon)					
(20) 喂,唔該Carmen呀	1	4	1		12
(Hi, Carmen please)					
(21) 喂, 你係咪Carmen 呀	4	4	1		8
(Hi, are you Carmen?)					
(22) 喂, 請問你Carmen□到嘛	1	9	1		7
(Hi, is Carmen there?)					

the earliest known in the literature. As one of the earliest, it is of historical interest since it can help us observe the evolution of mobile phone use from the late 1990s until now. In particular, I will like to draw attention to the way SMS has evolved by highlighting the following finding seen earlier:

Besides speech data, this research is also concerned with text data of SMS. Therefore, for those who claimed that SMS was available on their mobile phone, they were also asked how frequently they actually used SMS. Interestingly, even though SMS is available on most of the respondents' mobile phone, they hardly use it as a communication tool. The majority said they use SMS less than once a week. Only 8% of respondents claimed that they always use it (i.e. more than 10 times a week).

It is interesting to note that as of 2002 only 8% of respondents were in the habit of always using SMS. Seven or eight years later in 2009-2010, we have witnessed a massive explosion in the use of SMS. Almost every person, especially young people, now uses text-messaging systems on their mobile phones. Indeed, it may be claimed that some young people use their mobile phones to send and receive text messages more than voice messages.

Besides a historical insight, an implication that we can draw from this early study of mobile phone usage is within education: the outcry that came from parents, teachers and other stakeholders in the education of young people seems to be dying down as we realize that the mobile phone can indeed be a useful tool in the learning process since it can increase connectivity and interaction between students and their teachers and, more importantly, their peers in the learning process. While most parents seem to accept the computer as useful and would like to include it in the school curriculum a lot of them don't want their children spending too much time on the phone. But as Prensky (2004) has pointed out, little do they understand that the mobile phone is indeed now a computer, and some mobile phones may even be more powerful than some computers. These phones can be used to access the internet and thus to gain access to valuable learning resources that would otherwise not have been possible from remote distances.

In the next chapter, I will still deal with mobile phones as an instance of CMC but I will concentrate more on the linguistics aspects, especially how the constraints of time and space in the environment of text-messaging technology affect or shape the linguistic structures produced.

REFERENCES

Abdullah, M. H. (1998). *Electronic discourse: Evolving conventions in online academic environments.* (Available from Eric Document Reproduction Service No. ED 422593).

Abdullah, M. H. (2003). *The Impact of Electronic Communication on Writing.* (Eric Document Reproduction Service No. ED 477614).

ACNielsen. (2001). *Home internet access dominates; Asian mobile penetration booms; Aussies and Kiwis keen online shoppers.* http://asiapacific.acnielsen.com.au/news.asp?newsID=38, June 12, 2001.

Akers-Jordan, C. (2002). *The Implications of E-mail for the College Writing Instructor.* Retrieved on April 15, 2007, from http://spruce.flint.umich.edu/~cakers/research.htm

Baron, N. S. (2001). *Why Email Looks Like Speech. Proofreading, Pedagogy, and Public Face.* Retrieved on May 1, 2007, from http://www.american.edu/lfs/tesol/2003%20Paper-- Why%20Email%20Looks%20Like%20Speech.pdf

Bodomo, A. (2003). *Communicating in the Age of Information Technology: New Forms of Language and their Educational Implications.* Retrieved on June 1, 2007, from http://www.hku.hk/linguist/research/bodomo/QEF/index.htm

Bodomo, A., & Lee, C. (2002). Changing forms of language and literacy: technobabble and mobile phone communication. *Literacy and Numeracy Studies: An International Journal in Education and Training of Adults, 12*(1), 23–44.

Bodomo, A. B. (2007). Mobile phone texting in Hong Kong. In D. Taniar (Ed.), *Encyclopedia of Mobile Computing and Commerce*. Information Science Reference. Idea Group Inc.

Crystal, D. (2001). *Language and the Internet.* Cambridge: Cambridge University Press.

Healey, D. (2007). *The Internet: Helping create 'New English' or reinforcing old dominance?* Retrieved on April 20, 2007, from http://oregonstate.edu/~healeyd/thaitesol_healey.pdf

Prensky, M. (2004). *What Can You Learn from a Cell Phone? – Almost Anything.* Retrieved on April 29, 2007: http://www.ojc.edu/content/facultyStaff/What%20Can%20You%20Learn%20from%20a%20Cell%20Phone_%20Almost%20Anything!pdf

Thurlow, C. (2003). Generation txt? The sociolinguistics of young people's text-messaging. *Discourse Analysis Online*. Retrieved on May 18, 2007, from http://extra.shu.ac.uk/daol/articles/v1/n1/a3/thurlow2002003-01.html

ENDNOTES

1. For further details, readers may go to the project website at: http://www.hku.hk/linguist/research/bodomo/MPC/
2. Hong Kong Trade Development Council (HKTDC) (1997). *Transportation & Communications.* Product Magazines: Product trends. http://www.tdctrade.com/prodmag/transport/r9701e.htm, August 1997.
3. Mobile ownership rockets up to 98pc. (March 1, 2002). *South China Morning Post (SCMP).*

Chapter 7
The Grammar of Mobile Phone Written Language

INTRODUCTION: TECHNOLOGY AND LANGUAGE CHANGE

In the last chapter, I concentrated mainly on mobile phone *voice* communication. In this chapter, I will focus on mobile phone *text* communication. Mobile phone texting or communication through short message service (SMS) started slowly, as we saw in chapter 6, but has quickly emerged as a frequent daily linguistic, literacy or general communicative practice in which two or more people exchange messages by coding and decoding texts received and sent from their cell phones. Mobile phone texting is almost now as pervasive and as ubiquitous as mobile phone voice communication, if not more among some segments of users like young people. This communication process can be witnessed in buses, at homes, in offices, in restaurants, out in the woods, on the high seas, and even in the air! Hong Kong's main English language

DOI: 10.4018/978-1-60566-868-0.ch007

newspaper, the South China Morning Post (SCMP) edition of April 11, 2004 indicates that as huge a volume of 200 million SMS messages are exchanged monthly. SMS has become a multi-million dollar business for service providers.

Along with other kinds of computer, the Internet or other digital technology-mediated communication, SMS seems to be causing a silent evolution or even a revolution with regards to the linguistic and communication habits of people all over the world. This is especially so among the youth where one can safely say that more than 80% of people between the ages of 12 and 25 frequently use SMS as a mode of communication with their peers.

Given such a huge impact that this mode of communication has on the population, researchers and policy makers ought to turn their attention to this topic and attempt to find answers to questions about the consequences of SMS on various issues including language, communication and our general social behaviours.

In this chapter, I focus on the relationship between communications technology and language change, with particular reference to mobile phone texting. Does the emergence of these new communications technologies affect our language and communication habits on mobile phones and beyond? Does it change our language, bringing in new words and structures of expressions, and does it alter our general communication patterns? In short, is technology changing our language?

In examining these questions, based on observation and analysis of issues of language, literacy, and communications technology, I have proposed a model called Technology-conditioned approach to Language Change and Use (TeLCU) in chapter 4. As stated in that chapter, this approach projects the view that there is a causal relationship between the emergence of new tools and media of communication and the creation of new forms of language and communication. New tools and media of communication demand the creation of new forms and ways of communication. These new forms compete with existing forms and ways of communication, leading to changes in the way we use language in its various forms, including spoken and written forms.

This chapter builds on this fruitful discussion in the literature on the relationship between new technologies and the way language is used within these technologies and argues that there is indeed a significant causal relationship between communications technology and new language and communication practices or more specifically the evolution of new ways of using language. As Adams (1996) puts it, 'the new technologies are themselves dramatically changing the nature of the language we use'. Such an approach is also supported by Baron (1984) which concludes that '[n]o one in the computer industry has any hidden agenda for using hardware or software development to alter human language. Yet technology can indeed drive linguistic and social change' (p. 139).

Indeed, as argued in earlier parts of the book, new practices of language and communication may be attributed to a set of unique properties in new communications technology. Modern digital communications technology is characterized by flexibility, connectivity, and interactivity (Blurton 1999) that traditional forms of technology like radio and TV lack. In other words, it is possible to have many-to-many, many-to-one, one-to-many, and one-to-one modes of communication with modern digital information communications technology (ICT). These features of digital ICTs enable them to have a more pervasive influence on forms and uses of language.

In pursuing these questions and research agenda, this chapter is organized as follows:

First, after giving background statistics and other kinds of information about the evolution of SMS, I introduce and describe my data which comprises a corpus of about 500 SMS messages collected from Hong Kong youth in April 2002. Next, I look closely at the linguistic features of the corpus, noting peculiar instances of structure and usage. Following from this I present a detailed case study of mobile phone texting based on a class project on SMS (Alfreih 2007) supervised by me. The chapter concludes with an outline of the implications for these linguistic and communicative changes in our society with particular reference to social and educational habits, and consequences for the design of new tools of communication.

This chapter contains an important appendix that documents what is considered the first corpus on SMS. The corpus thus has historical signifance and its inclusion here in the book will provide a vital reference point for studies on the evolution of the mobile phone text communication.

THE DATA

What is SMS?

SMS (Short message service), first introduced commercially in 1995, refers to the transmission of short text messages between mobile phone users. The first SMS message was a Christmas greeting sent out in Britain in 1992. Today, SMS has emerged as one of the major digital communication media, with an estimation of over one billion messages exchanged per day around the world. SMS had a slow beginning in Hong Kong, mainly because of the inconvenience of inputting Chinese characters with mobile phone keypads. In addition to that, text messages could only be exchanged between subscribers of the same service provider. However, the SMS market has grown rapidly in Hong Kong. In December 2001, the 6 major service providers opened their networks to allow message exchanges across networks

(SCMP 2002). As of April 2004, over 200 million messages are exchanged every month in the territory.

Data Collection

The collection of textual data began with a questionnaire survey conducted in April 2002. The target subjects of the survey were Cantonese-speaking youngsters using the English language in Hong Kong. Since text messages involve private correspondences, prior permission was sought from the data providers. At the end of the questionnaire, the respondent was asked whether he/she would be willing to make his/her SMS texts available for the study. The study (Bodomo 2001-2002, Bodomo 2003- 2005, and Bodomo 2007) successfully collected 487 messages from 87 out of the 92 respondents who participated in the questionnaire survey. Data providers were asked to write down at least five messages that were already stored on their mobile phone message in-/out-box on a form provided by the investigator (Figure 1). Since the texts were collected from end-users, the corpus may also include messages sent from a non-mobile device/tool, such as ICQ.

LINGUISTIC FEATURES OF MOBILE PHONE TEXTING IN HONG KONG

As its name suggests, an SMS text message is indeed very short. It has to be short because users are constrained to input only up to 70 Chinese characters or 160 alphabetic letters. This also imposes some constraints on the basic structure of a message composed with the mobile phone. This section explores the various linguistic features identified in the corpus. The crisp nature of SMS language necessarily raises questions about its morphological and syntactic nature. Given a limited writing space and limited production time, there is logically more need to be as more economical as possible than to be as more elaborate as possible. In other words, a principle of economy outranks a principle of expressivity. The question then is: how is this balanced out? What aspects of the morphology and syntax are retained and what are given up? I here propose a general principle of economy and try to show how it generally manifests itself in SMS:

Technologically-conditioned Economy of Expression: *Words, phrases and sentences should be coded with as few symbols as possible without giving up comprehensibility.*

- Morphology: *produce words as short as possible.*
- Syntax: *produce strings as short as possible.*

Figure 1. Data collection form for data providers

Collection of SMS texts
Instructions to participant:
Please write out at least FIVE SMS texts which have already been stored in your mobile phone.
Please write clearly in the spaces provided below
The texts should be rewritten ACCURATELY, especially
Upper and lower case letters
Punctuation marks

Morpholexical Features: Shortening

I shall look at some morpholexical features in this sub-section. While there are many morpholexical features of mobile phone texting that distinguish it from normal writing, I concentrate on the more salient feature of shortening. Shortening in ICT texts has been widely studied in the literature of computer-mediated communication (CMC). Yet, scholarly research into SMS is still relatively rare and thus I have identified a set of shortened expressions in the corpus. I use the term here as a general cover word for the various strategies for shortening words, including acronymy, abbreviation and so on. The table below (Table 1) is a list of common ways for shortening and abbreviating lexical items and the number of times this happens in the corpus.

The following example from the corpus illustrates the use of shortenings in the real messaging context:

(1) r u 3 2nite? izit possible 2 hv dinner 2getda? call me

Table 1. Shortened expressions in the Corpus

Shortening	Target expression	Occurrences in corpus	Shortening	Target expression	Occurrences in corpus
U	You	120	Dun / dunno	Don't /don't know	11
R	Are	21	ur	Your	13
Pls	Please	13	B4	Before	5
Tmr	Tomorrow	5	2	to	17
Gd	Good	8	Tonite	Tonight	5
Wanna	Want to	10	Hv	have	5
y	Why	2			

Most (though not all) morpholexical items, as seen above, are thus produced in accordance with the economy principle: produce words as short as possible. I now turn to the syntactic aspects of the economy principle.

Syntax: Economy of Expressions versus Expressivity

Although there is a wealth of studies on the linguistic features of ICT texts, the aspect of syntax is rarely mentioned. This subsection is an attempt to take into account the syntactic nature of shortened expressions in SMS. The assumption here is that, in shortening a sentence, functional items such as tense and aspect are dropped, while lexical categories such as nouns and verbs are allowed to stay; in fact, this is more of a tendency than absolute adherence to the principle. So while in the morphology lexical items are more susceptible to reduction, in the syntax functional items are more susceptible to reduction or even dropping. I list below each group of items likely to be retained or dropped.

- What Is Retained?
- lexical Items
- The most essential piece of information:
    (2) lunch?
    (3) SLEPT?
    (4) QUIZ
    (5) exams finished la. now coming.
    (6) in ikea,thinking of u...
- What Is Left Out?
- Pro-drop

(7) miss u terribly! wanna c u & hold u so much. will hv quite a bz wk. =(
(8) MISS U VERY VERY MUCH DUNNO HOW TO TELL U

- Tense
  (9) YES I RECEIVE
  (10) My friend ask me to go taiwan

The above are ways in which mobile texters attempt to satisfy the principle of economy over and above that of elaborate expressivity.

Phonological Features: Homophony

Some interesting morphophonological and phonosemantic features also characterize the language of mobile texting. In this chapter, I will focus mainly on the concept of homophony. One of the creative practices of Mobile phone texters (Bodomo 2002) lies in their ability to substitute shorter (forms of) words for longer words that are homophonous or near-homophonous. I have found a very interesting one as shown below in (11).

- 3 for 'free' as in
  (11) r u 3 2nite?

Here, the homophonous word three (its shorter form being the numeral 3) is used to represent *free*.

The omission of tone: another aspect of the crisp nature of mobile phone text is the explicit omission of tone markings in the Cantonese romanization. The Linguistic Society of Hong Kong (LSHK) has an elaborate romanization system, including tone markings1 but this is hardly used in mobile texts. I list below a number of sentences showing various ways of romanization, especially in the rendition of final particles, which are very much a characteristic feature of the language of ICTs like mobile phone and ICQ texting (Lee 2002a, 2002b).

(12) exams finished la. now coming.
(13) Where R U worry ar me
(14) Heavy raining wor... but u still need to go rite? Be careful ar:p

Romanization of Common Cantonese Expressions

(15) Ai Ya... ho sun fu ar... the ref bk is very difficult... can't understand a word ar! @-@ ho mun!

(16) Hey bei sum gei la...
(17) Lo lic... everything will be fine again

Emoticonymy

Unlike face-to-face talk where emotions can be expressed visually (by facial expressions) or auditorily (by stress and loudness), text-based CMC mainly takes place without any face-to-face interactions between participants. Therefore, emotions are often expressed by graphical means. The introduction of the notion of 'smileys' or 'emoticons' has facilitated electronic communication to a large extent, and is an essential aspect of visual language and visual communication in the 21st century (Horn 1999).

By emoticonymy, I mean 'a subfield of CMC which involves the analysis and practices of employing smileys and related icons for conveying emotions and other linguistic and kinesic features intended by the author' (Bodomo and Lee, 2002, p.35). Emoticon (Emotional-icon) is also used to avoid misunderstanding in the interpretation of a message. For instance, a smiling face like:) is often used to indicate a joke or that the message sender is happy about something. Emoticons are usually placed at the end of a sentence, as in 'I am so sorry:-('. Most emoticons are created with different combinations of punctuation marks (colon and a close bracket form a smiling face). This is further illustrated in (18) and (19) below.

(18) Ladies hockey champion! Yeah >:D
(19) Tmr 記得 交 essay ^v^
- tmr gei3 dak1 gaau1 essay
- tomorrow remember submit essay (smiles)
- 'Don't forget to hand in your essay tomorrow.'

Mode-Mixing

Apart from emoticons, there also exist combinations of codes and symbolic systems in the messages. As Kress (2000) suggests, '[t]he appearance of modes other than language in the centre of the domain of public communication has several aspects: new, or newly prominent modes appear: texts, textual objects are more clearly seen to be multimodal, that is, to be constituted by a number of modes of representation' (pp. 183-184). So what is happening in SMS? In the corpus, we can identify a combination of words/letters and numbers, and graphic and alphabetic symbols. These are illustrated in (20), (21) and (22) below.

- Words/letters + Numbers
  (20) r u 3 2nite? izit possible 2 hv dinner 2getda?call me
- (Are you free tonight? Is it possible to have dinner together? Call me.)
- Graphical + alphabetic
  (21) ()"""()
- (▌'▌o▌'▌)")
- (.) /
- Happy New Year
- (Punctuation marks to visualize 'fireworks')
  (22) U:)= I:) U:(= I:(+ WORRY TAKE CARE
- (When you're happy, I am happy too. When you are sad, I also feel sad and worried. Take care.)

New Varieties of English

In this section I have examined specific features of written language on the mobile phone. The constraints of time and space force users to invent diverse and new ways of shortening or even leaving out entire words, and adding new signs and icons to make themselves comprehensible. I believe that the media of the mobile phone and other ICTs, such as the MSN that is popular worldwide, are changing our language use in profound ways, and making the youth in most parts of world evolve different ways of using language. New varieties of written English thus seem to be evolving. In Hong Kong, this new variety involves a mixture of English and Chinese signs, characters, words, and expressions. Even though some features may seem spontaneous and playful at first glance, some features and patterns are used quite consistently and repeatedly by different users. Indeed, these new varieties of written English are finding their way into school writing and are the source of worry for school teachers.

A CASE STUDY INSTANTIATING TeLCU

Introduction

In this part of the chapter I present a detailed study (Alfreih 2007) amplifying most of the features and practices associated with mobile texting that we have observed. As a departure from most of the case studies in the book which are taken from Hong Kong contexts where the author works, this particular case study is taken from California, USA where there author taught for one year in 2006/2007 while on sabbatical leave.[2]

Figure 2. Age distribution of instructors

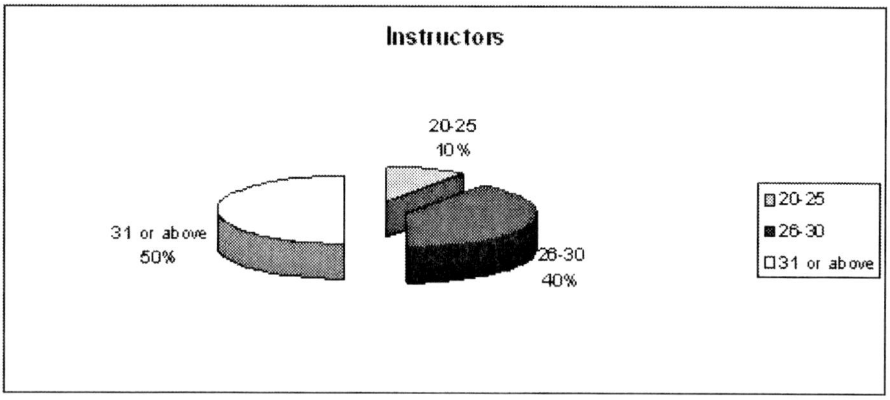

Methodologies and Data Collection

The methodologies applied in this study are discussed with a detailed description of the instrument used to collect and analyze the data. In this study, two methods were employed for data collection: (a) questionnaire survey, and (b) face-to-face interviews.

Questionnaire Survey

The first method for data collection was a questionnaire survey for instructors and students. Student's questionnaire consisted of 12 questions, and the Instructors questionnaire consisted of 14 questions. The questionnaires were divided into three main parts. The aim of the first part of the questionnaires was to gather some information about the participants' background and preferences regarding the use of new communication tools. The second part was mainly about the participants SMS writing behavior. The last part of the questionnaires was aimed to get an insight into the educational implication of this new language from participants' perspective. The students and instructors survey were identical except for the last part regarding the educational implications of SMS language on students writing behaviors in school. The instructors were asked additional questions in this part about their attitudes toward the use of this language in school assignments and the actions they took when encountering this language in students' writing. The questionnaires were distributed to 20 Santa Clara University students and 10 teachers who are graduate students in the Department of Education at Santa Clara University, and Santa Clara

Figure 3. Age distribution of students

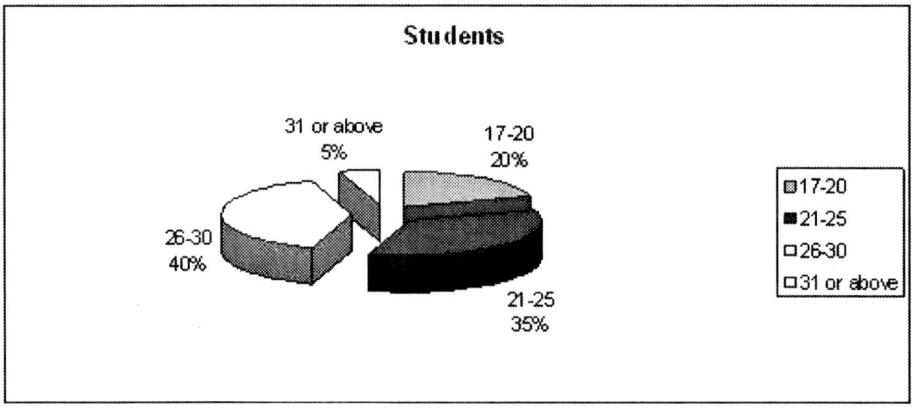

University Instructors. The distribution of the participants is shown in the following figures: Age in Figures 2 and 3, gender in Figures 4 and 5.

The instructors in this study teach different levels of students. The distribution of instructors based on their teaching experience is shown in Figure 6.

Face-To-Face Interviews

In addition to the questionnaire survey, two Santa Clara University students and two instructors participated in interviews to gather more information about their texting habits and attitudes toward this new form of language and its impact on students reading and writing behaviors. One of the instructors is a 48-year old English lan-

Figure 4. Gender distribution of instructors

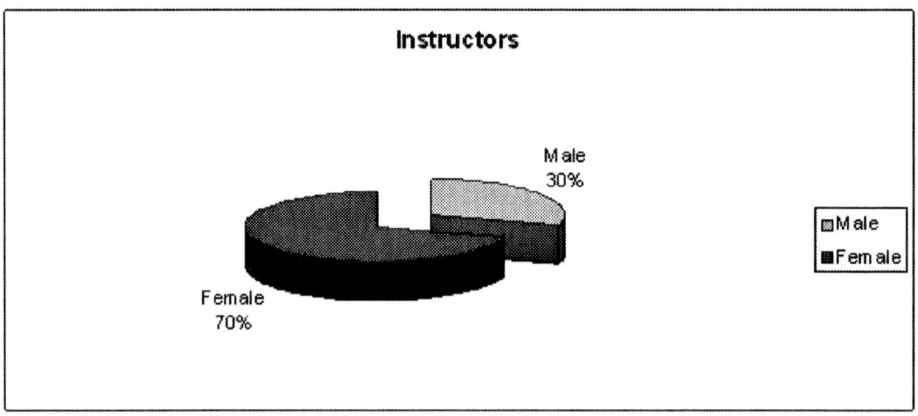

Figure 5. Gender distribution of students

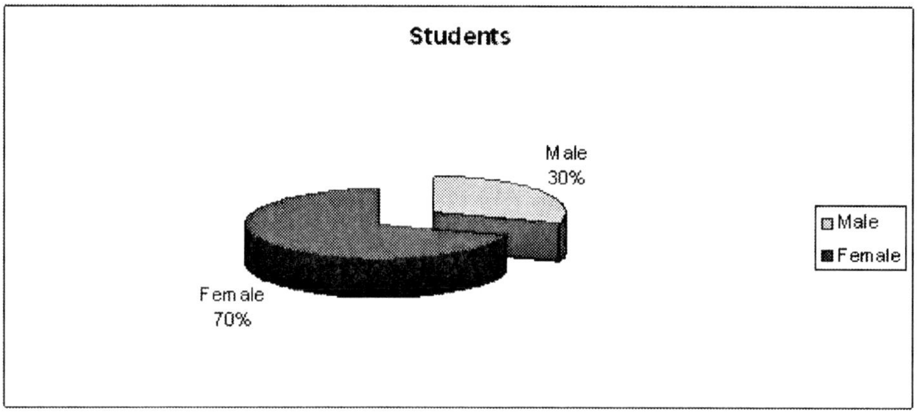

guage teacher while the other one is a 25-year old mathematics teacher. The reason for that was to see whether or not the subject matter and the age of instructor may have an impact on their attitudes toward this technology and the language use. Written notes were taken during the interviews that lasted for about 15-20 minutes. The questions had been set up before the interviews and were mainly about their mobile texting habits such as the frequency of texting, the language they use and their attitudes toward this form of language. They were also asked about their attitudes toward cell phones as an educational and communication tool. Questions regarding their views of the educational implications of this technology and the language used in text messages on students reading and writing behaviors were of special interest. Participants were asked to elaborate on their answers and explain

Figure 6. Instructors distribution based on their experience

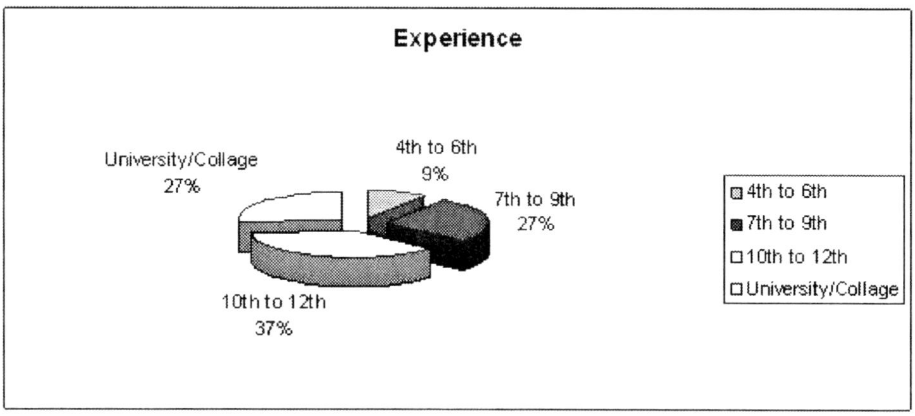

Figure 7. What is instructors' preferred tool for communication?

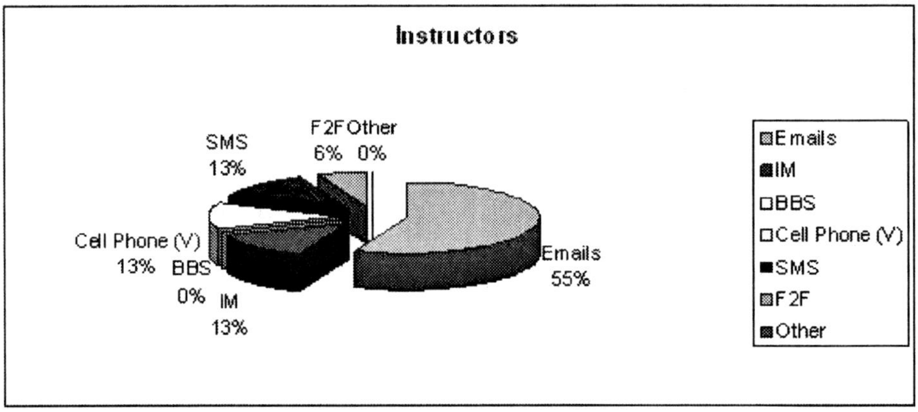

their responses. In general, the questionnaire survey focused on the "what" factor while the interviews were mainly concerned about the "why" factor.

Findings

In this section, report on the findings of this study will be presented based on the data collected from the questionnaire survey and interviews. This study adopts the TeLCU model that argues that there is a causal relationship between the introduction of new ICTs and new forms and uses of language. The findings of this study provide evidence that supports this argument and presents some unique features of SMS language.

The following section presents and compares text messaging habits and linguistic features of instructors and students texting including 1) methods of shortening, 2) grammatical awareness, and 3) use of emoticons.

Texting Habits

All the participants in this study have cell phones. The first question of the survey aimed to identify what kinds of communication tools the respondents used most often. The results are shown in Figures 7 and 8.

i. Preferred Tool for Communication

The majority of the instructors said they preferred Emails for their every day communication, with SMS, Cell phones (voice), and Instant messages in the second

Figure 8. What is students' preferred tool for communication?

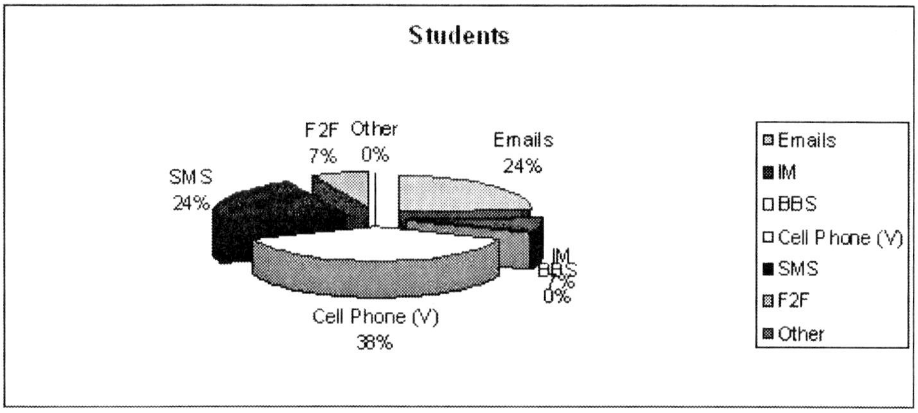

place. Only one instructor preferred fact-to-face communication and explained that by saying it is more personal and builds relationships. The most popular communication tool among students, according to the respondents, was voice cell phone, with SMS and emails in the second place.

In addition, one of the questions in the questionnaire survey was how frequently they used cell phone text messages to communicate. Figures 9 and 10 illustrate the frequency of instructors and students use of cell phone text messages.

Figure 9. How frequently do instructors use cell phone text messages to communicate?

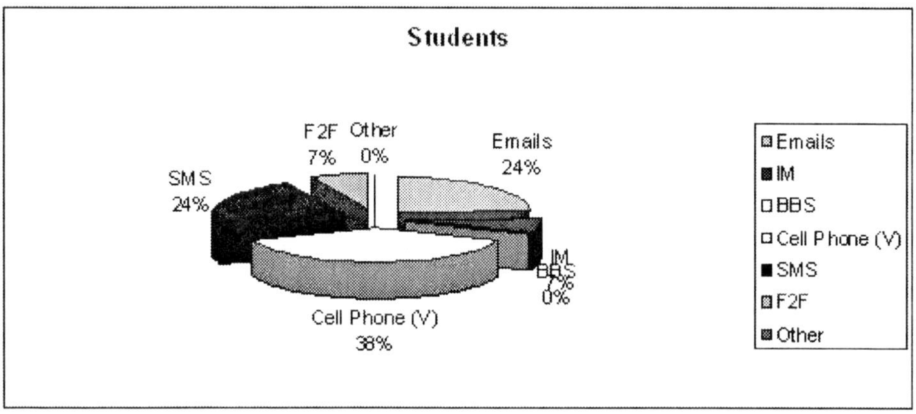

Figure 10. How frequently do students use cell phone text messages to communicate?

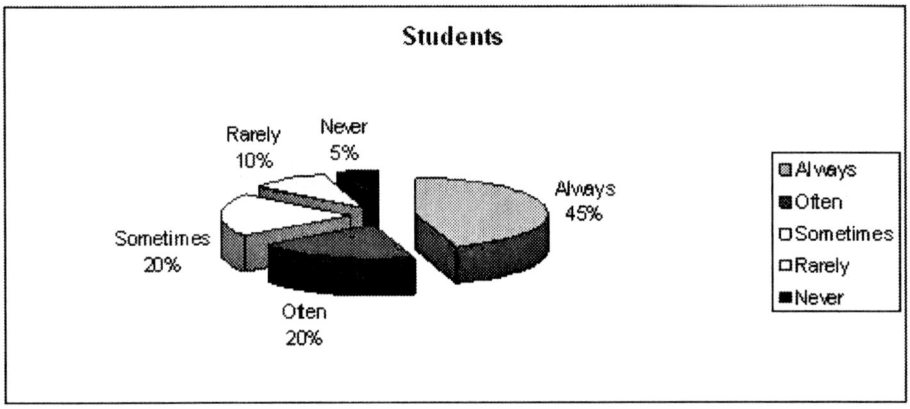

ii. Frequency

From the diagram above, it is clear that text messaging is more common among students than teachers. While 65% of the students from this survey use text messages frequently to communicate, only 30% of the instructors who participated in this study do.

iii. SMS Language

Participants were asked to rate their use of shortened words or phrases when writing text messages. As expected, Figures 11 and 12 shows that while 70% of the students reported that they always or often use this method, only 40% of the instructors said that they use shortened words when writing text messages.

iv. Shortening words in text messages

Participants were also asked to provide the five top acronyms of words or phrases that they use when writing text messages. The shortening provided by instructors and students are presented in Table 2 below.

Methods of Shortening

According to the data collected from the questionnaire survey, the most commonly used shortenings by instructors are cu for 'See You' and b4 for 'Before'. The com-

Figure 11. Do instructors use shortened words when writing cell phone text messages?

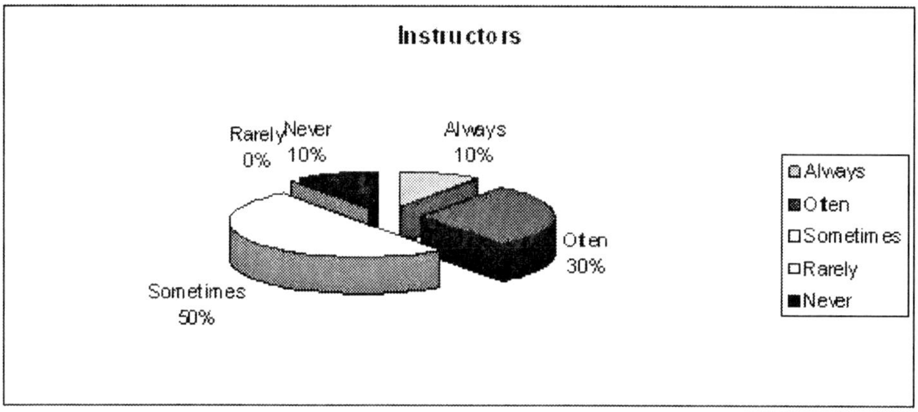

monly used shortenings by students are b4 for 'Before', lol for "Laughing Out Laud', u for 'You', and btw for 'By The Way'. Obviously, most of the shortenings found in text messages have unique features that can not be attributed to traditional methods of shortening. The shortening methods used in the data collected can be divided into two groups 1) phonetic-based methods, 2) spelling-based methods, 3) acronym of words and phrases, and 4) compression of words and phrases. These methods are presented in the following sections.

Figure 12. Do students use shortened words when writing cell phone text messages?

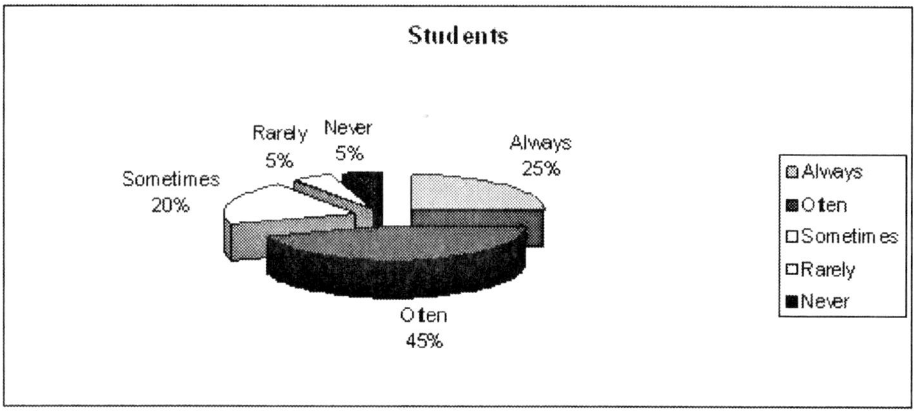

Table 2. The shortening provided by instructors and students

Shortening	*Target Expression*
Lol	Laughing Out Loud
Cu	See you
R	Are
u/ur	You / Your
K/ ok / kay	Okay
Sys	See You Soon
Nite	Night
Thx	Thanks
OMG	Oh My God
ASAP	As Soon As Possible
TC	Take Care
Plz/ pls	Please
Btw	By The Way
&	And
Tmrw/tmr	Tomorrow
Cuz	Because
IC	I See
BRB	Be Right Back
Frnd	Friend
Doin	Doing
Jk	Just Kidding
B4	Before
2nite/2day	Tonight/Today
2/4	To-Too/For

i. Phonetic-based methods

This method includes a) letter homophone, b) number homophone, and c) combination of letter and number homophone.

a). Letter homophone: In this method, users tend to employ letters to represent a word. The letters can either be a homophone to that word or similar to the pronunciation of the word. A number of examples are found in Table 2 including cu, u, r, ur, and k.

Figure 13. Do instructors pay attention to the correctness of their grammar and spelling when they write a cell phone text message?

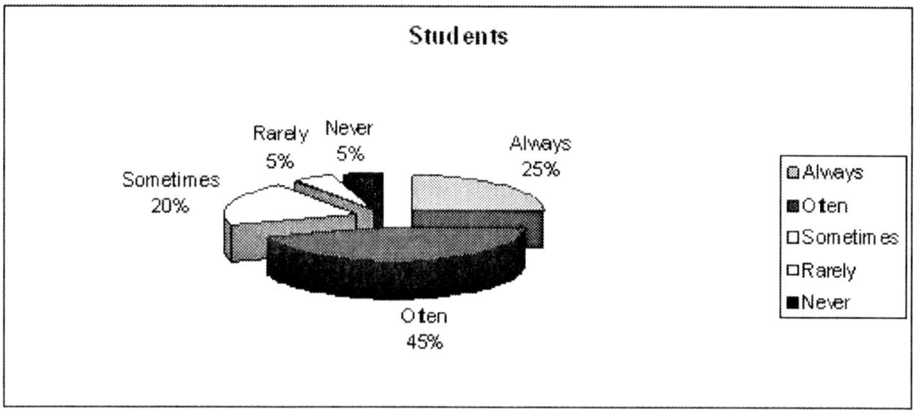

b). Number homophone: This method refers to using numbers to represent a word. This number can either be a homophone to the target word or it can be similar to the pronunciation of that word. Examples in this category from the data collected from the users are 2 and 4.

c). Combination of letter and number homophone: This method refers to using letters and numbers to represent a word. With reference to the data collected from the participants, a number of examples can be found such as b4, 2nite, and 2day.

ii. Spelling-based method

a). Omission of vowel-indicating letters: in this category, words are shortened by omitting vowel-indicating letters such as a, e, i, o, and u. A number of examples are found including frnd, tmrw, tmr, plz, and pls. This method is commonly used by students and no examples in this category were provided by instructors.

b). Acronym of phrases: In this category words are formed by the initial letters in a string of words. From the data collected, a number of examples in this category are found such as BRB, ic, btw, TC, ASAP, OMG, sys, and jk. This is the most commonly used type of shortening by instructors and teachers.

c). Compression of words: Examples in this category include cuz, luv, and thx/ths.

Figure 14. Do students pay attention to the correctness of their grammar and spelling when they write a cell phone text message?

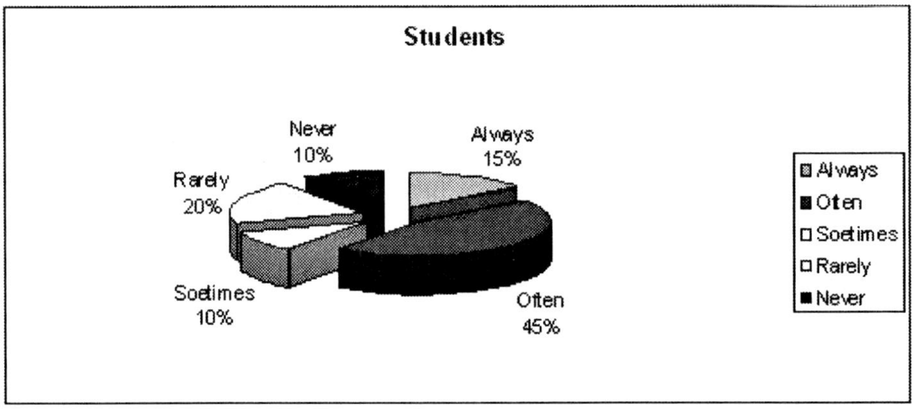

Grammatical and Spelling Awareness

An interesting finding came up when participants were asked to rate their grammatical and spelling awareness when writing text messages. The results are illustrated in Figures 13 and 14.

As the numbers in the diagram shows, 60% of the students said that they always or often check their grammar and spelling when writing a text messages. On the other hand, only 40% of the instructors said they often pay attention to their grammar and spelling in text messages.

Emoticons

In the context of text messaging, both senders and receivers are not able to see and listen to each other. For the senders to convey their feelings, they use a short sequence of keyboard characters to represent a facial expressions or emotions. In this study, the participants were asked to rate their use of emoticons in text messages as shown in Figure 15 and 16.

In general, students tend to use emoticons in text messages more often than instructors even though the difference is not significant as predicted.

It is important to note that during the interviews, all four participants said that using these distinctive features of SMS language depends mostly on the context in which the message is sent and the intended receiver. This indicates that users of SMS are aware of the informality of SMS language. The participants were also asked why they tend to use these elements when writing text messages. They said

Figure 15. Do instructors use emoticons expressions when writing cell phone text messages?

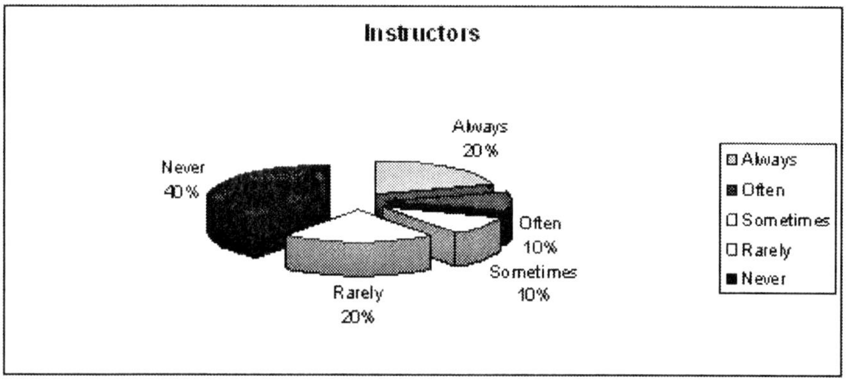

that minimizing the length of a text messages makes it easier to read and write as well. One of the instructors said that she usually replies to SMS as soon as she receives one and that she usually doesn't have the time to write a long message or proof-read it.

SMS Language in the Classroom

The rest of the survey was aimed to investigate the use of SMS language in school assignments and communication with instructors. Instructors were asked whether they encountered this type of language in students' writings in school and the major-

Figure 16. Do students use emoticons expressions when writing cell phone text messages?

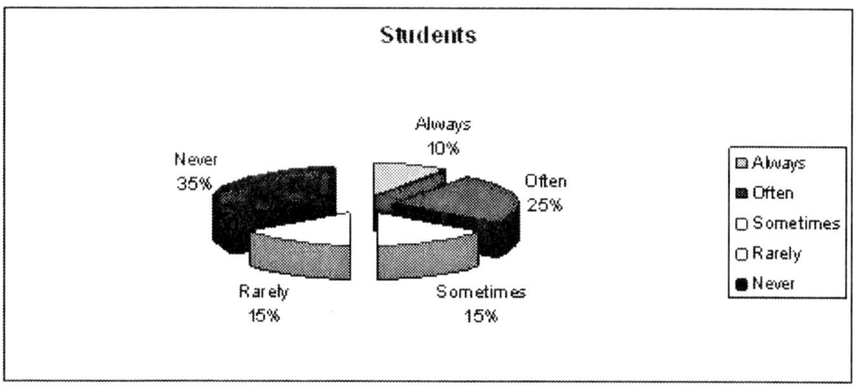

Figure 17. Do your students use this type of language when writing school assignments?

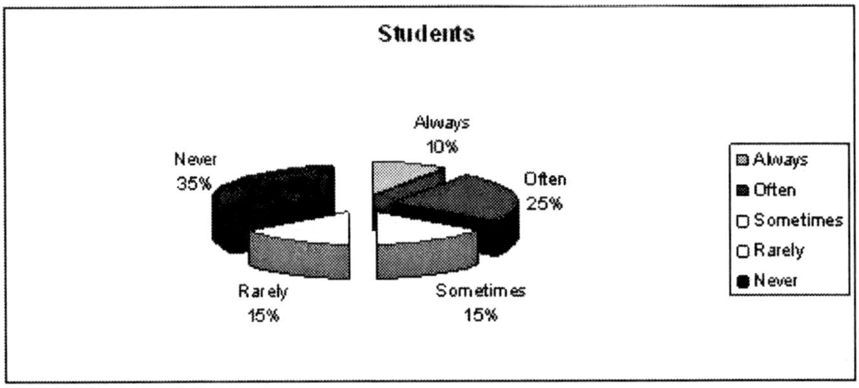

ity (70%) said that they never or rarely did as shown in Figure 17.

When Instructors were asked if students tend to use this language when communicating via emails, their responses were different. From instructors' responses, it is obvious that students tend to use this language when using ICT tools for communication as shown in Figure 18.

Students were also asked if they ever used this language for school assignments and all of them said that they never or rarely did as shown in Figure 19.

From the interviews, the students were asked if they would prefer to use this language in written assignments and they both said that they don't think that it is a good idea and that students need to understand the difference between informal writing and academic writing.

Figure 18. Do your students use this type of language when they communicate with you via emails?

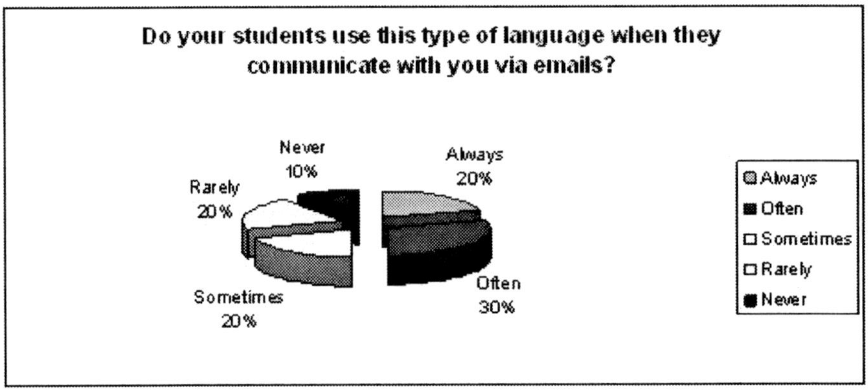

Figure 19. Have you ever used this form of language in your written assignments in school?

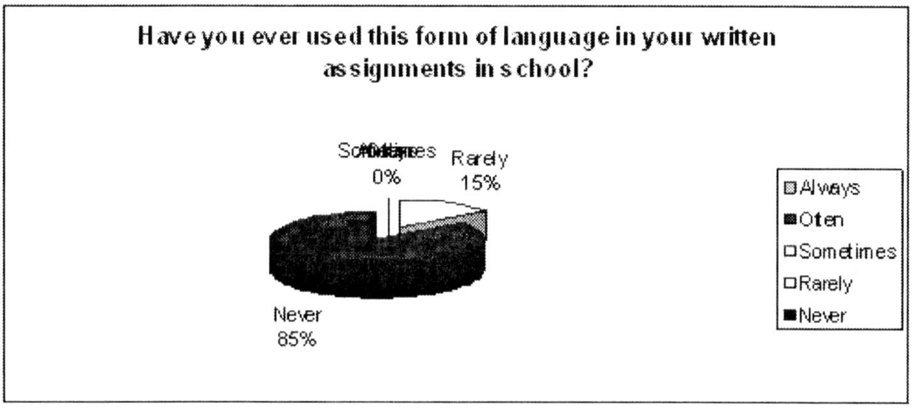

Students were also asked if they use this language when communicating with their instructors via emails. Students' responses to this question are illustrated in Figure 20.

One important question of this study was to see whether instructors' attitude toward the use of this language in students written assignments and emails are different. So far, it is clear that students tend to use this language more often when communicating with instructors via emails, but do instructors see this use of language as 'inappropriate'? From the data collected, the difference between the acceptance of this language in students written assignments and emails was significant. As shown in Figure 14, while all instructors never or rarely accept this language in written

Figure 20. Have you ever used this form of language in your written communication with your instructors via emails?

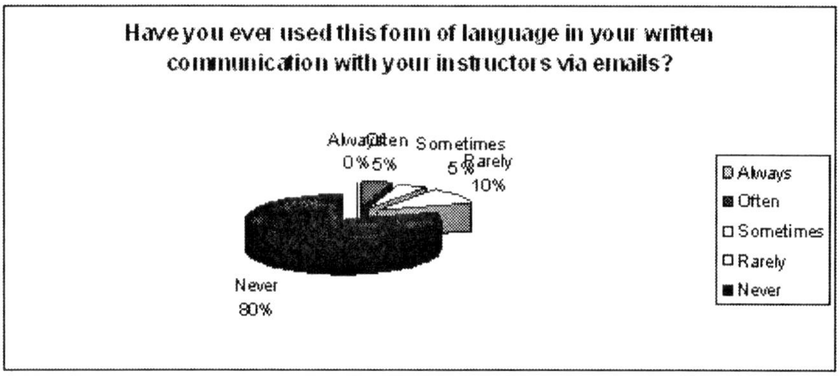

Figure 21. Do you accept this form of language in students written assignments?

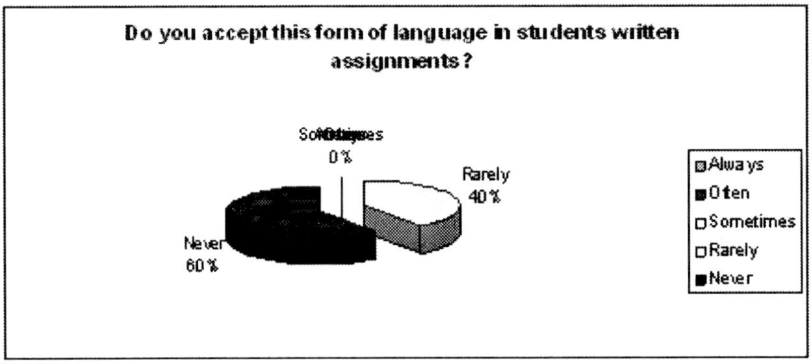

assignments, half of them said that they always or often do accept it in students' emails. The fact that teachers and other instructors are beginning to accept this form of language when students use ICT tools for communicating with them may be an indication that these instructors are coming round to understanding that there is a causal relationship between the introduction of new ICT tools and new forms of language and literacy that the model TeLCU is based on. In a sense, they are beginning to accept that their students' language will vary in accordance with the technological medium the students are communicating in (Figures 21 and 22).

The instructors from the interviews were asked the same questions and to explain their responses. One of the instructors said that the use f this language is not acceptable at all and that even though she does use this language sometimes, she believes that it is the teacher's responsibility to teach students formal writing and that if teachers don't appreciate and respect the English language so will students.

Figure 22. Do you accept this form of language in students' emails?

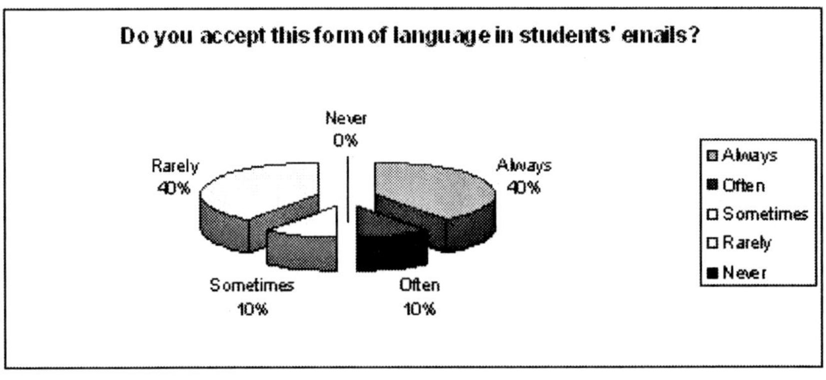

Figure 23. Instructors' responses

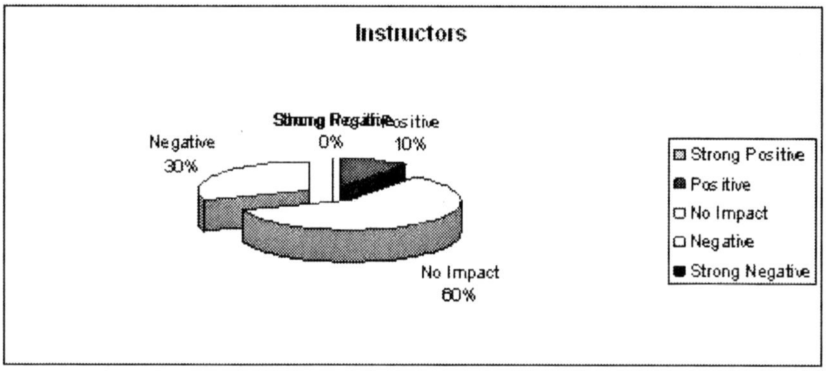

The other instructor was more open to the idea of accepting this new language in students' emails but totally rejected the idea of accepting this language in students written assignments and tests. The instructor explained his point of view by saying that he understands that this language is a part of a new culture and that teachers need to be lenient to an extent in order to be able to connect with students. Most of the instructors said that they would ask students to re-write their assignments or mark it when they encounter this type of writing in students' school assignments or tests.

The last question in the questionnaire survey was to see how strongly the participants believed that this new form of language affected students reading and writing behaviors. Instructors' and students' responses are shown in Figures 23 and 24.

From the diagrams above, one can see that most of the participants believe that new ICTs and the language used have no impact on students writing or read-

Figure 24. Students' responses

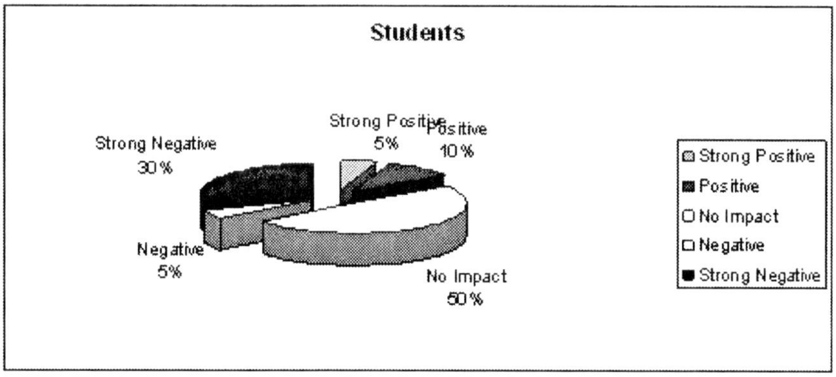

ing behaviors. From the interviews, one of the students said that with advances in technology and especially the internet, she spends a lot of time surfing the internet and reading about different topics and that helped her a lot in the area of reading and writing ability even though she thinks that using the internet only for chatting and playing games might have a negative impact on reading and writing abilities. The other student said that this new language is ruining the English language and that the more students use this language in on-line chats and text messages the more they will forget the basics of the English language.

Summary

This case study has examined the different linguistic features of SMS language and its educational implications based on the theoretical frame work of TeLCU as presented in chapter 4. Central to this model is that there is a causal relationship between the introduction of new ICTs and new forms and uses of language and literacy. A number of SMS language features have been introduced including shortening and emoticons. These unique features of SMS language are found to be related to the design and mobility of cell phones.

This study has examined different issues regarding students' use of SMS and their educational implications including instructors and students texting habits, the linguistic features of SMS that instructors and students provided, the role of new ICT tools on language and literacy change and use, the impact of this new form of language on students writing behaviors from instructors and teachers perspective, and finally, instructors attitude toward this new form of language.

Even though many claims have been made that the English language is deteriorating due to the use of new ICT tools, the data collected in this study prove otherwise. The majority of the participants in this study believe that ICT tools had no impact on students reading and writing behaviors. In addition, it seems that students understand the different contexts in which language is used and respect it.

With the rapid advances in communication technologies, it is expected that new ICT tools will be introduced and these advances will generate new opportunities for human communication. It is important to note the fact that this new language is a part of student and youth culture. For that reason, the educational implications of this new student and youth language deserve the attention of the education community, especially language teachers.

CONCLUSION

In this chapter, I have attempted to take stock of language structure and use within the technological environment of new media of communication such as the mobile phone. Drawing from different arguments and points of view within the literature I have highlighted the important debate as to whether the linguistic and communicative transformation that we are witnessing is one of technological determinism or social determinism. The position espoused in this chapter is that while technological determinism cannot be divorced from social determinism one has to tease out the two and look concretely and specifically at the immediate ways in which language functions and changes in the environment of new ICTs. I have discussed these issues based on my proposed technologically deterministic model known as TeLCU, details of which are in chapter 4. As has been proposed, users of language are forced to abide by a quite stringent economy principle in mobile phone texting. This has resulted in different ways of experimenting with language structure and use, phonologically, morphologically, and syntactically. The attendant innovations at these various levels of linguistic analysis seem to be creating new varieties of language and new ways of communicating.

REFERENCES

Adams, A. (1996). Language awareness and information technology. Curriculum and Teaching , 11(2), 69–76.

Alfreih, M. 2007. Student SMS Practices and their Educational Implications. Term Paper for the Course ED255 Special Topics Seminar: Languages, Literacies and Technologies. Spring Semester 2007, Santa Clara University, CA, USA.

Baron, N. S. (1984). Computer mediated communication as a force in language change. Visible Language , XVIII(2), 118–141.

Blurton, C. G. (1999). New directions of ICT-use in education. *UNESCO's World Communication and Information Report 1999*. URL: http://www.unesco.org/education/educprog/lwf/dl/edict.pdf.

Bodomo, Adams. (2000 to date). *The Hong Kong ICT Corpora*. University of Hong Kong.

Bodomo, Adams. (2002, May). *Communicating in the Age of Information Technology: New Forms of Language and Literacy and their Educational Implications*. Paper for the Interactive Seminar & Discussion, Ansted University, Penang, Malaysia.

Bodomo, A. (2003-2005). Communicating in the Age of Information Technology: New Forms of Language and Their Educational Implications. Research Project funded by Quality Education Fund (Ref: 2002/0384). URL:http://www.hku.hk/linguist/research/bodomo/QEF/

Bodomo, A. (2001-2002). *Linguistic Features of Mobile Phone Communication.* CRCG Research Project, HKU. URL: http://www.hku.hk/linguist/research/bodomo/MPC

Bodomo, A., and C. Lee (2002). Changing forms of language and literacy: technobabble and mobile phone communication. Literacy and Numeracy Studies: An International Journal in Education and Training of Adults , 12(1), 23–44.

Bodomo, A. B. 2007. Mobile phone texting in Hong Kong. In David Taniar (ed). Encyclopedia of Mobile Computing And Commerce. Information Science Reference. Idea Group Inc.

Horn, R. (1999) *Visual Language: Global communication in the 21st Century.* Macrovu Inc.

Kress, G. (2000). Multimodality. In Bill Cope and Mary Kalantzis (Eds.), *Multiliteracies: Literacy Learning and the Design of Social Futures* (pp. 182-202). London and New York: Routledge.

Lee, C. K. M. (2002a). Chinese and English Computer-mediated Communication in the Context of New Literacy Studies. Unpublished M.Phil. dissertation. Hong Kong: The University of Hong Kong.

Lee, C. K. M. (2002b). Literacy practices in computer-mediated communication in Hong Kong. *The Reading Matrix, Special Issue: Literacy and the Web. Reading Matrix.com.* Available online at: http://www.readingmatrix.com/articles/lee/article.pdf Council urges push on SMS. (October 17, 2002). *South China Morning Post.* Text maniacs find their calling. (April 11, 2004). *South China Morning Post.*

ENDNOTES

1 Tones in Cantonese are marked as follows: Tone 1: high level, Tone 2: high rising, Tone 3: mid level, Tone 4: low falling, Tone 5: low rising, and Tone 6: low level.

2 I gratefully acknowledge Maha Alfreih who as one of my students in an MEd course at Santa Clara University collected the data and analyses reported in this case study to test my TELCU model as part of her term paper.

[3] For example, Rachel Tabone sent me the following email on February 28, 2009 to request usage of the corpus: "I am a student doing BSc in Computing and Information Systems at the Department of Computing Goldsmiths College University of London as an external Student. As part of my final project I am building a prediction engine which would be deployed on mobile phones and would be able to aid the inputting of textual messages by predicting the next few words which the user will be about to write."

APPENDIX: THE FIRST SMS CORPUS

Introduction

In the more than eight years that it has taken me to research materials for this book, I have come up with what has been described by many people as one of the first and one of the most significant corpora on short message service that has been gathered from actual mobile phone usage and then tagged. These approximately 500 mobile phone messages are thus of historical significance. It is not safe to leave them on a website whose URL may not be accessible after a while. In order to ensure that they are well-documented for posterity, I have found it necessary to document the entire corpus in the form of an extensive appendix to this chapter. Several people continue to ask me to use it to test emerging technologies such as predictive SMS systems, including prediction engines.[3] It is hoped that the inclusion of this corpus in the book will provide the necessary historical source for discussing the evolution of mobile phone short message service.

SMS CORPUS. Compiled by: *Dr. Adams Bodomo*, Department of Linguistics, HKU **(Project Code: 10203814)**

Background Information of the Corpus

- Type of text: Text messages of SMS (Short Message Service)
- Time of data collection: April 2002
- Data collected by: Ms. Kitty Chan, Ms. Alvina Ho, Ms. Lam Yim, Ms. Karen Lau, and Mr. Tai Chung-pui
- Format specifications:
- Font size: 10
- Margin: 2.5 cm on each side of an A4 sheet
- Line Spacing: 15 pt (a blank line is left between two messages)
- Software: Microsoft Word 2000 / XP

Statistics

- No. of messages: 487 (each message is labeled with the format of <SMS 1> to <SMS 487>
- No. of Lines: approx. 633
- Lines per message: approx. 1.3
- Total no. of words: 5,236
- No. of characters with spaces: 19,712
- No. of characters without spaces: 16,381

- Approximate no. of words per message: approx. 11

Glossing Scheme*

Due to the complex linguistic nature of SMS texts, data in this corpus are glossed according the following five levels (though not all messages need to be glossed with all five levels):

(i) **CC**: Chinese/Cantonese Character Writing
(ii) **JP**: JyutPing Cantonese Romanization
(iii) **MPH**: Morpheme-for-Morpheme glossing
(iv) **FT**: Free Translation
(v) **NB**: Additional notes/explanation on the message / expression
  - Wherever applicable, complete or self-explanatory messages in English will not be glossed.

Labels and Abbreviations

1./2./3.SG/PL	1st/2nd/3rd person singular/plural pronoun		
COM	Comparative marker		
EMO	Emoticon		
EXP	Expressive		
GEN	Genitive marker		
HES	Hesitation		
INTENS	Intensifier		
INTRG	Interrogative marker		
LAUGH	Laughter		
NEG	Negation marker		
ORD	Ordinal numeral		
PART	Particle		
Personal Name	PN		
PREP	Preposition		
SFP	Sentence final particle		
??	Unknown item (s)		

The Corpus

<SMS 1>. I M GOING To HOME

FT: 'I'm going home.'

<SMS 2>. Choir agm

FT: 'I'm at the Annual General Meeting (AGM) of the Choir committee.'

<SMS 3>. Go BACK HOME

FT: 'I'm going (back) home.'

<SMS 4>. HAPPY VALENTINES DAY

<SMS 5>. Kung hei fat choi!

CC: 恭 喜 發 財
JP: *gung1 hei2 faat3 coi4*
FT: 'Wishing you prosperity and wealth.'
NB: This is a common Chinese New Year's greeting.

<SMS 6>. What r u doin?

FT: 'What are you doing?'
NB: r = 'are', u = 'you'

<SMS 7>. Haha, really? Hm, that's sweet. u r not on icq?

MPH: LAUGH, really? HES, that was sweet. You are not on ICQ?
FT: 'Haha...Really? Well, that was sweet. Aren't you on ICQ?'
NB: u = 'you', r = 'are'

<SMS 8>. 上 english 堂 好 悶 呀! 你 呢?

JP: *soeng5* english *tong4 hou2 mun6 aa3 nei5 ne3*
MPH: attend English lesson INTENS boring PART 2.SG PART
FT: 'English lessons are so boring! What do you think?'

<SMS 9>. 我 回 來 啦, 你 地 係 第 幾 行?

JP: *ngo5 wui4 loi4 laa3 nei5_dei6 hai2 dai6 gei2 hong4?*
MPH: 1.SG return come PART 2.PL PREP ORD how much row

FT: 'I am back. Which row are you in?'

<SMS 10>. 你 跟 我 *wear* 間條 三

JP: *nei5 gan1 ngo5* wear *gaan3tiu2 saam1*
MPH: 2.SG with 1.SG wear strip clothes (three)
FT: 'You and I should wear a strip shirt/T-shirt.'
NB: Here, the literal meaning of *saam1* 三 is actually 'three', which has nothing to do with the intended morpheme and its meaning, 衫 'clothes, shirt, or T-shirt'. However, since the syllable for 'clothes' and 'three' are homophones, and the character for 'three', 三 has fewer strokes, the message writer would prefer to use it as a substitute for 衫 'clothes'.

<SMS 11>. 喂, 你 今 朝 有 無 番*?* >_<

JP: wai3 nei5 gam1 ziu1 jau5 mou5 faan1
MPH: EXP, 2.SG this morning have NEG return ? EMO
FT: 'Hey, did you go (to school/work/home) this morning?'

<SMS 12>. 多 *D* 休息, 快 *D* 好 番 啦*!*

JP: do1 di1 jau1 sik1, faai3 di1 hou2 faan1 laa1
MPH: more COM rest, quick COM well return PART
FT: 'Take more rest so that you will get well as soon as possible.'

<SMS 13>. 你 *mud* 訓 "教" 呀*?* 我 係 度 你 都 訓*?*

JP: *nei5 mat1 fan3_gaau3 aa3 ? ngo5 hai2 dou6 nei5 dou1 fan3*
MPH: 2.SG why sleep PART ? 1.SG PREP here 2.SG still sleep
FT: 'Why are you still sleeping when I am here?'

<SMS 14>. What r u goin to do after school?

FT: 'What are you going to do after school?'
NB: u = 'you', r = 'are'

<SMS 15>. ? Don't know what u are talking about.

FT: 'I don't know what you are talking about.'

<SMS 16>. I know ar... But I have nothing else lar bor

FT: 'I know, but I have nothing else.'
NB: *ar* and *lar bor* are sentence final particles in Cantonese and they are romanized by the author here.

<SMS 17>. Hahaha... Very funny. very funny...

FT: 'Hahaha, it's very funny!'

<SMS 18>. Revision lor

FT: 'I am revising.'
NB: *lor* = Cantonese SFP. This message could be an answer to a question like 'What are you doing?'.

<SMS 19>. Haha... ofcourse la. You ppl are so...

FT: 'Haha, of course! You people are so...(elided)'
NB: *la* = Cantonese SFP, *ppl* = people

<SMS 20>. Happy 21 dearest! Afraid calling u will wake u up... can we meet on Monday?

FT: 'Happy 21st birthday, my dearest. I don't want to call you as I don't want to wake you up. Can we meet on Monday?'

<SMS 21>. So glad I didn't phone the hunk last nite

NB: *nite* = 'night'

<SMS 22>. Ladies hockey champion! Yeah >:D

NB: >:D = 'happy, laughingly'

<SMS 23>. Hey dear, had a wonderful talk wif sam. Thanks 4 ur time. Work hard and take good rest tmr.

FT: 'Hey, Dear, I had a wonderful chat with Sam. Thanks for your time. Work hard and take a good rest tomorrow.'

NB: *wif* = 'with', *4* = 'for', *ur* = 'your', *tmr* = tomorrow

<SMS 24>. Glad to receive ur call. We have not chatted for ages lar. Can I see u when u r back?

NB: *ur* = 'your', *lar* = Cantonese SFP, *u* = 'you', *r* = 'are'

<SMS 25>. Where R U worry ar me

FT: 'Where are you? I am worried.'
NB: *R* = 'are', *U* = 'you', *ar* = Cantonese SFP

<SMS 26>. I feel so sorry

<SMS 27>. Pls forgive me

NB: *pls* = 'please'

<SMS 28>. Gd Show

NB: *Gd* = 'good'

<SMS 29>. You too... wish you all the best in everything!

<SMS 30>. Superpass! elizabeth:D

NB: *Superpass*, directly translated from the Cantonese expression 勁過*ging6 gwo3* (superpass), an expression to wish somebody good luck in exams; this is common among university students.

<SMS 31>. wanna give u a kiss

FT: 'I want to give you a kiss.'
NB: *wanna* = want to, *u* = 'you'

<SMS 32>. I love you

<SMS 33>. Hehe very painful ar just now

FT: 'It's was very painful just now!'

NB: *Hehe* = laughter, *ar* = Cantonese SFP

<SMS 34>. AGM ar? ++oil ar!!

MPH: Annual General Meeting PART ? add oil PART!!
FT: 'You're going to have the AGM? Do your best and good luck!'

<SMS 35>. my part was already finished...yeah! u ar at home and Tping?

NB: *u* = 'you', *ar* = 'are'

<SMS 36>. so happy, can win the match. haha... champion!

FT: 'I'm so happy that we won the championship!'

<SMS 37>. u left ur pen in lab

FT: 'You left your pen in the laboratory.'
NB: *u* = 'you', *ur* = 'your'

<SMS 38>. 去唔去飲茶*?* 去就打番俾我

JP: *heoi3 m4 heoi3 jam2 caa4 ? heoi3 zau6 daa2 faan1 bei2 ngo5*
MPH: go NEG go drink tea ? go then call back give 1.SG
FT: 'Do you like to go to "yam cha"? Give me a call if you are interested.'

<SMS 39>. Ai Ya... ho sun fu ar... the ref bk is very difficult... can't understand a word ar! @-@ ho mun! just want to tell as a release... hai...

MPH: EXP INTENS unbearable PART DET reference book is very difficult can't understand a word PART EMO INTENS boring ! just want to tell as a release... SIGH

FT: 'Oh, I can't bear it! The reference book is very difficult. I can't even understand a word! It's really boring! Well...I just want to tell you as a way to get some relief.'

<SMS 40>. Finish your service mei ar? I still haven't eat ar... wanna eat together?

FT: 'Have you finished your service? I still haven't eaten anything. Do you want us to eat together?'

NB: *mei* = 未 *mei6* 'not yet', *ar* = Cantonese SFP, *wanna* = 'want to'

<SMS 41>. Haha, still 17 premarked question ar... may be u canget back before the session end...:p

FT: 'Haha…I still have 17 premarked questions to go through. Perhaps you can have them back before the session ends [EMO]'.

NB: *ar* = Cantonese SFP, *u* = 'you'

<SMS 42>. Ho chur ar... I really want to take a rest on Sat b4 campaign

FT: 'I am very exhausted. I really want to take a rest on Saturday before the campaign begins.'

NB: *ho chur* = *hou2 coe2*, 好□, where *hou2* being the INTES and *coe2* does not have a corresponding character in the language, meaning 'exhausted, tired'; Romanized Cantonese coined by the author.

<SMS 43>. Hohoho... lo lic ar! Me is also bok jonging... the stuff is so hard that I nearly wants to give up ar... But your call gives me energy back la, thx! Smile babe

FT: 'Work hard! I'm also working hard. The stuff is so hard that I almost want to give up! But your call brought me back to life again. Thanks! Smile, babe!

NB: *lo lic* = 努力 *nou5lik6* 'work hard', *bok jonging* = 搏盡 *bok3 zeon6* 'working extremely hard'. Here, the English progressive marker *–ing* is being attached to the Cantonese romanization.

<SMS 44>. Heavy raining wor... but u still need to go rite? Be careful ar:p

FT: 'It's raining heavily. But you still need to go, right? Take care.'

NB: *wor*, *ar* = Cantonese SFP, *rite* = 'right'

<SMS 45>. I have finished the tutor la... Have u leave your office? Ate dinner? Better take a little break b4 driving ar... Ho Keung has lost...

FT: 'I have finished the tutorial. Have you left your office? Have you taken dinner? You should better take a break before you drive. Ho Keung has lost.'

NB: *la, ar* = Cantonese SFP, The last sentence *Ho Keung has lost* is a new piece of information.

<SMS 46>. How's your foot? Does it hurt much? Anywhere got hurt again? Worry u all the day lei... Phone me when u are free so that I can know whether u are alrite la

FT: 'How's your foot? Does it hurt much? Have you hurt yourself anywhere else? I've been worrying for you for the whole day. Call me when you are free so that I will know whether you are alright.'

NB: *u* = 'you', *lei* = Cantonese SFP, *alrite* = alright

<SMS 47>. Hai... feeling bad now... seems no mood and want to vomit! Nothing la... just want to tell u... you work hard la!:p Should have a fast progress le

FT: 'Oh, I'm feeling bad now. I seem to have no mood to do anything and I also want to vomit. Well...nothing special indeed. I just wanted to tell you about my present condition. You should work hard. I am sure you will make significant progress.

<SMS 48>. Lo lic... everything will be fine again as u've pay your effort in it! Don'y blame yourself due to the missing check I'm always there to listen

FT: 'Work hard! Everything will be fine if you put much effort in it. Don't blame yourself for having missed the check. I am always there to listen to you.'

NB: *Lo lic* = work hard!

<SMS 49>. I have finished practice ball la:p How about u? Gan Ba Tei! Support u always

FT: 'I have finished my training session (for a kind of ball game). How about you? Work Hard! I'll always support you!'

<SMS 50>. TRAPPED between the stairs!

FT: 'I got trapped between the staircases!'

<SMS 51>. Sorry cannot have dinner with you, working

FT: 'Sorry, I cannot have dinner with you as I'm still working.'

<SMS 52>. Sorry I am not back stay home you work hard

FT: 'Sorry, I am not back yet. I'll stay at home and you should work hard!'

<SMS 53>. 想贏取標緻*206XT*房車? 即刻將你的電郵地址透過手機以*SMS*傳送至*3383383388* 給我們.即可參加抽獎*!4*月 *7*日截止·

NB: [Advertisement from mobile service provider]

<SMS 54>. *HAPPY BIRTHDAY To You*

<SMS 55>. *Here is your bill for thismoth O. Outstanding Bal: $0 Your Account No: 0406xxxx. View your bill at www.smartone.com.hk*

NB: Billing information from mobile service provider

<SMS 56>. *My phone is low batt... so we dine at pacific place? When and where do we meet?*

FT: 'My phone's battery is low. So shall we dine at Pacific Place? When and where do we meet?'

<SMS 57>. *Here is your bill for this month (02/04). Outstanding Bal: $78.4 Please pay now. Your Account No: 0406xxxx. View your bill at www.smartone.com.hk*

NB: [Billing information from mobile service provider]

<SMS 58>. *Chair bay sum gay, 505 spiritually support you!*

FT: 'Chairman/lady, try your best! 505 (room no.) is supporting you spiritually/morally.'

NB: *bay sam gei* = Cantonese expression 俾心機, 'to put effort, to try one's best'

<SMS 59>. *Happy Birthday! Celebrate later la:)*

FT: 'Happy Birthday! Let's celebrate later. [EMO = happy]'

NB: *la* = Cantonese SFP *laa1*

<SMS 60>. *I can't forgive them.*

<SMS 61>. Happy New Year!! 新年快樂*!!*

JP: Happy New Year!! *San1 nin4 faai3 lok6*
MPH: Happy New Year!! New year happy
FT: 'Happy New Year!! Happy New Year!!'

<SMS 62>. *幸福豬*

JP: *hang6 fuk1 zyu1*
MPH: happy pig
FT: 'Happy Pig'

*<SMS 63>. You have 1 New Voice Msg (s). 70 listen, pls press *55*

NB: [Information from mobile service provider]

<SMS 64>. Which check pt?

NB: pt = 'point'

<SMS 65>. yup.

FT: 'yes.'

<SMS 66>. have lunch yet?

FT: 'Have you had lunch yet?'

<SMS 67>. chi sin

FT: 'Non-sense! Crazy!'
NB: *chi sin* = Cantonese expression 黐線 *ci1sin3*, used to describe something or somebody being non-sensical, out of mind.

<SMS 68>. 請注意, 動物園家走失*o*左一隻馬騮同一隻豬西, 馬騮家*o*向沙灘曬緊太陽. 豬西則睇緊自己部手機*o*既短訊.

JP: *cing2 zyu3 ji3, dung6 mat6 jyun4 gaa1 zau2 sat1 zo2 jat1 zek3 maa5 lau1 tung4 jat1 zek3 zyu1 sai1, maa5 lau1 gaa1 hon2 saa1 taan1 saai3 gan2 taai3 joeng4, zyu1 sai1 zak1 tai2 gan2 zi6 gei2 bou6 sau2 gei1 ge3 dyun2 seon3*

FT: 'Your attention please, the zoo family has lost a monkey and a pig, the monkey is sun-bathing on a beach, the pig is checking her SMS.'

<SMS 69>. 生日快樂*! 20*歲喇. 大個女喇. 祝你有個開開心心*o*既*20*歲生日*!*

JP: *saang 1 jat6 faai3 lok6*! 20 *seoi3 laa3. zuk1 nei5 jau5 go3 hoi1 hoi1 sam1 sam1 ge3* 20 *seoi3 saang1 jat6*

MPH: birth day happy! 20 years PART. Big CL girl PART. Wish 2.SG have CL happy GEN 20 birth day!

FT: 'Happy Birthday! Now you're 20! You're already a grown-up. Wishing you a happy 20th birthday!'

<SMS 70>. 新年快樂. 龍馬精神. 身壯力健

JP: *san1 nin4 faai3 lok6. lung4 maa5 zing1 san4. san1 zong3 lik6 gin6*

MPH: new year happy. Dragon horse energetic. Body strong strength healthy

FT: 'Happy New Year! Wishing that you are energetic, healthy, and strong at all times.'

NB: A serious way of traditional Chinese New Year greetings.

<SMS 71>. n_n

NB: EMO = bored

<SMS 72>. U dun CARE ME!

FT: 'You don't care for me!'

NB: U = 'you', dun = 'don't'

<SMS 73>. I SLEEP LA. You TAKE MORE REST!

FT: 'I'm going to sleep. You should take more rest!'

NB: *LA* = Cantonese SFP *laa3*

<SMS 74>. GRACE, I AM IN K318 AR, HOW ARE YOU AR? COMPUTER IS IN STUDIO AR:P WILL WE HAVE DINNER?:P HE, HE.

FT: 'Grace, I'm in K318 now. How are you? The computer is in the studio. Shall we have dinner? [Smiling]

NB: K318 = a room number, ar = Cantonese SFP *laa3*

<SMS 75>. GRACE, I WAIT U IN RUNRUN SHAW AR:P HO MA? LUNCH TOGETHER:P

FT: 'Grace, shall I wait for you in Run Run Shaw? [EMO = happy / smiling]? Let's have lunch together. [EMO = happy / smiling]'

NB: u = 'you', RunRun Shaw = name of a building in HKU, *ar, ma* = Cantonese SFPs *aa3* and *maa3*.

<SMS 76>. G, I AM NOW SLEEPING IN HALL. LUNCH OK AR. 12:15 AR:P U CAN PHONE MY HALL GA:P

FT: 'G, I'm now sleeping in the hall. Lunch will be fine. [EMO = happy]. Let's say 12:15 pm. You can call my hall number [EMO = happy].

NB: *ar, ga* = Cantonese SFPs *aa3* and *gaa3*, u = 'you'

<SMS 77>. CAN WE HAVE LUNCH TOGETHER? WHEN U OK? >.<

FT: 'Can we have lunch together? When will you be free? [EMO = kissing]'

<SMS 78>. HO MO LIU AR~ WHERE ARE U AR?

FT: 'I'm getting bored. Where are you?'

NB: *HO MO LIU* = Romanized form of the Cantonese expression *hou2 mou4 liu4*, 'feeling very bored'

<SMS 79>. HAPPY NEW YEAR! Zebada! +<01>

<SMS 80>. Wei- wei, TigeR LazY aH! haha!

<SMS 81>. 早晨, 男朋友仔*!*

JP: *zou2 san4 naam4 pang4 jau5 zai2*

FT: 'Good morning, little boy friend'

<SMS 82>. 我開完會喇... *n_n*

JP: *ngo5 hoi1 jyun3 wui2 laa3*

MPH: 1.SG open PERF meeting PART [emo = bored]

FT; 'I've just finished a meeting.'

<SMS 83>. 你訓O左未呀... ?

JP: *nei5 fan3 zo2 mei6 aa3*
MPH: 2.SG sleep PERF yet PART
FT: 'Are you already asleep?'

<SMS 84>. My physics DIE! Can you teach me after mock? MISS YOU SO MUCH

NB: die = metaphorically means that he/she didn't do well in an exam

<SMS 85>. hey, keith. Are we going to have a meeting today? when is it? - louis (management)

<SMS 86>. HI

<SMS 87>. 生日快樂!:D 仲有新年快樂! 希望你有一個開開心心o既20歲!

JP: *saang1 jat6 faai3 lok6 zung6 jau5 san1nin4 faai3 lok6 hei1mong6 nei5 jau5 jat1 go3 hoi1 hoi1 sam1 sam1 ge3 20 seoi3*
MPH: born day happy EMO still have new year happy wish 2.SG have one CL happy GEN 20 years
FT: 'Happy Birthday (happily) and also Happy new year. Wishing you a joyful 20th year.'

<SMS 88>. QUIZ

<SMS 89>. Pk Merry X mas

<SMS 90>. Wa, thanks ar! same to u la. i still have exam on thu ar

NB: wa = EXP, u = 'you', ar, la = Cantonese SFPs *aa3* and *laa1*

<SMS 91>. happy new year out club soon

NB: out club = 'I'm going to have a date', hall jargon

<SMS 92>. Happy new year! Wish you all the best:p

NB::p = happy, laughingly, joyful

<*SMS 93*>*. i am so stupid that forgot bringing pencil bag...*

FT: 'I'm so stupid to have forgotten to bring my pencil case'

<*SMS 94*>*. i think no need la i still have 1 can use... ai...*

FT: 'I think there is no need. I still have one that works...<sigh>'
NB: la = Cantonese SFP *laa3*

<*SMS 95*>*. can hand in on time?*

FT: 'Can you hand it (possibly an assignment) on time?'

<*SMS 96*>*. Still worj'g hard or already given up? sorry 4 be'g 2 playful 2 most of da time.. should i learn 2 b more like a grown-up or be whom i m?*

FT: 'Are you still working hard or have you already given up? I'm sorry for being too playful for most of the time. Should I learn to be more like a grown-up or should I be who I am?'
NB: worj'g (typo, should be work'g) = 'working', 4 = 'for', be'g = 'being', 2 = 'too', da = 'the', b = 'be'

<*SMS 97*>*. how r u feeling?*

FT: 'How are you feeling?'
NB: r = 'are', u = 'you'

<*SMS 98*>*. luv u 24/7!*

NB: luv = 'love', u = 'you', 24/7 = July 24th

<*SMS 99*>*. miss u terribly! wanna c u & hold u so much. will hv quite a bz wk. =(*

FT: 'I'm missing you terribly! I want to see you and hold you so much. I'll have quite a busy week (emo = sad)
NB: u = 'you', wanna = 'want to', c = 'see', u = 'you', hv = 'have', bz = 'busy', wk = 'week'

<*SMS 100*>. 唔好*ANGRY with me*啦.

JP: *m4 hou2* angry with me *laa1*
MPH: NEG good angry with me PART
FT: 'Don't be angry with me!'

<SMS 101>. wat cha do'g? i m still @ P nite. had alot 2nite..! >.< call u when i m home. miss u.

FT: 'What are you doing? I'm still at Presentation night. I had a lot tonight. (emo = kissing) I'll call you when I get home. Miss you.'
NB: wat cha = 'what are', do'g = 'doing', @ = 'at', P nite = 'Presentation Night', a hall event, u = 'you'

*<SMS 102>. *kiss**

<SMS 103>. done w tt ages ago. wear r u now?

FT: 'I've been done with it for ten years now. Where are you now?'
NB: w = 'with', wear = 'where', r = 'are', u = 'you'

<SMS 104>. sorry.i dun mean 2 b so mean 2 u.

FT: 'Sorry! I don't mean to be so mean to you.'
NB: dun = 'don't', 2 = 'to', b = 'be', u = 'you'

<SMS 105>. actualli wanna talk 2 u face 2 face. wanna c u so much. miss u desperately. >.<

FT: 'Actually, I want to talk to you face-to-face. I really want to see you. I'm missing you desperately. (kiss)'
NB: actually = 'actually', wanna = 'want to', 2 = 'to', c = 'see', >.< = emoticon, kissing

<SMS 106>. r u 3 2nite? izit possible 2 hv dinner 2getda? call me

FT: 'Are you free tonight? Is it possible to have dinner together? Call me.'
NB: r = 'are', u = 'you', 3 = 'free', 2nite = 'tonight', hv = 'have', 2getda = 'together'

<SMS 107>. CAMPAIGN

SUSCESSAH;)

LOVE U

FT: 'The campaign was successful. I love you'
NB: ;) = 'Happy, with one eye blinking', u = 'you'

<SMS 108>. LAU LUEN HUNG

BAY SOME GAY

NB: *Lau Luen Hung* = PN, *bay sam gei* = Cantonese expression 俾心機, 'to put effort, to try one's best'

<SMS 109>. LIP STILL HURT A BIT..ENJOY UR MEAL N WORK HARD ON UR HW TONITE:)

FT: 'My lips still hurt a bit. Enjoy your meal and work hard on your homework tonight. (smiling)'
NB: *UR* = 'your', N = 'and', HW = 'homework', tonite = 'tonight',:) = smiling, happily

<SMS 110>. U:)= I:) U:(= I:(+ WORRY TAKE CARE

MPH: You happy = I happy You unhappy = I unhappy + worry Take Care
FT: 'If you are happy, I will be happy. If you're sad, I'll be sad and worried. Take Care!'
NB: mix of emoticons and words: U = 'you',:) = happy,:(= sad, unhappy

<SMS 111>. haha..

hoho..sorry didn't phone u b4 the match to cheer u up, i was busy.. sorry about that. anyway... take most rest la! c u in HK in june!

FT: '(laughingly) Sorry I didn't call to cheer you up before the match. I was busy. Sorry about that. Anyway, take a lot of rest! See you in Hong Kong in June!'
NB: u = 'you', b4 = 'before', la = Cantonese SFP *laa3*, c = 'see'

<SMS 112>. MISS U VERY VERY MUCH DUNNO HOW TO TELL U

FT: Miss you very much. I don't know how to tell you.
NB: u = 'you', dunno = 'don't know'

<SMS 113>. Wei...no rain la i guess! Yeah! I have lesson till 3pm.can go back together ...

FT: 'Hey, I guess it has stopped raining! Yeah! I will have lesson until 3pm. We can go back together.'
NB: la = Cantonese SFP *laa3*

<SMS 114>. when u get up, pls give me a call la!

FT: 'Please call me when you get up!'
NB: u = 'you', pls = please, la = Cantonese SFP *laa3*

<SMS 115>. 喂!俾心機做功課呀!不過不要做得太夜啦, 早些休息啦!

JP: *wai3, bei2 sam1 gei1 zou6 gung1 fo3 aa3! Bat1 gwo3 bat1 jiu3 zou6 dak1 taai3 je6, zou2 se1 jau1 sik1 laa1*
MPH: Hey put effort do homework PART! But NEG need do DAK1 INTENS late PART, early COM rest PART
FT: 'Hey, put more effort in your homework! But don't work till too late. You should rest earlier!'

<SMS 116>. Sorry...I fail everyone...cant make it at last...but this is not an end.. maybe do sth for next year..we're partner till the end!

FT: 'Sorry. I've failed everyone. I couldn't make it at last. But this is not the end. Perhaps we can do something for next year. We are partners till the end!'

<SMS 117>. 完了!無論結果怎樣也完了。我也懺悔而回。俾心機讀書吧!讓所有成為回憶好了!我們一起也努力過……

JP: *jyun4 liu5! Mou4 leon6 git3 gwo2 zam2 joeng6 jaa5 jyun4 liu5. ngo5 jaa5 caam3 fui3 ji4 wui4. bei2 sam1gei1 duk6 syu1 baa6! Joeng6 so2 jau5 sing4 wai4 wui4 jik1 hou2 liu5! Ngo5 mun4 jat1 hei2 jaa5 lou5 lik6 gwo3*
MPH: finish PART! However result how also finish PART. 1.SG also regret to come. Put effort read book PART! Let all be memory good PART! 2.PL together also work hard PERF
FT: 'It's over! However it is, it's over! I returned with regrets. Let's study hard.

Leave everything behind in our memories. We have tried our best...'

<SMS 118>. 真的想贏!但已經再沒辦法了...

JP: *zan1 dik1 soeng2 jeng4! Daan6 ji5 ging1 zoi3 mut6 baan6 faat3 liu5...*
MPH: really want win! But already again NEG method PART
FT: 'I really wanted to win. But there was no way out...'

<SMS 119>. its vicky. heard about result of game. dont be too upset, sure ur team has tried ur best. sometimes u ve 2 take into a/c of circumstancial elements & luck..

FT: 'It's Vicky. I just heard about the result of the game. Don't be too upset. I'm sure that your team has tried its best. Sometimes you have to take into account circumstantial elements and luck.'

NB: ur = 'your', u = 'you', ve = have, a/c = account, circumstantial = (typo: circumstantial)

<SMS 120>. Pity u. What a long meeting.

NB: u = 'you'

<SMS 121>. See u 3:35pm at Dai pai fong a.

FT: 'See you at 3:35 pm at the main entrance of HKU.'

NB: u = 'you', Dai pai fong = jargon of HKU, meaning the main entrance of the University of Hong Kong. a = Cantonese SFP *aa3*.

<SMS 122>. thanks for the t-shirt, it's grate!

NB: grate = 'great'

<SMS 123>. HAPPY BIRTHDAY! SWEET TWENTY-TWO!

<SMS 124>. SORRY! FORGIVE ME OK? DINNER NEXT TIME

<SMS 125>. THX

NB: THX = 'thanks'

<SMS 126>. 我一直想對你**說**這幾個字,我怕一旦**說**出,我們連朋友都沒法做了,但我控制不住感情,鼓起勇氣對你**說**:益力多,你今日飲佐未呀!哈哈

JP: *ngo5 jat1 zik6 dou1 soeng2 deoi3 nei5 syut3 ze5 gei2 go3 zi6, ngo5 paa3 jat1 daan6 syut3 ceot1, ngo5 mun4 lin4 pang4 jau5 dou1 mou5 dak1 zou6 liu5, daan6 ngo5 hung3 zai3 bat1 zyu6 gam2 cing4, gu2 hei2 jung5 hei3 deoi3 nei5 syut3: lik1 lik6 do1, nei5 gam1 jat6 jam2 zo2 mei6 aa3! Haa1haa1*

MPH: 1.SG always want to 2.SG say DET few CL word, 1.SG fear once say out, 1.PL even friend NEG way make PART, but 1.SG control NEG emotion, arouse courage to 2.SG say: Yakurt, 2.SG today drink PERF NEG PART! [LAUGH]

FT: 'I have been wanting to say these few words to you. I fear that once I say them out, we can't even be friends. But I can't control my emotion. So, I am saying this courageously to you: "Yakurt, have you drunk today?"'

<SMS 127>. JUST WANT TO SAY GDNITE TO U

FT: 'I just want to say good night to you'

NB: GDNITE = 'good night', U = 'you'

<SMS 128>. thank you for your present............

<SMS 129>. Wait you 230 at kowloon bay 車頭 *to Lam tin*

FT: 'I will wait for you at 2:30 at the front area of the platform towards Lam Tin in Kowloon Bay Station'

NB: kowloon bay = a district in HK, here it refers to the MTR station. 車頭*ce1 tau4*, car-head, referring to the front area of the platform.

<SMS 130>. weiwei- irene ar...testing testing ar

if u receive this msg on your mobile..tell me by leaving a msg in icq la- thx

FT: 'Hey, Irene, testing! Testing! If you can receive this message on your mobile, tell me by leaving an ICQ message. Thanks.'

NB: ar, la = Cantonese SFPs *aa3* and *laa1*, u = 'you', msg = 'message', thx = 'Thanks'

<SMS 131>. YES LA! BUT MY MA SAID HER COM. HASN,T KEPT THIS RECENTLY.SO SORRY AR!

FT: 'Yes. But my mother said her company hasn't kept this recently. Sorry!'
NB: la, ar = Cantonese SFP *laa3*, *aa3*, ma = 'mother', com. = 'company'

<SMS 132>. HEY i am back,free tomolo?

FT: 'Hey, I'm back. Will you be free tomorrow?'
NB: tomolo = 'tomorrow'

<SMS 133>. Bill date:07/04

Used Airtime:983min

Current balance:$159.00

Due Date:21/04

NB: [Billing information from mobile service provider]

<SMS 134>. TABLE TENNIS TEAM WILL HAVE PRACTISE TMR (8-10) PLZ DON'T BE LATE! CALL ME IF ANY PROBLEM!

THX!

NB: plz = 'please'

<SMS 135>. 今天還未跟你說生日快樂呢...不要溫書太夜,早點休息吧*!=>*

JP: *gam1 tin1 waan4 mei6 gan1 nei5 syut3 saang1 jat6 faai3 lok6 ne1..bat1 jiu3 wan1 syu1 taai3 je6, zou2 dim2 jau1 sik1 baa6!*
MPH: today still not yet to 2.SG say birth day happy PART..NEG need revise book INTENS late, early COM rest PART [EMO]
FT: 'I still haven't said Happy Birthday to you today…Don't revise till too late. Rest earlier! (smiling)'

<SMS 136>. 兼兼,恭喜你的新書出版。封面的設計有好新的感覺,白色唔錯呀*!*今天我去沙田大眾時你的新書已經買斷了。好好成績*!*繼續努力*!*收到我的信嗎*? Yen*上

JP: *him1 him1, gung1 hei2 nei5 san1 syu1 ceot1 baan2. fung1 min2 dik1 cit3 gai3 jau5 han2 hou2 dik1 gam2 gok3, baak6 sik1 m4 co3 aa1! Gam1 tin1 ngo5 heoi3 saa1 tin4 daai6 zung3 si4 nei5 dik1 san1 syu1 ji5 ging1 maai6 tyun5 liu5. hou2 hou2 sing4 zik1! Gai3 zuk6 lou5 lik6! Sau1 dou2 ngo5 dik1 seon3 maa1? YEN soeng.*

<SMS 137>. WaiWai

*HAPPY BDAY**

yeah:)

FT: 'Hey
Happy Birthday
Yeah (smiling, happily)'

<SMS 138>. 你係咪惱我,*SORRY*我都想你□,不過□係唔得,你唔好惱,至多我應承你唔*CALL*其他女,亦唔比其他女*CALL*我,好無?你唔惱我*o*羊都應承你。

JP: *nei5 hai6 mai6 nou5 ngo5, SORRY ngo5 dou1 seong2 nei5, bat1 gwo3 hai6 m4 dak1, nei5 m4 hou2 nou5, zi3 do1 ngo5 jing1 sing4 nei5 m4 CALL kei4 taa1 neoi2, jik6 m4 bei2 kei4 taa1 neoi2 CALL ngo5 hou1 mou2? Nei5 m4 nou5 ngo5 me1 dou1 jing1 sing4 nei5*

MPH: 2.SG be NEG angry 1.SG, SORRY 1.SG also want 2.SG □, but □ be NEG fine, 2.SG NEG good angry, COM many 1.SG promise 2.SG NEG CALL other girl, also NEG let other girl CALL 1.SG, good NEG? 2.SG NEG angry 1.SG whatever also promise 2.SG

FT: 'Are you angry with me? Sorry. I also want you xx but I couldn't make it. Please don't be mad. I promise that I won't call any other girls and I won't let other girls call me. OK? If you are not angry, I can promise you anything.'

<SMS 139>. Are you here?

*<SMS 140>. SUNDAY*賬單

截數日

03/03/02

應繳款項

$113.25

將於截數日十四日內經自動轉賬扣除

NB: [billing information from mobile service provider]

<SMS 141>. happy one month!:p

NB: 'one month' = having been lovers for one month

<SMS 142>. U are really mo liu wor,haha...u miss me lei,haha:p

FT: 'You are talking nonsense.You must be missing me! (laughingly)'
NB: u = 'you', mo liu = Cantonese expression *mou4 liu4*, meaning 'bored, nonsensical', *wor, lei* = Cantonese SFPs *wo3, le5*

<SMS 143>. in ikea,thinking of u...

FT: 'I'm in IKEA thinking of you…'
NB: u = 'you'

<SMS 144>. I wanna find u 2...

FT: 'I'm also looking for you.'
NB: wanna = 'want to', u = 'you', 2 = 'too'

<SMS 145>. Goodnight ar

NB: ar = Cantonese SFP *aa3*

<SMS 146>. Happy birthday !!! Wish u a very very good ex_co year!

NB: u = 'you', ex_co = 'executive committee'

<SMS 147>. add oil!

FT: 'Do your best. Work Hard!'

<SMS 148>. It really makes me upset to hear what u r wondering.

NB: u = 'you', r = are'

<SMS 149>. CINDY2000257XXX JADE2000086XXX

FT: 'Cindy's student no. is 2000257XXX, Jade's is 2000086XXX.'

<SMS 150>. can u put 2000 into 082-3-040837?

FT: 'Can you transfer 2000 dollars to [no. of an account]?'

<SMS 151>. pppp k k

p p k k

pppp kkk

pp k k

pp k k

NB: PK is a person's name.

<SMS 152>. Hey bei sum gei la... Happpy up jong.(from QQ)

FT: 'Hey, do your best! I wish you all the best in your work on the executive committee'

NB: bei sum gei = 'do one's best', up jong = university jargon, meaning being part of the executive committee of a student society

<SMS 153>. You 2.Wish u every success!

FT: 'You too. I wish you every success!'

NB: 2 = 'too', u = 'you'

<SMS 154>. GOODNIGHT CHI CHUN, SLEEP WELL AR!

SIU MUI '...'!

NB: CHI CHUN, SIU MUI = PN, AR = Cantonese SFP *aa3*

<SMS 155>. CHI CHUN, PLS DON'T ANGRY WITH ME LA, I KNOW HOW TO DO LA

FT: 'Chi Chun, please don't get angry with me. I know what to do.'
NB: CHI CHUN = PN, PLS = 'please', la = Cantonese SFP *laa1*

<SMS 156>. Forgot to say tim..CONGRATULATIONS for having a successful inaugration ar!Sorry about the flash.....also hope you are not sleeping in soc rm tonight la!

FT: 'I forgot to say "Congratulations for having a successful inauguration. Sorry about the flash. I hope that you are not going to sleep in our society's office tonight'

NB: *tim, ar, la* = Cantonese SFPs, soc rm = the office place of a student society

<SMS 157>. Hello ar!Long time no come to 10/F lar!I'm in Caterina's room to revise tmr's panda test ar.You must be very busy with the inaug stuff lar,dun get sick bor!

FT: 'Hello! I haven't been to 10/F for a long time. I'm now in Caterina's room to revise for tomorrow's test. You mush be very busy with the inauguration. Don't break down!'

NB: ar, lar, bor = Cantonese SFPs, tmr = 'tomorrow', inaug = 'inauguration'

<SMS 158>. Where are u?

Are u ok?
NB: u = 'you'

<SMS 159>. Wei... How are you ar?Long time on see la!Busy with jong stuff, the Stanford exchange?

FT: 'Hey, how are you? Long time no see! Are you still busy with your executive committee affairs? Or are you busy with the Stanford exchange stuff?

NB: la, ar = Cantonese SFPs *laa3* and *aa3*

<SMS 160>. Last night so good la! u ar so nice. man!

NB: la = Cantonese SFP *laa3, u* = 'you'

<SMS 161>. Terri, Je te souhauite des bons chances!

MPH: Terri, 1.SG 2.SG wish of good luck
FT: 'Terri, I wish you good luck!'
NB: This message is written in French.

<SMS 162>. I am sitting on the plane now

<SMS 163>. I am so tirre. ll have to sleep, I call u later and date u dinner to-night!

FT: 'I'm so tired. I have to sleep. I'll call you for dinner tonight!'
NB: tirre = (typo) tired, u = 'you'

<SMS 164>. 神看見你渴, 他創造了水, 神看見你餓, 他創造了米, 神看見你沒有朋友, 所以他創造了我, 然而神也看見這世上沒有傻瓜, 順便也創造了你(開玩笑, 別生氣)

JP: *san4 hon3 gin3 nei5 hot3, taa1 cong3 zou6 liu5 seoi2, san4 hon3 gin3 nei5 ngo6, taa1 cong3 zou6 liu5 mai5, san4 hon3 gin3 nei5 mut6 jau5 pang4 jau5, so2 ji5 taa1 cong3 zou6 liu5 ngo5, jin4 ji4 san4 jaa5 hon3 gin3 ze5 sai3 soeng6 mut6 jau5 so4 gwaa1, seon6 bin2 jaa5 cong3 zou6 liu5 nei5 (hoi1 waan4 siu3, bit6 sang1 hei3)*

MPH: god look see 2.SG thirsty, 3.SG create PERF water, god look see 2.SG hungry, 3.SG create PERF rice, god look see 2.SG NEG have friend, so 3.SG create PERF 1.SG, but god also look see DET world up NEG have idiot, by-the-way also create PERF 2.SG (open play laugh, NEG angry)

FT: 'God saw that you were thirsty, so He created water. God saw that you were hungry, so He created rice. God saw that you didn't have any friends, so He created me. God also saw that there were no fools on earth, so He created you. (Just kidding, don't get mad!)'

<SMS 165>. HI TERI BONJOUR

ME WOKE UP LA.

AS BUSY AS A BEE AR.

FT: 'Hi, Teri, good morning. I've already got up. I'm as busy as a bee now!'
NB: bonjour = good morning (French), la, ar = Cantonese SFP *laa3, aa3*

<SMS 166>. 如果靚仔係一種罪, 我已經罪犯滔天;如果有型係一種錯, 我已一錯再錯;如果聰明要受懲罰, 我豈非要千刀萬割?如果謙虛都要受折磨, 我又點能逃得過

JP: *jyu4 gwo2 leng3 zai2 hai6 jat1 zung2 zeoi6, ngo5 ji5 ging1 zeoi6 faan6 tou1 tin1; jyu4 gwo2 jau5 jing4 hai6 jat1 zung2 co3, ngo5 ji5 jat1 co3 zoi3 co3; jyu4 gwo2 cung1 ming4 jiu3 sau6 cing4 fat6, ngo5 hei2 fei1 jiu3 cin1 dou1 maan6 got3? Jyu4 gwo2 him1 heoi1 dou1 jiu3 sau6 zit3 mo4, ngo5 jau6 dim2 nang4 tou4 dak1 gwo3?*

MPH: if beautiful boy be one CL crime, 1.SG already crime commit fill heaven; if have style be one CL mistake, 1.SG already one wrong again wrong; if intelligent need accept punishment, 1.SG INTRG NEG need thousand knifes ten thousand cuts

FT: If being handsome were a crime, I had already committed a very heavy one. If being stylish were a mistake, I had already committed this twice. If being intelligent had to be punished, I would have to receive thousands of cuts. If being humble had to be tortured, how could I escape from it?

<SMS 167>. Hey TERRI GoOd LuCk.

<SMS 168>. THX 4 THE CARD YESTERDAY.IT IS A GREAT ENCOURANGMENT. VERY SORRY FOR LETTING U DOWN THOUGHOUT THE YEAR.

NB: THX = 'thanks', 4 = 'for', U = 'you'

<SMS 169>. i just return from that:) will u ar? so many ppl wor

FT: I just returned from that (event). Will you join me? There were so many people.'

<SMS 170>. where are u?

CHA KIN

NB: u = 'you', CHA KIN = PN

<SMS 171>. u with somebody or alone?

NB: u = 'you'

<SMS 172>. Just trying if i can send msg. Bojin in ur last subject la.

NB: msg = 'message', Bojin = 搏盡 *bok3 zeon6* 'working extremely hard'.

<SMS 173>. yue loi yue lang sum sheung see sing blue ;>

FT: 'Wishing that you get more and more beautiful and that you can achieve whatever you want'

NB: yue loi yue lang = Romanized form of the Cantonese expression: to become more and more beautiful, sum sheung see sing = Romanized form of the Cantonese expression: *sam1 soeng2 si6 sing4*: to achieve whatever you want

<SMS 174>. Can't help. He has exams on the coming Saturdays.

<SMS 175>. He's busy these days... Probably won't make it this week.

<SMS 176>. Don't know...

Have to ask, but he porbably has to accompany Sandy. Maybe ask him if he's free this Sat or Mon, or just have him back tomorrow.

NB: probably = (typo) 'probably'

<SMS 177>. i meet u at 1:30 to hv lunch ok?

NB: u = 'you', hv = 'have'

<SMS 178>. I Wanna go w u to buy new clothes i try my best to be quick pls wait for me

NB: wanna = 'want to', w = 'with', u = 'you', pls = please

<SMS 179>. honey u r bored?

NB: u = 'you', r = 'are'

<SMS 180>. I WILL GO HOME LATER

<SMS 181>. FEELING BETTER?

<SMS 182>. BUT GOIN 2 PARIS AFTER XSAM

NB: GOIN = 'going', 2 = 'to', XSAM = 'exam'

<SMS 183>. AGREE STILL GOT A LOT TO WORK ON LAST D OF SAM IS 7.6

FT: 'I agree. There is still a lot to work on. Sam's last date of work is July 6th.
NB: D = 'day'

<SMS 184>. GO SOMETIN' FUNNY TO TELL?

<SMS 185>. Hows ant? He seems tired... And will he change the entry ar?

NB: ant = ?, ar = Cantonese SFP *ar*

<SMS 186>. May i go to sogo to see playmobil?

<SMS 187>. Wai Happy New <Year ar_XX= Hope U Super Pass_XX=:)

From Felix

FT: 'Hey, Happy New Year! I hope that you will pass all of your exams. (smiling) From Felix.'

<SMS 188>. I ve put your key back to your room

NB: ve = 'have'

<SMS 189>. Can u write a short essay for the spirit of a_team in the web page?

NB: u = 'you'

<SMS 190>. 8:00@minibus station ok?

NB: @ = 'at'

<SMS 191>. I will go to mong kok to buy sth will u go there?

NB: sth = 'something', u = 'you'

<SMS 192>. Will you wait for me?

<SMS 193>. Within 30min

<SMS 194>. Yes

<SMS 195>. OK

<SMS 196>. when will it finish?

<SMS 197>. When will you go? Now?

<SMS 198>. Will you go to eat 'siu' with me? I invite you wo

NB: eat 'siu' = to have late night snacks / meal, wo = Cantonese SFP *wo3*

<SMS 199>. No

<SMS 200>. i don't know!

<SMS 201>. You are in my room right?

<SMS 202>. I have let gary to have a look of my homework but he let the other to copy it!

FT: 'I just let Gary have a look at my homework but he let other people copy it!'

<SMS 203>. Very angry!

<SMS 204>. My mother bought me 4x4 ar!

NB: 4x4 = ? (a product?), ar = Cantonese SFP *aa3*

<SMS 205>. I will finish the lecture at 5:30

<SMS 206>. Not finished yet i think its about 1130

<SMS 207>. Have u finish your tution? Will u go to anywhere else or u will back hall?

NB: u = ‘you’

<SMS 208>. My friend ask me to go taiwan

<SMS 209>. Please call me when u finish your lesson

NB: u = ‘you’

<SMS 210>. i can't wake u up ar,I have lesson today,sorry!

NB: u = ‘you’, ar = Cantonese SFP *aa3*

<SMS 211>. i sleep la

NB: la = Cantonese SFP *laa3*

<SMS 212>. i wanna go now,s.o use key come in,can't sleep!

NB: u = ‘you’

<SMS 213>. DIU LEI

NB: Cantonese swearing

<SMS 214>. WHAT DO U SAY AT LAST AR?

NB: U = ‘you’, ar = Cantonese SFP *aa3*

<SMS 215>. WEI DIM AR LI

WHERE R U?

FT: ‘Hey, how are you? Where are you?’
NB: wei = ‘hey’, DIM AR LI = romanized Cantonese expression *dim2 aa3 nei5*, meaning ‘how are you?’, R = ‘are’, U = ‘you’

<SMS 216>. CHIU OK AR

NB: chiu = could mean a person’s name or an expression of disapproval, ar = Cantonese SFP *aa3*

<SMS 217>. WEI TOMOLO YUM CHA 1200 OK MA?

FT: 'Hey, shall we have Yum Cha at 12 noon tomorrow?'
NB: wei = 'hey', tomolo = 'tomorrow', yum cha = 'going to Chinese restaurant to take Chinese dim sum', ma = INTRG *maa3/1* 嗎

<SMS 218>. FINISHED? CAN U DINNER TONITE?

FT: 'Have you finished (your work)? Can you have dinner with me tonight?'
NB: U = 'you', TONITE = 'tonight'

<SMS 219>. Pls Call back "92XXXXXX"

FT: 'Please return a call at [tel.no.]'
NB: Pls = 'please',

<SMS 220>. bay sum gay la CM =)

FT: 'Try your best, CM (smilingly)'
NB: bay sum gay = 'try your best, work hard', la = Cantonese SFP *laa3/1*

<SMS 221>. Hi my name is johny boring in library

FT: 'Hi, my name is Johnny and I'm feeling bored in the library'

<SMS 222>. tomolo can't c movie yum cha ho ma?

FT: 'I can't see movie with you tomorrow. Shall we have Yum Cha instead?'
NB: tomolo = 'tomorrow', c = 'see', yum cha = 'going to Chinese restaurant to take Chinese dim sum', ho ma = INTRG, *hou2 maa3/1*好嗎?

<SMS 223>. HI are u free, would u like to have tea?

NB: u = 'you'

<SMS 224>. 912XX525

:-)

NB: This message contains a telephone number and a smiling emoticon.

<*SMS 225*>. *wei where r u ar c movie ma?*

FT: 'Hey, where are you? Will you like to see movie?'
NB: wei = 'hey', r = 'are', u = 'you', c = 'see', ma = INTRG *maa3 / 1* 嗎

<*SMS 226*>. *YES I RECEIVE*

FT: 'Yes, I have received (something).'

<*SMS 227*>. 二十歲生日快樂*!*

JP: *ji6 sap6 seoi3 saang1 jat6 faai3 lok6*
MPH: twenty years birth day happy
FT: 'Happy 20th Birthday'

<*SMS 228*>. 珮珮*:*今日*retake*要俾心機*!*一定會過*!*

JP: *pui3 pui3, gam1 jat6* RETAKE *jiu3 bei2 sam1 gei1! Jat1 ding6 wui5 gwo3*
MPH: PN today RETAKE must put efforts! Sure will pass
FT: 'Pui Pui, you must do your best in today's re-examination. I'm sure you'll pass!'

<*SMS 229*>. *I am back! we are going to have lunch in sai wan, call me after chong meeting la!*

NB: chong meeting = university jargon, meaning a student society executive meeting

<*SMS 230*>. *HAPPY BIRTHDAY!WHSH U A WONDERFUL DAY.*

NB: U = 'you'

<*SMS 231*>. 生日快樂呀壽星女*!*你日點慶祝呢*?*

JP: *saang1 jat6 faai3 lok6 aa3 sau6 sing1 neoi2! Nei5 dim2 hing3 zuk1 ne1?*
MPH: birth day happy PART birthday star girl, 2.SG how celebrate PART?
FT: 'Happy Birthday, the birthday girl! How are you going to celebrate it?'

<*SMS 232*>. *happy birthday:>*

<SMS 233>. MISS U SUDDENLY!

NB: U = 'you'

<SMS 234>. 你找我嗎*?***我有去上堂呀***!*

JP: *nei5 zaau2 ngo5 maa1? Ngo5 jau5 heoi3 soeng5 tong4 aa3!*
MPH: 2.SG find 1.SG PART? 1.SG have go up lesson PART
FT: 'Were you looking for me? I did attend the lecture!'

<SMS 235>. 唔駛擔心□,我唔會俾頭先個人影響到.驚我食得飽過頭仲好*:p*

JP: *m4 sai2 daam1 sam1 ?? ngo5 m4 wui5 bei2 tau4 sin1 go3 jan4 jing2 hon2 dou2. geng1 ngo5 sik6 dak1 baau2 gwo3 tau4 zung6 hou2 [EMO]*
MPH: NEG need worry ?? 1.SG NEG will let head first CL person affect DOU, fear 1.SG eat COMP full too even good
FT: 'Don't worry! I would not let that person influence me. You should rather fear that I will eat too much. (laughingly)'

<SMS 236>. 亞康*:*如果你有時間, 請你致電給. 他已經取消了飛線

JP: *aa3 hong1: jyu4 gwo2 nei6 jau6 si4 gaan3, cing2 nei5 zi3 din6 kap1 taa1 ji5 ging1 ceoi2 siu1 liu5 fei1 sin3*
MPH: PN: if 2.SG have time, please 2.SG give call 3.SG already cancel PERF call divert
FT: Aa Hong (the message writer): If you get time, please call him. He has already cancelled the call-divert function (on his mobile phone)'

<SMS 237>. Come on! It's supposed to be fun. Cheer up. I'll do whatever you want me to...

<SMS 238>. Leaving now. Do have a nice day and be faithful.

Love you...

<SMS 239>. Relax, and don't get yourself stressed up. You're doing fine. In my prayer... Take care.

<SMS 240>. I didn't get any message... Just get off now. I think I'll be training at Sandy Bay with the sunshine. Are you okay there?

<SMS 241>. I'm going to swim now. Be confident and only if then your dream comes true. You'll be fine.:)

<SMS 242>. What if you back hall to get changed first, and I go to find you out later when I finish studying here.

FT: 'What if you go back to the hall and get changed first, and then I will meet you later when I have finished studying here?'

<SMS 243>. Oh, right. Forgot that... Well, let's meet tomorrow then. Don't really want you to rush around.

<SMS 244>. 楊振文律師行何律師 *"25249266"*

JP: *joeng4 zan3 man4 leot6 si1 hong2 ho4 leot6 si1* '25249XXX'
MPH: PN lawyer firm PN lawyer [telephone no.]
FT: 'Lawyer Ho from Yeung Chan Man Law Firm: tel: 25249XXX'

<SMS 245>. I basically don't mind getting them tomorrow, but I leave it to your decision, okay?

<SMS 246>. Should got that right probably. Is it painful this time?

FT: 'I should probably be able to get that right. Is it painful this time?'

<SMS 247>. I call you later when I'm done then.:)

<SMS 248>. Forgot to divert the line... I'm having the cookies and not finishing, trying to have you taste them. Really nice!:)

FT: 'I forgot to divert the calls...I still have the cookies and haven't finished them yet. I will get you to taste them. They are really nice. (smilingly)'

<SMS 249>. sei poh, wish u prettier and prettier, happy forever, love u always ar!

FT: 'You stupid thing, I wish that you'll become prettier and prettier. Happy always and I love you always!'

NB: sei poh = *sei2 po2* in Cantonese, which is normally used to address close friends, u = 'you', ar = Cantonese SFP *aa3*.

<SMS 250>. hv gd meals & gd rest; wish u gd health & gd luck! less snacks, more water too ah! beating air 4u! call me when "letting dogs out" (fong 9) ah! luv u all!

FT: 'Take good meals and good rest. Wishing you good health and good luck. Eat less junk food and drink more water. I will support you at all times. Call me when you go out to walk your dog. I love you all!'

NB: hv = 'have', gd = 'good', u = 'you', ah = Cantonese SFP *aa3*, beating air = romanized Cantonese expression 打氣 *daa2 hei3*, meaning 'to cheer somebody up', fong 9 = romanized Cantonese expression 放狗*fong3 gau2*, meaning 'to walk a dog'

<SMS 251>. KHFC ah! arrived home safely la, thx again for tonite! call me to go to see monsters or anything or hear more stories la! health & happiness with you always!

NB: KHFC = Kung Hei Fat Choy, traditional Chinese new year greeting, la = Cantonese SFP *laa3*, thx = 'thanks', tonite = 'tonight'

<SMS 252>. MERRY CHRISTMAS! MARRY POSTER XMAS SPELL: I LOVE JAY QUICKLY TELL OTHERS! HAHA. MISS U LA:)

<SMS 253>. WHERE ARE U? PHONE ME ASAP. FR. ISABELLA

NB: U = 'you', ASAP = 'as soon as possible', FR. = 'from'

<SMS 254>. BE QUICK LA! VERY EMBARRASSING HA!

NB: la = Cantonese SFP *laa1*

<SMS 255>. HI! BEAUTIFUL GIRL! COFFEE AT MY HOUSE? HAH?

<SMS 256>. i am having lecture now and i find this class very boring. miss u all.

NB: u = 'you'

<SMS 257>. exams finished la. now coming.

FT: 'I've just finished the exam. I'm coming now'

NB: la = Cantonese SFP *laa3*

<SMS 258>. coach, i am finding u urgently la. i need your chinese and eng full name. pls reply me asap.

FT: 'Coach, I am looking for you desperately. I need your Chinese and English full name. Please reply as soon as possible.'

NB: u = you, la = Cantonese SFP *laa3*, pls = 'please', asap = 'as soon as possible'

<SMS 259>. have u finished your lessons?

NB: u = 'you'

<SMS 260>. how come u dun come to class today ah? u said u'd come and so i brought the heavy notes...

FT: 'How come you didn't come to class today? You said you would come and so I brought the heavy notes for you'

NB: u = 'you', dun = 'don't', ah = Cantonese SFP *aa3*, u'd = 'you would'

<SMS 261>. At 15:23 on 02/04/28 You have 1 new Voice-mail message(s) and 0 new Fax message(s).

NB: [Information message]

<SMS 262>. Price slash on International SMS! Send SMS to 35 countries/destinations for only $2/message, or '1 SMS + $1' or 'SMS Package' customers'.

NB: [Information message from mobile service provider]

<SMS 263>. HAVE A NICE DAY:)

<SMS 264>. Are you ok?

<SMS 265>. check yr email for poster

NB: yr = 'your'

<SMS 266>. HAPPY NEW YEAR

<SMS 267>. Merry X'mas!

<SMS 268>. Get up soon pl for presen

FT" 'Please get up soon! You are going for a presentation.'
NB: pl = 'please', presen = 'presentation'

<SMS 269>. Confirmed with my school that we r to teach. time could be arranged

FT: 'I have confirmed with my school that we are going to teach. Time could be arranged.'
NB: r = 'are'

<SMS 270>. Just wanna tell ya i ve brought ur shirt and short if u r to be there

FT: 'Just want to tell you that I have brought your shirt and shorts, in case you will be there.'
NB: wanna = 'want to', ya = 'you', ve = 'have', ur = 'your', u = 'you', r = 'are'

<SMS 271>. www.altavista.com babefish translator

<SMS 272>. Hope u r having a gd time.

NB: u = 'you', r = 'are', gd = 'good'

<SMS 273>. Time to get up to hand in all the forms and documents needed

<SMS 274>. ssDo u have to ignore me like that?! fine...u'll see

NB: u = 'you'

<SMS 275>. where the heck are you?

<SMS 276>. well, her english is not that gd

NB: gd = 'good'

<SMS 277>. in fact, her english sucks! Haha

<SMS 278>. pl bring my charger. i left it in ur rm.

FT: 'Please bring my charger. I left it in your room.'
NB: pl = 'please', ur = 'your', rm = 'room'

<SMS 279>. *y r u not answering the fone?*

FT: 'Why are you not answering the phone?'
NB: y = 'why', r = 'are', you = 'you', fone = 'phone'

<SMS 280>. *r u free this sat? i am having an easter party at my place. pls come and call me if u can. take care!*

NB: r = 'are', you = 'you', pls = 'please', u = 'you'

<SMS 281>. *easter party is cancelled. call me for further details.*

<SMS 282>. *ARE WE GOING FOR LUNCH? WHERE ARE U? CALL ME NOW PLS CUZ I WILL HAVE CLASS AT 2.*

NB: U = 'you', pls = 'please', cuz = ''cause' = 'because'

<SMS 283>. *sorrie that i can't make it today. cya next time pal.*

NB: sorrie = 'sorry', cya = 'see you'

<SMS 284>. *i cant make it to class today. pls take the lecture notes for me. Th\x.*

NB: cant = 'can't', pls = 'please', thx = 'thanks'

<SMS 285>. *wei wei...sui yah! Still not here yet? Where are u ah? call me now la pls!*

FT: 'Hey, stupid thing! You are not in yet? Where are you? Please call me now!'
NB: wei wei = 'hey', u = 'you', ah, la = Cantonese SFPs *aa3* and *laa3*, pls = 'please'

<SMS 286>. *wei...free for tea today? call me ah:)*

FT: 'Hey, are you free for tea today? Call me. (smilingly)'

NB: wei = 'hey', ah = Cantonese SFP *aa1*

<SMS 287>. i'm in the lib now 6 fl. i've got a pl for u la

FT: 'I'm on the 6th floor in the library. I've already reserved a place for you.'
NB: lib = 'library', 6 fl = '6th floor', pl = 'place', possible a study place in the library, u = 'you'

<SMS 288>. take care la, dun be so down la. cheer up! dunno what to say la...

NB: la = Cantonese SFP *laa3*, dun = 'don't', dunno = 'don't know'

<SMS 289>. call bro up now

NB: bro = 'brother'

<SMS 290>. HAPPY BIRTHDAY!!!:)

NB: :) = happy, smilingly

<SMS 291>. Bill date: 26/04 used Airtime: 1013 mins.

NB: [Billing information from mobile service provider]

<SMS 292>. At 19:54 on 02/04/10 You have 1 new Voice-mail message(s).

NB: [information message]

<SMS 293>. At 22:01 on 02/04/10 You have 2 new Voice-mail message(s).

NB: [information message]

<SMS 294>. lesson time ar, man!

NB: ar = Cantonese SFP *laa3*

<SMS 295>. Haha...I'm still in tst lor...

NB: tst = Tsim Sha Tsui, a district in Hong Kong, lor = Cantonese SFP *lo1*

<SMS 296>. Lei ho boring ar? Gum charm?

CC: 你好 BORING 呀? 咁 慘
JP: *nei5 hou2* BORING *aa4? Gam3 caam2*
MPH: 2.SG INTENS BORING PART? INTENS miserable
FT: 'Are you feeling bored? Poor you!'

<SMS 297>. That's ok la, jack pui lei ma. Anyway...miss you much

FT: 'That's ok. Jack will accompany you. Anyway, I miss you so much.'
NB: la, ma = Cantonese SFPs *laa1*, *maa3*, pui lei = romanized Cantonese expression *pui4 nei5* 陪你, meaning 'to accompany you'.

<SMS 298>. Bill date: 27/03 used Airtime: 1418 mins.

NB: [Billing information from mobile service provider]

<SMS 299>. DEAREST JESSICA, HAPPY BIRTHDAY! LOVE FROM SABRINA

<SMS 300>. Jessica, good morning!

<SMS 301>. Good morning Jessica, see you tonight in the high table dinner, remember to wear the yellow hall t-shirt and jeans! Take care! Love, Sabrina

<SMS 302>. Hi Jessica, sorry i can't wait for u la...i thought i could donate ga ma...anyway, next time go with u la.

NB: u = 'you', la, ga ma = Cantonese SFPs *laa3, gaa1 maa3*

<SMS 303>. Hi where r u?

NB: r = 'are', u = 'you'

<SMS 304>. whered u go?

FT: 'Where did (would?) you go?'
NB: u = 'you'

<SMS 305>. im leavn tdy 2 singapor

FT: 'I'm leaving today for Singapore'
NB: levn = 'leaving', tdy = 'today'

<SMS 306>. y?

FT: 'Why?'

<SMS 307>. can u email wat u did to me?

FT: 'Can you email what you did to me?'
NB: u = 'you', wat = 'what'

<SMS 308>. u done histry?

FT: 'Have you done the history (assignment?)?'
NB: u = 'you'

<SMS 309>. duno how come u aint sleepn?

FT: 'I don't know why you are not sleeping.'
NB: duno = 'don't know', u = 'you', sleepn = 'sleeping'

<SMS 310>. what r u doin?

FT: 'What are you doing?'
NB: r = are, u = 'you', doin = 'doing'

<SMS 311>. gold sing

<SMS 312>. tiger lazy

<SMS 313>. merry x'mas! Holla'

<SMS 314>. 想要*hallmark*賀年屏幕圖像?請即回覆, 收費*$3*。只適用於諾基亞手機, 如不需要請按刪除, 即上*www.wow.net.hk*仲有更多選擇。

NB: [Information message from mobile service provider]

<SMS 315>. 你個攤位收得工未?跟住去邊?

JP: *nei5 go3 taan1 wai2 sau1 dak1 gung1 mei6? Gan1 zyu6 heoi3 bin1?*
MPH: 2.SG CL assign place collect DAK work yet? Follow PERF go where?
FT: 'Has your counter been closed already? Where will you go afterwards?'

<SMS 316>. ()""()

(▌'▌o▌'▌)")

(.) /

Happy New Year

NB: The upper part forms a picture of fireworks.

<SMS 317>. hi hi

<SMS 318>. So boring ar!

FT: 'I'm feeling bored!'
NB: ar = Cantonese SFP *aa3*

*<SMS 319>. PEOPLES*流動*Email,*隨時用手機收發電郵免月費仲送*5MB*郵箱及電郵提示!用手機上*Wapeoples*或打*29458888*查詢

NB: [Promotional information message from mobile service provider]

<SMS320>. 想免金一次過, 登記晒*IDD*及漫遊服務*?*請填妥三月份的*PEOPLES*月結單內*Easy Call*組合申請表, 並於*4*月*4*日或以前傳真回本公司。

NB: [Promotional information message from mobile service provider]

*<SMS 321>. PEOPLES*客戶尊享手機优惠:諾基亞*5210*新貨再度應市, 憑此短訊可以*$1488*購買, 數量有限, 請即搶購。

NB: [Promotional information message from mobile service provider]

<SMS 322>. ngo jo yeah yat chai hup ngan funn @_@

JP: *ngo5 jou6 je5 jat1 cai4 hap1_ngaan5_fan3*

CC: 我 做 野, 一 齊 瞌眼訓
MPH: 1.SG do things one together sleepy [EMO]
FT: 'I'm working. Both of us are feeling drowsy at the same time (sleepily).'

<SMS 323>. have you fin the meeting yet? I knew that you may be hungry now...

NB: fin = 'finished'

<SMS 324>. I am @ genki waiting call me =>

FT: 'I'm at Genki queuing up for a table. Call me! (smilingly)'
NB: @ = 'at', genki = a Japanese sushi restaurant

<SMS 325>. weve got good time today =>

NB: weve = 'we've'

<SMS 326>. afraid of wake u up, whisper a good morning

FT: 'I didn't want to wake you up. So I'm whispering a good morning to you'
NB: u = 'you'

<SMS 327>. work hard =>

<SMS 328>. Sorry, I need be a little bit late, call me after lesson.

<SMS 329>. i arrive, call me.

NB: 'I've arrived. Call me!'

<SMS 330>. cannot contact you, call me back.

<SMS 331>. in lesson, call you back later.

FT: 'I'm attending a lesson. I'll call you later.'

<SMS 332>. 生日快樂 *<george.wong)*

JP: *saang1 jat6 faai3 lok6*
MPH: birth day happy PN

FT: ‘Happy Birthday, from George Wong’

<*SMS 333*>. 喂!您□邊呀?

JP: *wai3 nei5 hai2 bin1 aa3*
MPH: hey 1.SG PREP where PART
FT: ‘Hey, where are you?’

<*SMS 334*>. 我好掛住你!

JP: *ngo5 hou2 gwaa3 zyu6 nei5*
MPH: 1.SG INTENS miss PERF 2.SG
FT: ‘I’m missing you very much!’

<*SMS 335*>. 我愛你

JP: *ngo5 ngoi3 nei5*
MPH: 1.SG love 2.SG
FT: ‘I love you’

<*SMS 336*>. *Surprise? We both use CSL. Get yr phone card yet?*

FT: ‘Are you surprised that both of us are using CSL (a mobile service provider). Have you got your phone card?’
NB: yr = ‘your’

<*SMS 337*>. *Just finished lesson, hungry & tired.*

<*SMS 338*>. *Still sick? Or better now*

FT: ‘Are you still feeling sick? Or are you getting better now?’

<*SMS 339*>. *Got your parents yet?*

<*SMS 340*>. *In Cheung Chau now?*

FT: ‘Are you at Cheung Chau (an Island in Hong Kong) now?’

<*SMS 341*>. *Remember today is the last day of flea+cens’ sale? Please go if you can spare time.*

<SMS 342>. Happy New Year

<SMS 343>. Kung Hei Fat Choi ah! Wish u all the best in the coming yr. n' most importantly.... Get more 'lai si ah!!!'

NB: Kung Hei Fat Choi = traditional Chinese new year greeting, 恭喜發財 *gung1 hei2 faat3 coi4*, to wish someone wealthy, ah = Cantonese SFP *aa3*, u = 'you', yr = 'year', n' = 'and', lai si = red packets

<SMS 344>. 新年快樂

JP: *san1 nin4 faai3 lok6*
MPH: new year happy
FT: 'Happy new year'

<SMS 345>. Happy Valentine's! Je t'aime. =)

FT: 'Happy Valentines day'! I love you (smilingly)'
NB: Je t'aime = 'I love you' (French)

<SMS 346>. Giving lesson? I'm so cheap that I fell asleep on the sofa without shower n washing away my makeup! Very tired, but I hope it wouldn't happen next time!

FT: 'Are you teaching now? I was so bad that I fell asleep on the sofa without taking a shower and removing my make-up! I was very tired. But I hope that it will never happen again!'
NB: n = 'and'

<SMS 347>. Hey, how are u ah, very boring at office now... let's have tea or dinner later on lar. Hope to see you soon.

FT: 'Hey, how are you? I'm feeling bored in the office now. Let's have tea or dinner later on. I hope to see you soon.'
NB: u = 'you', ah, lar = Cantonese SFPs *aa3* and *laa3*

<SMS 348>. HAPPY EASTER!

<SMS 349>. Work hard together la!

NB: la = Cantonese SFP *laa1*

<SMS 350>. Be Happy!

<SMS 351>. HAPPY BIRTHDAY la!

<SMS 352>. Will u go to the library today?

NB: u = 'you'

<SMS 353>. What's the matter?

<SMS 354>. happy new year, V

<SMS 355>. pls call back

NB: pls = 'please'

<SMS 356>. ROGER CALL BACK.

FT: 'Roger, please call back!'

<SMS 357>. Happy birthday

<SMS 358>. BB豬， 您吾HAPPY係好正常, 我知錯啦,I愛U豬仔 !!

JP: BB *zyu1, nei5 m4* HAPPY *hai6 hou2 zing3 soeng4, ngo5 zi1 co3 laa3,* I *ngoi3* U *zyu1 zai2*

MPH: BB pig, 3.SG NEG HAPPY be INTENS right usual, 1.SG know wrong PART, 1.SG love 3.SG pig boy

FT: 'Baby pig, it's understandable that you're not happy. I know that I'm wrong, I love you, Piggy!'

<SMS 359>. Call me =p

NB: =p = smilingly (EMO)

<SMS 360>. I have called u but can't connect don't angry baby

FT: 'I called you but the line was engaged. Don't get angry, baby.'
NB: u = 'you'

<SMS 361>. BB豬 CALL ME B4 2 THANKS

FT: 'Little Piggy, call me before 2 o'clock. Thanks'
NB: BB豬 = 'piggy', B4 = 'before', 2 = 'to'

<SMS 362>. B, WHERE ARE U< CALL ME

NB: U = 'you'

<SMS 363>. I'm FREE call me

<SMS 364>. Baby I LOVE YOUR LOVELY AND CUTY SOUND

NB: CUTY = 'cute', SOUND, here, it means 'voice'

<SMS 365>. BB PLEASE WAKE ME UP AT 2:00 AND I SAY SORRY FOR MY TEMPER

<SMS 366>. 數碼通客戶可以*10%*折扣出*MotorolaV70*或以特价*$1630*出*Nokia5210*伏惠期至*10/3*

NB: [promotional information message from mobile service provider]

<SMS 367>. 你最新的賬單*(9/3)*已發出應繳款項*$98*請即繳付

NB: [promotional information message from mobile service provider]

<SMS 368>. Also ar.... Anxiety in study... I think... eg 1) thesis la, not enough time ma, 21 supervisor/teaching staff, cos dunno how they think abt us me etc.

FT: 'Anxiety from my study is also a factor, I guess. (i) I haven't got enough time for my thesis (ii) I don't know how my supervisor and other teaching staff would think about me....etc.'

NB: This message is possibly a part of a 'conversation'. ar, la, ma = Cantonese SFPs *aa3*, *laa3*, *maa3*, cos = ''coz = because', abt = 'about'

<SMS 369>. 想要*Twins*最新上榜砍曲大浪漫主義做手機鈴聲, 請選擇回覆, 毋需輸入文字, 再選擇發送, 收費只通用於諾基亞, 如不需要請按刪除。

NB: [promotional information message from mobile service provider]

<SMS 370>. But, y u ask this?

FT: 'But why do you ask about this?'
NB: y = 'why', u = 'you'

<SMS 371>. Bill: 10/03 Used:612 min Bal:$176 Due: 24/03

NB: [Billing information from mobile service provider]

*<SMS 372>. Dial *096111 to get 25% off for F & B at Kln hotel w/ticket stub of Beauty & the Beast & 15% off at Disney Store $ 3/download*

NB: [promotional information message from mobile service provider]

<SMS 373>. Dial #888 for NewYear's fortune telling, jokes, blessing, & ringtone. Send more SMS & you may get a Nokia 8310! Chi lang only.

NB: [promotional information message from mobile service provider]

*<SMS 374>. Enjoy over 20 m-coupon offers in shopping, food and entertainment by dialing *092 to register with Orange Club*

NB: [promotional information message from mobile service provider]

<SMS 375>. Dial #1212 to enjoy priority concert booking service for !'Leon Live Is Live! (from 26/10 to 08/11. Service charge: $4/ticket)

NB: [promotional information message from mobile service provider]

<SMS 376>. You have received 1 voice message

NB: [Information message from mobile service provider]

<SMS 377>. hi hi

*<SMS 378>. GameFreer*首創流動互動 '武林風雲傳' 俾你同多個英雄對戰, 掀起連場激鬥! 每星期的武林盟主有機會贏*PS2*一部。即上*WAP*或打*#111*開戰

NB: [promotional information message from mobile service provider]

<*SMS 379*>. 由現在到四月八日,*Nokia6510*淨手機只需*$2450,*詳情請親臨本公司門市查詢。

NB: [promotional information message from mobile service provider]

<*SMS 380*>. 聖□快樂, *wish u have a happy and lucky - from George Wong*

FT: 'Merry Christmas. Wishing you a happy and lucky new year – From George Wong'
NB: 聖□快樂 *sing3 daan3 faai3 lok6* = 'Merry Christmas'

<*SMS 381*>. *You have 1 New Voice Msg(s) To listen, pls press *55*

NB: [Information message from mobile service provider]

<*SMS 382*>. *To (Rosa): Don't give up, I will support you all the time.*

<*SMS 383*>. *HAHA,* 我現在睡覺了, *bb.*

JP: HAHA, *ngo5 jin6 zoi6 seoi6 gaau3 liu5*, BB
MPH: LAUGH 1.SG now sleep PART, bye bye
FT: '(laughingly) I'm going to sleep now, Bye!'

<*SMS 384*>. 小心身子呀*!*

JP: *siu2 sam1 san1 zi2 aa3*
MPH: small heart body little PART
FT: 'Take care'

<*SMS 385*>. 我正在上*lesson*

JP: *ngo5 zing3 zoi6 soeng5 tong4*
MPH: 1.SG right

<*SMS 386*>. *+ oil* 呀

MPH: add OIL PART
FT: 'Try your best!'

<*SMS 387*>. 陣間先*Phone*你

JP: *zan6 gaan1 sin1* PHONE *nei5*
MPH: later first PHONE 2.SG
FT: 'I'll phone you later.'

<*SMS 388*>. *Merry X'mas!*

<*SMS 389*>. 掛住你

JP: *gwaa3 zyu5 nei5*
MPH: hang PRG 2.SG
FT: 'I'm missing you'

<*SMS 390*>. *Happy B'day.*

NB: B'day = 'birthday'

<*SMS 391*>. *How r u?*

NB: r = 'are', u = 'you'

<*SMS 392*>. *Reply ASAP.*

NB: ASAP = 'as soon as possible'

<*SMS 393*>. *Happy Birthday!*

<*SMS 394*>. *I Love U!*

NB: U = 'you'

<*SMS 395*>. *I miss U!*

NB: U = 'you'

<*SMS 396*>. *What's going on?*

<*SMS 397*>. *What ar?*

NB: ar = Cantonese SFP *aa3*

<SMS 398>. Hello!

<SMS 399>. Have a tea, ok?

<SMS 400>. Good luck

<SMS 401>. bye! CU later.

NB: CU = 'See you'

<SMS 402>. 黃小姐......夠鐘起身喇*!!* "咯.. 咯.. 咯.. 咯"

JP: *wong4 siu2 ze2, gau3 zung1 hei2 san1 laa3! Gok4 gok4 gok4 gok4*
MPH: WONG little sister……enough clock rise body PART! Knock Knock Knock Knock
FT: 'Miss Wong, it's time to get up! Knock knock…'

<SMS 403>. hi hi 我係*florance:)*

FT: 'Hi, I am Florence (smilingly)'
NB: 我係 = *ngo5 hai6* 'I am'

<SMS 404>. 我都係*LA,* 好*BUSY. I M IN TOILET NOW, HAHA*

FT: 'Me too. I'm also very busy. I'm in the toilet now! (laughing)'
NB: 我都係 = 'me too', LA =Cantonese SFP *laa1*

<SMS 405>. haha.... 因為我訓唔著，同埋今朝*9:00*有比賽*lor,* 但我依家返到屋企*lu,* 好眼訓, 但訓唔到, 慘. . . 慘. . . 慘 >.<

JP: HAHA…*jan1 wai6 ngo5 fan3 m4 zeok6, tung4 maai4 gam1 ziu1* 9:00 *jau5 bei2 coi3 lo1, daan6 ngo5 ji4 gaa1 faan1 dou3 uk1 kei2 lu3, hou2 ngaan5 fan3, daan6 fan3 m4 dou2, caam2 caam2 caam2*
MPH: LAUGH…because 1.SG sleep NEG ZEOK, and this morning 9:00 have competition PART, but 1.SG now return arrive home PART, INTENS eye sleep, but sleep NEG DOU, miserable, miserable, miserable [EMO]
FT: '(laughing) Because I couldn't sleep, and I had a contest at 9am this morning. But I have returned home now. I'm feeling very tired but I can't fall asleep!

Poor me!'

<SMS 406>. weiwei.....still in whampo?? Take care ar, and quite cool now!

FT: 'Hey, are you still in Whampo? Take care. It's quite cool now!'
NB: wei wei = 'hey', ar = Cantonese SFP *aa3*, Whampo = place in Hong Kong

<SMS 407>. lunch? call back.

<SMS 408>. I will dine out with Dabby tonight. Call Gerald after 4pm

<SMS 409>. lunch? hehe

<SMS 410>. Congratulations Well done Proud of you:)

<SMS 411>. I am at Fung Wong Shan now!!!

NB: Fung Wong Shan = Hill in Hong Kong

<SMS 412>. Good show ar. Ohoh Yeah

NB: ar = Cantonese SFP *aa3*

<SMS 413>. 恭喜發財 *FROM VIVI*

JP: *gung1 hei2 faat3 coi4* FROM VIVI
FT: 'Wishing you wealthy. From Vivi.'
NB: traditional Chinese new year greeting

<SMS 414>. You have received 1 voice mail

NB: [information message from mobile service provider]

<SMS 415>. Hihi

<SMS 416>. 九龍酒店*9*折優惠

JP: *gau2 lung4 zau2 dim3 gau2 zit3 jau1 wai6*
MPH: Kowloon hotel 9 reduction good offer

NB: 'Kowloon hotel 10% off special offer'

<SMS 417>. Go to rock

<SMS 418>. Thanks for ur hard work... Buy u a dinner later. Siu Lai

FT: 'Thanks for your hard work. I'll buy you dinner later. Siu Lai (the message writer)'
NB: ur = 'your', u = 'you', Siu Lai = PN

<SMS 419>. Okay I goin now wait 4 u at times

FT: 'Okay, I'm going now. See you at Times Square.'
NB: goin = 'going', 4 = 'for', u = 'you'

<SMS 420>. Best wishes for your coming exam! Add Oil!! From Sabrina:)

NB: *Add Oil* = direct translation of the Cantonese expression *gaa1jau2*, which is used to encourage somebody to work hard.

<SMS 421>. How is the interview? I am online in ICQ.... Give me a call if you have finished at 91233XXX Siu Lai

<SMS 422>. Happy new year

<SMS 423>. buy me sth to eat

<SMS 424>. I hv a nice day

NB: hv = 'have'

<SMS 425>. I'll treat u vvvv good

FT: 'I'll treat you very well.'
NB: u = 'you', vvvv = iteration of INTENS

<SMS 426>. C U tmr

FT: 'See you tomorrow'
NB: C = 'see', U = 'you', tmr = 'tomorrow'

<SMS 427>. happy new year

<SMS 428>. wei wei?? Dun be unhappy la. Let's add oil lei!

FT: 'Hey, don't be sad. Let's do our best!'

NB: wei wei = 'Hey', dun = 'don't', la, lei = Cantonese SFPs, *add oil* = direct translation of the Cantonese expression *gaa1jau2*, which is used to encourage somebody to work hard.

<SMS 429>. Dun worry Be happy la:)

NB: dun = 'don't', la = Cantonese SFP *laa1*

<SMS 430>. 真係掛*0*架*!* 你尋**晚**好夜先睡*?*

JP: *zan1 hai6 gwaa3 gaa3! Nei5 cam4 maan5 hou2 je6 sin1 seoi6?*

MPH: really be miss PART! 2.SG last night INTENS late first sleep?

FT: 'I really miss you! Did you sleep very late last night?'

<SMS 431>. 我都叫你脾*D HEART*機**溫**書*!* 算啦*.* 今**晚**再努力*!*

JP: *ngo5 dou1 giu3 nei5 bei2 di1 HEART gei1 wan1 syu1! Syun3 laa1! Gam1 maan5 zoi3 lou5 lik6*

MPH: 1.SG already tell 2.SG put CL effort review book! Never_mind PART, today night again endeavor

FT: 'I already told you to study hard! Never mind. Try your best again to-night!'

<SMS 432>. Don't think too much la! Support U!

NB: la = Cantonese SFP *laa3*, U = 'you'

<SMS 433>. SLEPT?

<SMS 434>. Call me when you are free please! ^ _ ^

<SMS 435>. hi. having lesson ar, what's matter?

NB: ar = Cantonese SFP *aa3*

<SMS 436>. *I sleep la call you tom afternoon*

FT: 'I'm going to sleep. I'll call you tomorrow afternoon.'
NB: la = Cantonese SFP *laa3,* tom = 'tomorrow'

<SMS 437>. *May I meet you online later tonight*

<SMS 438>. *Edith you can call me how.*

<SMS 439>. *MISSING U NOW! WHERE ARE U?*

NB: U = 'you'

<SMS 440>. *HAPPY LUNAR NEW YEAR! WISH YOU MORE DUCK YEE THAN BEFORE! KAKAKA!*

NB: DUCK YEE = 'cute', KAKAKA = laughters

<SMS 441>. *I will be okay. Don't worry so much!*

<SMS 442>. 勁過*!* 加油*!*

JP: *ging6 gwo3! Gaa1 jau2!*
MPH: super pass! Add oil!
FT: 'I wish that you'll pass all the exams! Work hard!'
NB: ging6 gwo3 = Hall jargon, meaning being able to pass all examinations with good results

<SMS 443>. *AT HALL?*

<SMS 444>. *WHAT'S UR VERSION OF SPSS?*

NB: SPSS = a computer software for statistical analysis

<SMS 445>. *HEHE*

NB: HEHE = (giggling)

<SMS 446>. *Gong hei Fat Choi! Happy new year!*

NB: Gong hei fat choi = traditional Chinese new year greeting.

<*SMS 447*>. *OK la, u take care la.*

FT: 'That's ok. Take care!'
NB: la = Cantonese SFP *laa1*, u = 'you'

<*SMS 448*>. *THIS IS SO BORING*

<*SMS 449*>. *OH MY GOD, I HOPE THIS FINISHES SOON.*

<*SMS 450*>. *WHO CALLED? NO BATTERY AT*

NB: This message is incomplete.

<*SMS 451*>. *CARD RECHARGED ON.... REMAINING BALANCE IS ....*

NB: [information message]

<*SMS 452*>. *MERRY CHRISTMAS!*

<*SMS 453*>. *Happy BDay!!!*

FT: 'Happy Birthday!!!'

<*SMS 454*>. *Add Oil! Call me back!*

NB: *Add Oil* = direct translation of the Cantonese expression *gaa1jau2*, which is used to encourage somebody to work hard

<*SMS 455*>. *How's Sch today?*

NB: *sch* = 'school'

<*SMS 456*>. *Available? Wanna to c u!*

NB: *wanna* = 'want to', *c* = see, u = 'you'

<*SMS 457*>. *Miss u! Miss u! Miss u!*

NB: *u* = ‘you’

<SMS 458>. Waiting for you!

<SMS 459>. Hello!

<SMS 460>. Good Day!

<SMS 461>. ^_^

NB: emoticon = happy, smiling

<SMS 462>. MISS YOU A!

NB: *A* = Cantonese SFP

<SMS 463>. 家姐 你又做咩呀? 我岩岩考完*econ*囉! 你睥心機上課啦! 得閒再傾

JP: *gaa1ze1 nei5 jau6 zou6 me1 aa3? Ngo5 ngaam1 ngaam1 haau2 jyun4* econ *lo1. nei5 bei2 sam1 gei1 soeng5 fo3 laa1, dak1 haan4 zoi3 king1*

MPH: family sister 2.SG again do what PART? 1.SG just examine PERF economics PART ! 2.SG give effort attend lesson PART. Have time again chat

FT: ‘Sister, what are you doing again? I just finished my economics exam. You should pay attention to the lecture. Let’s chat again when we get time.’

<SMS 464>. 你細妹透過*hkonline.com*同你講: 去*camp*你就好啦迷! 幾時先返呀? 我好掛住你!你都申請個*send*野睥我啦 咁得意

JP: *nei5 sai3 mui2 tau3 gwo3* hkonline.com *tung4 nei5 gong2*: *heoi3* camp *nei5 zau6 hou2 laa1 mai4 gei2 si4 faan1 aa3? Ngo5 hou2 gwaa3 zyu6 nei5! Nei5 dou1 san1 cing2 go3* send *je5 bei2 ngo5 laa1 gam3 dak1 ji3*

MPH: 2.SG small sister through hkonline.com with 2.SG say: go camp 2.SG so good PART ?? when first return PART 1.SG INTENS miss PROG 2.SG 2.SG too apply CL send things give 1.SG PART INTENS funny

FT: ‘Your sister would like to say this to you through hkonline.com: It’s good that you are going camping. When will you return? I’m missing you very much! You should also apply for one in order to send things to me. It’s fun!’

<SMS 465>. Boring lecturer

<SMS 466>. Happy Birthday

<SMS 467>. I love you forever!

<SMS 468>. 早晨, 你再做甚麼*?*

JP: *zou2san4 nei5 zoi3 zou6 sam6 mo1*
MPH: good morning, 2SG again do what
FT: 'Good Morning. What are you doing?'

<SMS 469>. **Wait at CYM.**

FT: 'Wait for me at CYM.'
NB: CYM = an amenities centre at the University of Hong Kong.

<SMS 470>. **Tmr** 記 得 交 **essay ^v^**

JP: tmr *gei3 dak1 gaau1* essay
MPH: tmr remember submit essay EMO
FT: 'Don't forget to submit your essay tomorrow. (smiling).'
NB: *tmr* = 'tomorrow'

<SMS 471>. **TST 7:00pm** 元氣

FT: Let's meet at 7pm at元氣, Tsim Sha Tsui
NB: TST = 'Tsim Sha Tsui', 元氣 = name of a Japanese Sushi restaurant

<SMS 472>. **Happy Birthday To You!** 又大一**歲**了**.**

JP: Happy birthday to you! *Jau6 daai6 jat1 seoi3 liu5*
MPH: Happy birthday to you! Again big one year FP
FT: 'Happy Birthday to you! You're one year older again!'

<SMS 473>. 我已經上了巴士啦**!** 半個鐘就到**.**

JP: *ngo5 ji5 gin1 soeng5 zo2 baa1 si1 laa3 bun3 go3 zung1 zau6 dou3*
MPH: 1.SG already up-PERF bus PART half CL clock soon arrive
FT: 'I've already got on the bus. I'll be there in half an hour's time.'

<SMS 474>. **Sorry, I'm going to be late for 5 minutes. yours, alice**

<*SMS 475*>. **Add oil with your studying, hang on there, just a few more hours to go. yours, alice.**

NB: *add oil* = direct translation of the Cantonese expression *gaa1jau2*, which is used to encourage somebody to work hard.

<*SMS 476*>. **Get well soon, drink more water and rest more. take care! yours, alice.**

<*SMS 477*>. **Good luck with your presentation tonite, all the best. yours, alice.**

NB: *tonite* = 'tonight'

<*SMS 478*>. **Happy new year ar, wish u won't go to wrong classroom this yr la, hehe**

NB: *ar*, *la*: Cantonese SFPs, *u* = 'you', *yr* = 'year', *hehe* = laughingly, giggling

<*SMS 479*>. **u2 la, beauty nurse**

NB: u2 = 'you too'

<*SMS 480*>. **The phy is too difficult....**

<*SMS 481*>. **Add oil ar! U can do this ga!**

FT: Work hard! You can do this!
NB: *add oil* = direct translation of the Cantonese expression *gaa1jau2*, which is used to encourage somebody to work hard; *U* = 'you'; *ar, ga* = Cantonese SFPs.

<*SMS 482*>. **Happy new year ar!**

NB: ar = Cantonese SFP

<*SMS 483*>. **Happy Birthday,哥!**

NB: 哥 = 'brother'

<*SMS 484*>. **i'm having lesson, call u later!**

NB: u = 'you'

<*SMS 485*>. **yesterday: went to haagen daz and eat a big big cone with 2 ice-cream, SOOO FAT**

<*SMS 486*>. **MISS U**

NB: *u* = 'you'

<*SMS 487*>. **i'm in yuen long now, call u tonite.**

FT: 'I'm in Yuen Long now. I'll call you tonight.'
NB: *u* = 'you', *tonite* = 'tonight'

Chapter 8
New Forms of Reading and Writing
Cell Phone Novels

INTRODUCTION

"This is the story of Rita, who carries a vital piece of information on which depends the future of India. It is written by RoGue, but controlled by you. The author keeps landing her in trouble and only you can save her. After each chapter, predict what RoGue's going to do next and upset his plans. Suggest an alternate course for the story, via comments, and fight the author. Download Chapters I-XV and start playing."

The above quotation is an online introduction to a novel titled *Cloakroom* at the following website: http://cloakroom.blogspot.com/2004/11/cloakroom-chapter-11.html (retrieved, July 7, 2008). *Cloakroom* is widely regarded as India's first cell phone novel or mobile phone novel, a kind of writing that seems to have been largely

DOI: 10.4018/978-1-60566-868-0.ch008

ignored by the established literary world but that has been gaining widespread popularity among the youth, especially young users of the internet in places like Japan, China, and India.

In previous chapters, particularly in Chapter 3, I discussed new forms of reading and writing, such as practiced by students in their everyday academic studies, where I documented a growing preference for online course materials over hardcopy materials, e-books as against print books, etc.

In this chapter I discuss an even more radical consequence of these new forms of reading and writing on the literacy and literary world. As shown in the quotation above, instant messaging, such as text messaging, is taking the idea of a novel to new heights. While the traditional novel as we know it, is a product of, presumably, many years of laborious writing and imagination by a single individual, the author, and delivered to the reading public as a carefully edited and published product by editors and publishers, the "novel" as produced through computer-mediated instant messaging is a radically different sub-genre or even a different genre altogether. It is interactive, instantaneous, as seen by the call to readers to get involved in the developing plot, and it is thus read as it is written!

In this chapter I present the outlines of the cell phone novel, how it started, and what its main features are. I then do a case study of how prevalent this phenomenon of cell phone novel is among students. The next section concentrates on the outline features of the Cell phone novel. This is then followed by the survey. I finally conclude the chapter with some implications and consequences of the cell phone on established literary writing and on the general reading habits of young people.

OUTLINE FEATURES OF THE CELL PHONE NOVEL

Background

The anatomy of the cell phone or mobile phone itself has transformed greatly, from heavy brick-shaped mobile phones with a small screen showing only the dialing number, a few years ago, to the minuscule objects that we know them today to be[1]. Today, mobile phones are more than just tools to make phone calls, indeed the BlackBerry today is a whole mobile office database! They are portable multifunctional assistants which can play music, take pictures, surf websites and even read books, and yes, *novels*, cell phone novels, the topic of this chapter.

The cell phone novel or mobile phone novel is a relatively new and increasingly popular trend, mainly in Japan but spreading to many places like China and India. One may argue that reading books on a mobile phone is not that peculiar. With our mobile phones or PDAs (personal digital assistants), we can use the pre-installed software to view articles transferred from computers or any other sources. However, the mobile phone novel refers to a new type of literature relying on a new medium, the mobile phone, and generating a new business which deals with the mobile phone novel specifically. It is defined as a type of novel which is specially designed for reading on a mobile phone and transmitted through text messages or e-mails from websites running such businesses.

The mobile phone novel is a new product of the 21st century. In 2000, a previously unknown author named Yoshi set up his own website, which provided a mobile phone novel delivery service. He wrote his serialized novel *Deep Love* and made it available to those who subscribed to it through mobile phone. To make his website known to the public, Yoshi launched a promotion campaign by sending out business cards to 2000 High School girls in front of Tokyo's Shibuya Station, the center of Tokyo youth culture. (Japan Echo Inc, 2004) Yoshi's *Deep Love* became extremely popular among young Japanese as the news was spread by word of mouth and within three years his website received more than 20 million hits.

With the great success of *Deep Love*, various websites offering similar services emerged and the mobile phone novel business suddenly became a prosperous industry in Japan. Readers pay a certain amount of monthly fees and can have unlimited access to a selection of titles, some exclusively online and others from print books. (Wilson, 2007) The websites then send daily fresh chapters to the readers' mobile phone via email but the readers can also simply download them directly from the websites onto their phones. Normally, a mobile phone novel has 200 to 500 pages, with each page containing 500 Japanese characters. (Katayama, 2007) Each time, about 1600 characters are packaged onto the readers to be run on certain applications. While this number limit challenges most literary writings, mobile phone novel writers such as Yoshi take advantage of that and create a new style which is straightforward, touching and easy to read, a scenario that attracts a lot of young people who previously seldom read novels. (Japan Echo Inc, 2004). Moreover, with the development of mobile phone technology, as the screen gets bigger and sharper, and as the broadband makes the transmitting speed much faster, and the reading aid applications are more advanced, the mobile phone novel is steadily becoming more and more popular among people in Japan. (Katayama, 2007)

Currently, the mobile phone novel has been recognized as successful, at least, in the Japanese market. In 2007, five of the top ten best-selling novels were originally mobile phone novels which were later distributed as print copies (Onishi, 2008). These works are also adapted into other forms of entertainment, such as TV-drama, movie, and manga (Japanese comics) and they have been hugely successful in these adopted forms, as well. I will now focus on a closer, albeit brief, examination of one mobile phone novel which came out in Japan.

Koizora (Sky of Love)

Koizora is a popular mobile phone novel based on a real story - that of the writer, a young amateur writer called Mika. It is a love story about a teenage girl named Mika.

In terms of statistics, the viewership of this mobile phone novel, *Koizora*, has reached beyond 12 million hits. Later, it was published as print copies and the sales volume went up to more than 1.7 million copies. In 2007, the story was adapted into a blockbuster, *Koizora*. The domestic box office amounted to almost 4 billion yen. (appr. One U.S. dollar= 100 yen) (Sina, 2007)

Plotline

In one summer, Mika, an ordinary High School girl met with a boy of her age, Hiro. They fell in love. However, as they began to settle down into a deep love affair, Mika gets raped. With the love and care from Hiro, she gradually leaves the disaster behind and gets ready for a whole new life.

Right at this time, Mika encounters another huge challenge. She is pregnant with Hiro's child. When she tells Hiro about the baby, Hiro tells her to give birth to it, smiling. He comforts her that he will take care of them both. But misfortune comes once again. Mika loses the fetus. However, with the help of each other and their families, they overcome the difficulties together and the love is even deeper.

Unfortunately though, in the second year of their affair, Hiro suddenly decides to break up with her and thus abandons her. Mika falls into deep sorrow and sadness. That is when she meets another College boy, Yu. Yu likes her a lot. In order to forget Hiro, Mika decides to date him. In this way, Mika starts a new relationship and decides to enroll in Yu's College for further study. However, when they decide to spend the first Christmas together, Mika learns the truth about why Hiro left her. That is because he suffers from a serious disease and does not want to affect Mika and her life.

Excerpt from original novel: (Unofficial Translation)

From that day on, Mika never replies to Hiro's messages and never answers his phones.

Hiro still comes to her classroom for her, but…

Mika always avoids him, apparently.

However, Hiro never stops the phone calls and the messages. [Mika, how do you think of Hiro?]

One day, just after Mika's single-sided break-up with Hiro, Aya asks Mika this question around the corner in the classroom, with a beautiful smile on her face.

[It doesn't matter how I think. He has a girlfriend anyway.] Mika has had enough about Hiro and his girlfriend. [But Nozomu told me that they broke up.]

[What?]

[Hiro said he met the one he really liked and wanted to split with his girlfriend. It seems that he really likes that girl a lot. Do you know who she is? That is you, Mika.]

As a matter of fact, Mika has given up on Hiro. However upon hearing this kind of words, she wavers.

Resolution is really useless to her! Aya keeps on talking.

[Hiro said that he wanted a good chat with Mika. So he will wait for you in the library today after school.]

[O.K. I see.] Dang~Dang~Dang

The bell rings, indicating that the last class of the day is over.

When packing her bag, suddenly someone taps her shoulder from the back.

She turns around.

It is Aya. [Let's make efforts together~ I will go tell Nozomu that I like him and I will certainly be her girlfriend. Call me later!] Aya blurts the words and blinks at Mika. Before Mika can respond, Aya walks away. Aya will reveal her affection to Nozomu.

[Ah~ how come I became nervous for her.] Mika felt sweat in her palm of the hand, walking towards the library. On arrival, she opens the door, biting her lower lip.

~Gara Gara~ Hiro is there, sitting on the ground, looking tired.

[Hi.] [Long time no see. You don't really return my messages, do you?]

[I am sorry.]

Long silence remains.

Then, it seems that Hiro wants to break the ice and he starts to talk.

[About the previous things, I am sorry.] The words uttered from his mouth are so sudden that they confuse Mika.

[What?] [When you come to my house, I do that to you so fast. I am really an ass. I am sorry that I am really a bad guy.] [Actually, I also liked you, Hiro, back then... but it is shocking that you have a girlfriend. It is not that I hate you.] [So, what about now? Hate me?]

[Now... a little unconvinced and untrustworthy.]

[I am really single now. In order to let you believe my heart, I will try my best. I like you so much...hope you can be my girlfriend.] The answer has already been made. Needless to consider! Mika has made up her mind. [Yes...]

Just for one more time. Mika only trusts Hiro for one more time.

She decides to accept Hiro's love. Maybe, deep in Mika's heart, this is what she has already been expecting.

From now on, be sure to reply to my messages. Hiro smiles and caresses Mika's hair. When Mika first met Hiro, she could never imagine they will have a relationship. From this time, Mika's life will be different. The plain life which Mika is supposed to have comes to an end. After arriving home, Mika called Aya to update her on the latest information.

Just like her manifesto, Aya succeeded in becoming Nozomu's girlfriend by confessing to him about her love for him. The two girls who both begin a relationship on the same day are so excited that they keep chatting all night long and forget to sleep.

Discussion

In the section above, basic facts such as the story and its style are introduced briefly. In the following passages, several dimensions of analysis will be made on this novel.

Content

The content of the story has been introduced in the plotline part. Judging from its content alone, it can be inferred that the story is targeting young people, especially young school girls. In the novel, several important issues are mentioned, such as sex offence, teenage love, pregnancy and miscarriage, which are closely connected with the life of ordinary young Japanese girls. Therefore, a kind of empathy can be generated to draw the viewers into the story and the inner-world of the characters. (Marx, 2007)

Besides love affairs, the novel also talks about issues of friendship and family relationships such as divorce, which resonates with the audience.

Style

As from the excerpt, although resulting from unofficial translation, it is obvious that the writing style is colloquial and easy to read. Generally speaking, the novel is composed of conversations and soliloquy. In order to be compatible with the new reading style - the story is serialized and delivered to readers one episode at a time - it makes use of cliff-hanging endings and lively tone to capture the viewers (Associated Press, 2008).

From the episode above, we can also find that the progression of the plots is rather rapid, compared with other traditional novels. This can also be due to the constraint of the medium – the mobile phone screen can only show one page of no more than 500 characters (Katayama, 2007) and each time only a limited number of words can be transmitted. From another aspect, it also satisfies readers' longing to see more character and plot development.

There is also an interesting point with the style of mobile phone novels. Many mobile phone "slangs" appear in the novel, such as the use of punctuation marks "~~" and ellipsis "…". This phenomenon reflects changes brought about along with CMC tools. Since the mobile phone novel is carried by mobile phones, the language it uses may more or less be influenced by what is used on the phones. This can also fit in with the TelCU model introduced in earlier parts of the book, the model of Technology-conditioned approach to language Change and Use. (Bodomo & Lee, 2002) This model emphasizes the causal relationship between new ICT (Information Communications Technology) and new forms of language and literacy. Mobile phone novels are products of the new mobile technology, which in turn affects the language used for the mobile phone novel.

Public Reaction

As revealed by statistics, the Japanese public showed great enthusiasm for *Koizora*, which can be seen from the great success of the original novel, print copies and the corresponding movie.

The *Koizora* fever spread to places other than Japan, especially to Asian regions, such as Taiwan, Korea and mainland China. Its related mobile phone novel phenomenon also attracted attention from different countries, domains and industries. If one performs an internet search with the key words "mobile phone novel Japan" or "Koizora", hundreds of thousands of pages can be found, among which topics, languages and countries vary.

Interactivity

An innovative characteristic of *Koizora* or other mobile phone novels, as seen above with *Cloakroom*, is that it allows readers to comment on the story they just read and gives ideas to the writers about the future of the story (Onishi, 2008). In this way, it provides a channel for readers to participate in the novel development, which makes it more attractive and dynamic. This is a significant difference between the Mobile Phone Novel and the traditional print novel.

Another aspect of this issue of interactivity involves the participation of other industries. In the case of *Koizora*, it eventually got published as print copies and, further still, it has been adapted into movies. This high-degree of integration or interactivity between different genres is another hallmark of the mobile phone novel, in general. Other examples of successful adaptations from Mobile Phone Novels include Yoshi's *Deep Love* which was further adapted into comic books and TV drama which became very successful, and his "*Wingless Angels*" series depicting different troubled girls, which was also made into TV drama later.

To conclude this outline of the general features of the mobile phone novel, one might pose the question as to why this innovative literary phenomenon has been so successful, especially in Japan.

The short sentences, basic dialogues, inviting plots and lively characters are all contributors to the success of mobile phone novels. In addition, cultural background also matters. Japan is a country with more than 1000 years of literature. People love reading books. Furthermore, the technological infrastructure like the network, phone functions and other related issues all facilitate the popularity. Last but not least, people's tendency toward using phones also has influence over the market. In a community, if people only see mobile phones as tools to make phone calls, the mobile phone novel service most likely cannot be popularized.

A related issue to the question of why the mobile phone novel has been successful is who are behind this success. In other words, what segment of the population are the most avid readers of the mobile phone novel? It has been shown that most readers of mobile phone novels are young people, especially females, according to a research by Bandai (Associated Press, 2005). From the popular mobile stories themselves, we can gather some clues about this question. Most stories are love stories and depict the current world and the problems in society from an incisive point of view. Many sensitive issues are usually covered. In *Koizora*, we can encounter themes of gang rapes, teenage pregnancy and mid-age crisis, among others. To give another example, the first and most famous representative of mobile phone novels, *Deep Love*, reflects the phenomenon of "enjo kousai" (teenage girl prostitution) which is a unique social problem within the Japanese society.

A SURVEY OF CELL PHONE USE AMONG STUDENTS

In the foregoing section, I have presented the background and outline features of mobile phone or cell phone novels. Like other computer-mediated practices, we see that the participants in this peculiar communication event are mostly young people, particularly young females. To obtain some empirical evidence of the extent to which young people, especially students, know about and participate in this novel novel reading habit I decided to do a survey of a class of about 30 students taking a course on Language and Information Technology[2].

Survey Background

The survey took place on the WebCT of a linguistic course, *Language and Information Technology*, in April 2008. Twenty four questionnaires were filled in by twenty four informants.

Findings

The Respondents

Native language: All of them speak Cantonese.

Other languages spoken

Figure 1. Other languages spoken by the respondents

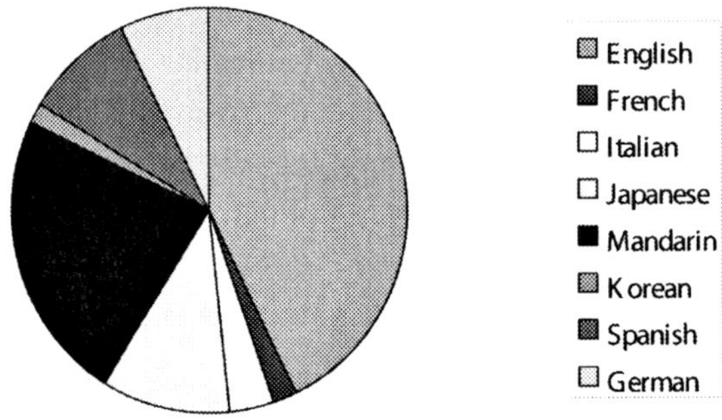

Among the 24 respondents, apart from their native language, English, French, Italian, Japanese, Korean, Spanish and Mandarin are also spoken (see Figure 1).

Age

One didn't answer; while the others are between 21-25.

Education

All of them are undergraduates.

Gender

There are 14 females and 10 males.

How much do the Respondents know about the Following Communication Tools?

The answers to questions 6 to 23 were summarized in Table 1.

How Often do the Respondents use the Mode(s) of Communication in Q.1?

The answers to questions 24 to 41 were summarized in Table 2.

Language Used on Line

Question 42 asked the respondents what language they usually use in online communication, in general. The result shows that 75% of them use a mixture of Chinese and English (see Figure 2).

The Term Cell Phone Novel/Mobile Phone Novel

Question 43 asked the respondents if they have heard of the term Cell Phone Novel/ Mobile Phone Novel. The result is shown in Table 3.

Have you ever Read a Cell Phone Novel?

Question 44 is a follow-up question on question 43. If the answer to question 44 is yes, then the respondent was asked to tell if he/she had ever read a Cell Phone Novel. Although most of the respondents knew of the existence of Cell Phone Novel, only three of them had read one.

Can Reading and Writing a Cell Phone Novel Improve Language Skills?

Question 45 is designed to see if the respondents agree that students can improve their language skills by reading and writing Cell Phone Novels. The result indicates that most of them have a neutral opinion with regards to this issue.

Justification

The respondents were asked to justify their choices in question 45, regarding reading and writing a Cell Phone Novel and its correlation to the improvement of language skills. Their justifications are listed in Tables 4, 5 and 6.

As can be seen from the above tables, opinions vary on whether or not literacy practices with cell phone novels can improve linguistic skills. Even though a related question as to whether or not practices with cell phone novels might indeed decrease language skills was not asked, one would imagine that people who disagreed with the question at stake might indeed be answering in the affirmative on this. On the whole people are skeptical because they think that these novels are not written by "professional" writers and that indeed the language used may be rather simple and too colloquial to reach the level of literacy that we may refer to as literary.

Never the less, as cell phone enthusiasts show, it is a medium that gives its practitioners a chance to improve on their reading, and indeed they can acquire a lot of vocabulary and expressions that go a long way to improving their literacy and literary skills. This will remain an open debate for sometime as we witness the development of this new form of literature.

Table 1.

	A. Very familiar and have used it	B. Quite familiar and have used it	C. Only heard of it but never used it	F. Never heard of it
AOL Instant Messenger (AIM)		2	17	5
Bulletin Board	3	15	4	2
Email	21	2	1	
ICQ	16	7	1	
Chatroom (e.g. IRC)	8	8	8	
Land-line phone	12	4	2	6
Letter (hand-written)	15	8	1	
Mobile Phone (voice)	18	5		
Fax	7	17		
MMS	5	5	12	2
MSN Messenger	22	2		
Newsgroup	3	10	8	3
Online Forum	12	11	1	
SMS	19	5		
MUDs and MOOs	1		1	22
Video Conferencing	2	14	7	1
Yahoo Messenger	1	6	16	1
Others (name them below:				

Others used by the respondents: QQ, facebook, skype, xanga, webshots, blogs, google messenger, telegram, webCT, walkie talkie and pager.

Is the Cell Phone Novel a New/Different Genre of Writing?

A **genre** according to the Oxford Compact English Dictionary is a kind or style of art or literature. Question 47 aims to find out if the respondents consider the Cell Phone Novel to be a new or different genre of writing. As can be seen from the responses almost half of respondents think that it is a new genre (see Table 9).

Justification

Similar to question 45, question 47 also required the respondents to justify their choices. Tables 10, 11, and 12 show their justification. Three respondents didn't

Table 2.

	A. Always (at least once a day)	B.	C. Quite Often (3-5 times a week)		E. Never	F. Never heard of it
AOL Instant Messenger (AIM)				3	19	2
Bulletin Board		2	3	12	6	1
Email	19	4		1		
ICQ	1			16	7	
Chatroom (e.g. IRC)		1	2	10	11	
Land-line phone	4	5	4	4	2	5
Letter (hand-written)	1		3	11	3	
Mobile Phone (voice)	20	3		1		
Fax	1	1	3	18	1	
MMS		2	1	9	9	3
MSN Messenger	17	5	2			
Newsgroup		2	5	8	8	1
Online Forum	4	6	4	9	1	
SMS	15	6		3		
MUDs and MOOs	1			1	3	19
Video Conferencing	1		3	13	6	1
Yahoo Messenger		3		5	15	1

Figure 2. What language do you usually use in online communication, in general?

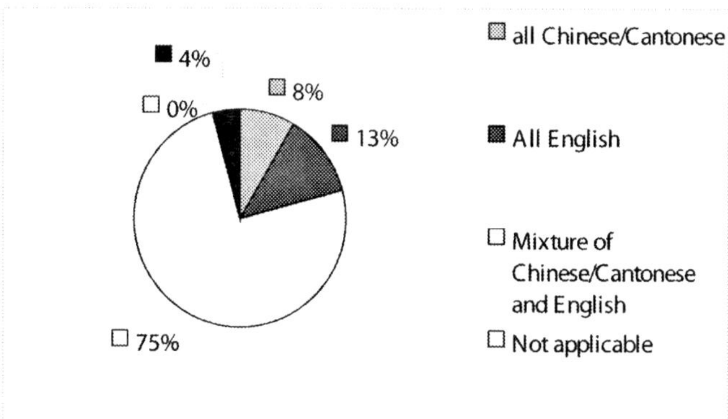

Table 3. The term Cell Phone Novel/Mobile Phone Novel

Answers□	No. of Respondents
Yes	23
No	1

Figure 3. Have you ever heard of the term Cell Phone Novel/Mobile Phone Novel?

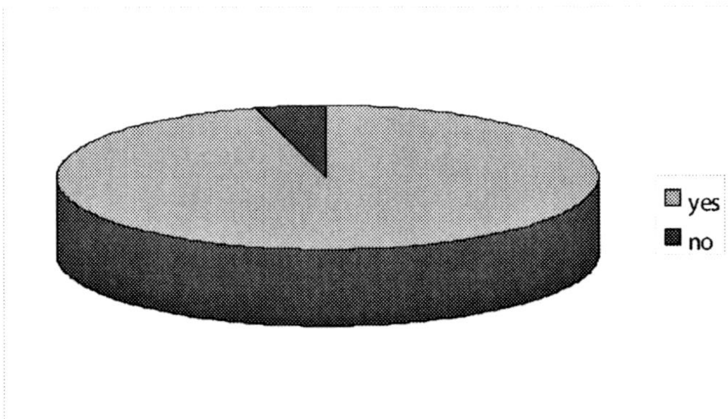

justify their choices.

As with the previous debate, the question of whether or not cell phone novels are a new, different genre of literature or they are just a sub-genre of the novel will continue to be with us for sometime. Ultimately our understanding of this issue will

Figure 4. Have you ever read a Cell Phone Novel?

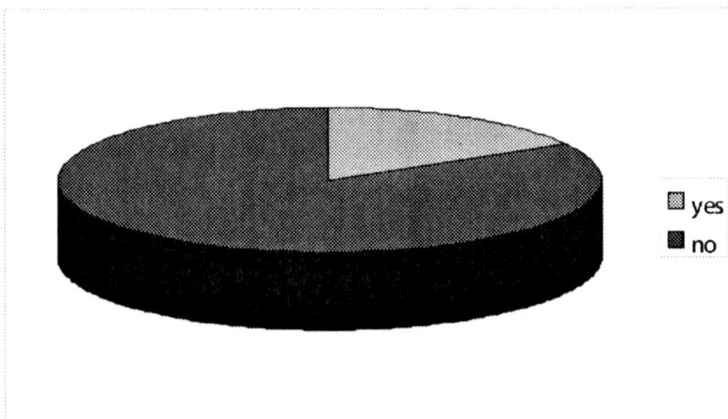

Figure 5. Do you agree that reading and writing a Cell Phone Novel can improve language skills?

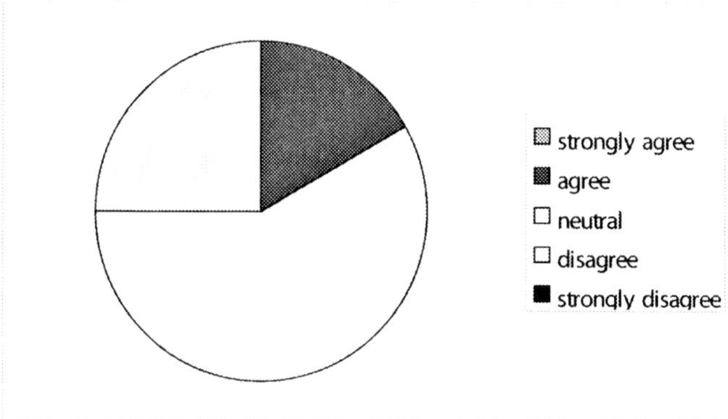

Table 4. Have you ever read a Cell Phone Novel?

Answers	No. of Respondents
Yes	4
No	20

Table 5. Do you agree that reading and writing a Cell Phone Novel can improve language skills?

Answers	No. of Respondents
Strongly agree	
Agree	4
Neutral	14
Disagree	6
Strongly disagree	

Table 6. Justification for 'agree'

As the language used in cellphone novels are as short as possible while expressing its meaning, I think it will help people to develop a skill to express themselves in the shortest way.
They may learn new vocabs and apply them in their writing in real life.
Increase in chance/time for time in reading would benefits the reading skill.
by reading and writig novels they can improve language skills

Table 7. Justification for 'neutral'

Some of the big vocabularies cannot be used and acquired there
I think if the Cell Phone Novels are well organized and written, just like the printed ones that found in the library, the reading ability may enhance.
It is depends on their choice of novel, also depends on their reading and writing style
I don't think that helps to improve our reading skills...i think it's the same for us reading real books.
The language use in SMS may harms linguistic development
I 'd rather treat that as a two-sided sword.
They could improve language skills by reading print books.
I have never read a cell phone novel, so I remain neutral.
Depends on the languagues used.
Havn't read before, so no comment
I am not sure if the grammar used in those cell phone novels are correct or not.
Cell Phone Novel is a medium of communication. Depends on how to manipulate it
forms of language used in mobile phone novels may not be the standard form of that particular language which is recognized in formal situations or academic environment

rest with our interpretation of the term 'genre', on the one hand, and how we view the style and content of the cell phone novel, on the other.

Awareness of any Online Websites or Books or Newspaper Reports

Question 49 was meant to find out if the respondents were aware of any online websites or books or newspaper reports about the new reading phenomenon of cell phone novels and their impact (see Table 13).

Table 8. Justification for 'disagree'

The wirtting style of cell phone mobile is too bad.
I do not disagree reading classics in the form of cell phone novels, but I wonder how many new cell phone novels are written by a good writer decently now.
The novels were written by less professional authors.
The cell phone novels are usually informal that are not able to help students improve their language skills.
what are written in cell phone novels are onlt slang and informal language which cannot reach the stardard of fornal langauge
Because of the limited space in SMS, language used is very simple and colloquial, and grammar is also neglected

Table 9. Is the Cell Phone Novel a new/different genre of writing?

No. of Respondents	Answers
Yes	12
Neutral	9
No	3

Figure 6. Is the Cell Phone Novel a new/different genre of writing?

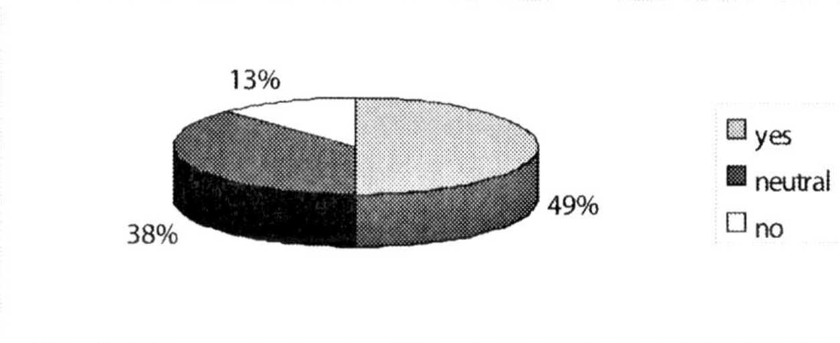

Online Websites or Books or Newspaper Reports

Question 50 asked the respondents to list as many websites, books, or newspaper reports as possible, if their answer to question 49 is yes. The following are the answers they gave (see Table 14).

Table 10. Justification for 'yes'

As it is written and read entirely different to a traditional book.
Its writing style varies from that of traditional writing
It creates creativity in writing.
It was born as an "e version" while all other genres were born as a paper version.
It is just that novel SMS is a new thing which has not yet been accepted by others. But as days gone by, it will gradually be a wide-accepted genre.
It's different from ordinary novel.
ppl seldom use it before
It contains a specifis kinds of content
as forms of language used in mobile phone novels are different, I consider this as a new kind or style of novels.
the definition of literature is very broad, even a photo or a building is studied in comparative literature study, so, this kind of writing style, can be a focus of Complit

Table 11. Justification for 'neutral'

I'm not sure, but generally I think it will definitely affect the ways we use language and may invent some new words
becasue the techology change very fast.
The cell phone novel can encourage more people to write. Though the languages used may not be too formal, it still starts a new genre of writing. People can write whatever and wherever they want.
I have never read a cell phone novel, so I remain neutral.
the wordings used in those novels may be informal
it would demonsrate a "forum/blog-like" style of novel at the beginning, but this may change within short period of time due the differences between those platform
It is a special type of literature.
Because I have not read any mobile novel, therefore I cannot be very sure whether the characteristics of a mobile novel can make it become a new genre

Knowledge of specific authors and titles of cell phone novels

The respondents were asked if they know any of specific authors and titles of cell phone novels in Chinese, English, Japanese or any language they know in question 51 (see Table 15).

Authors and Titles of Cell Phone Novels

Like question 50, question 52 intended to know whose novels and what the novel titles the respondents know if their answer to question 51 is yes. Their answers are listed in Table 16.

Is it Possible or Desirable to Adopt Cell Phone Novels as Language Teaching Material?

Question 53 was to investigate the respondents' opinions on the adoption of Cell Phone Novels as language teaching material. According to Table 17 and Figure 9, no one thinks that it is absolutely impossible to adopt Cell Phone Novels as language

Table 12. Justification for 'no'

a novel is still a novel.
I would consider cell phone as a platform rather than a new form of literature.
It follows normal syntactic rules.

Table 13. Awareness of any online websites or books or newspaper reports

No. of Respondents	Answers
Yes	6
No	18

Figure 7. Awareness of any online websites or books or newspaper reports

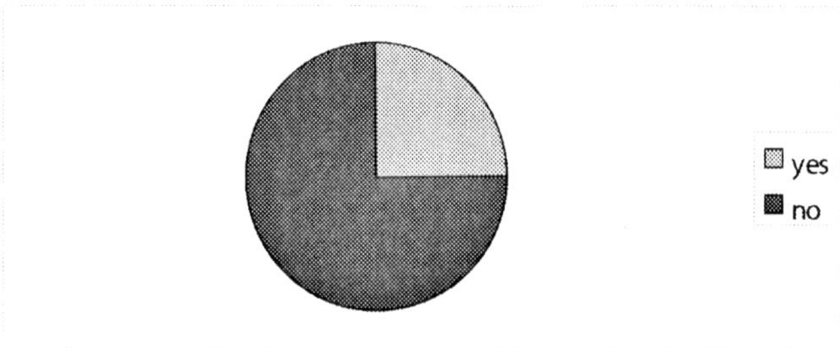

teaching material, and only one respondent thinks that the adoption is very likely.

Do you have any other Comments Regarding the Topic of Cell Phone Novels?

Ten of the respondents gave their comments (see Table 18).

In this section, I set out to find out how widespread knowledge there existed among students, at least, those in the author's class, about the concept and practice of the mobile or cell phone novel. To obtain some empirical evidence of the extent to which young people, especially students, know about and participate in this novel reading habit I decided to do a survey of a class of about 30 students taking a course taught by me. What we have uncovered is a plethora of ideas about the cell phone

Table 14. Online websites or books or newspaper reports

http://en.wikipedia.org/wiki/Cell_phone_novel
text message novels shake literary (A10 SCMP on 21 Jan 2008)
BBC, zdnet.com, Technology Weekly
Rin, "If you"
Heard it from TV news
http://www.nytimes.com/2008/01/20/world/asia/20japan.html?pagewanted=1&_r=1&ei=5087&em&en=9275f067f59eb69c&ex=1200978000

Table 15. Knowledge of specific authors and titles of cell phone novels

No. of Respondents	Answers
Yes	5
No	19

Figure 8. Knowledge of specific authors and titles of cell phone novels

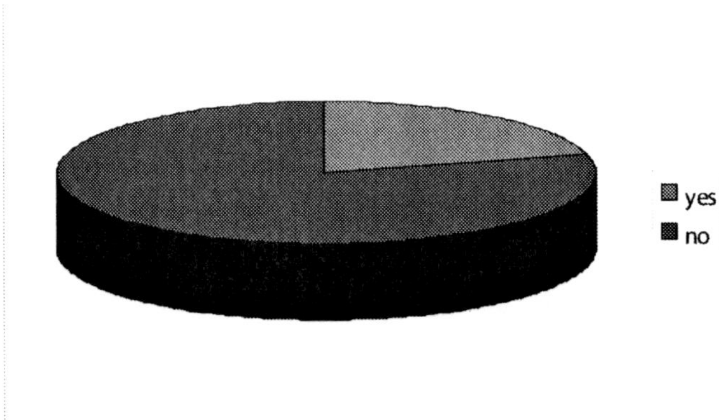

novel. While it is not as practiced as extensively as it is in places like Japan and India, the participants are indeed well aware of this literacy and literary practice and have provided a myriad of opinions ranging from whether or not it creates opportunities for improving one's language skills, whether or not it can be incorporated into formal school curricula to whether or not it is a new genre of literature.

CONCLUSION

To conclude this chapter, in addition to speculating about the future of the mobile or cell phone novel, I will like to take a final look at the issue of whether or not the

Table 16. Authors and titles of cell phone novels

Deep Love, the city besiged,etc
鍗江曆澶?
" MOSHIMO KIMI GA" "If You" (by RIN), "KOIZORA" "Love Sky" (by MIKA)
Rin, 'If you'
Yoshi <<Deep Love>>

Table 17. Opinions on the adoption of Cell Phone Novels

No. of Respondents	Answers
Absolutely impossible	0
Unlikely	7
Quite likely	7
Likely	4
Very likely	1
Neutral	5

cell phone is a new literary genre by itself or a sub-genre of the novel.

We do this by drawing ideas from one of the greatest novel specialists, M. M. Bakhtin, about the changing nature of the novel.[3] Bakhtin acknowledges the difficulty or impossibility to grasp the novel in one coherent and comprehensive theory (of genre); the novel, he argues has historically been considered the rebel among the genres, under-estimated and neglected, constantly fighting to be understood, but, at the end of the day, it is this openness, this struggle to incorporate new developments in history, society and literature, and this ungraspable character which *defines* the novel and shows its great achievement and literary value: the novel is the ever-developing genre which can keep up with change and innovation; it is 'a genre-in-the-making' (Holquist 1981).

It seems clearly that the advent of the mobile or cell phone novel is testimony to Bakhtin's perceptive comments about the ever-changing nature of the novel. This view from Bakhtin may indeed provide important clues to the idea that the cell phone is indeed a sub-genre of the novel: the novel, because of its ever-changing

Figure 9. Opinions on the adoption of Cell Phone Novels

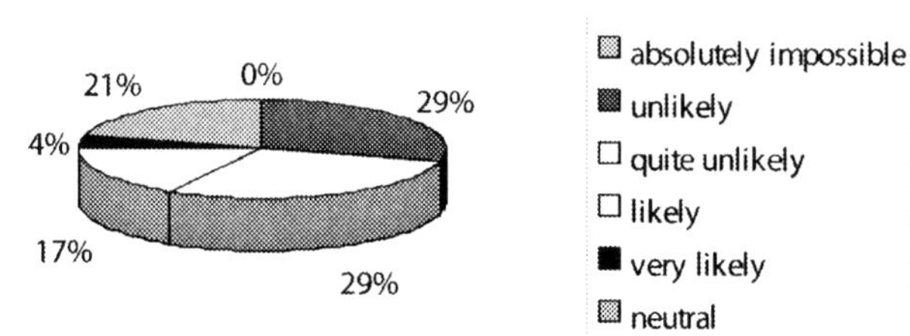

Table 18. Comments regarding the topic

Suggested new question: Do you think cell phone novel is more easily accepted in Asia than other places of the world?,,Suggested question: In your opinion, what makes a cell phone novel successful?,Suggested question: Would you consider cell phone novel a literature form or simply a commercialized activity?
I think Cell Phone Novels will lower our writing quality because it tends to use sentences and words as simple as possible, which will narrow down the range of vocabularies.,,,I think Cell Phone Novels are not attractive to me because it demands high concentration to read
its effect on readers may vary depending on the languages concerned., not common here perhaps due to the fact that there isnt much of a reading culture., seems not to be popular in hong kong yet. dont think a lot of people have heard about it here.,mobile phone novels encourages people to read and write because they are convenient,they may choose to play with their gadgets/chat on the phone in their leisure time eg when travelling on the train. travelling time may not be not enough for reading as in japan.
even it is a new form of writing, but seems not many people know about it,,,it is only for the youngsters, the elderlies doesn't have any idea about it,
The Cell Phone Novels is a trend that suit the need of the advance of technology development,,,,
Cell Phone Novels attract lots of Japanese schoolgirls to read, Cell Phone Novels create a new type of business in Japan, Cell Phone Novels may lead to a problem of copyright, Most of the topics of the novels are love, but tragedy loves,
for adopting it as teaching materials, it is still under developed to be adopted in the syllabus.,,,,
It is a good idea to adopt Cell Phone Novels as language teaching material. The use of cell phone is so widespread that there should be more to be further developed based on the existing technology. Skype is also a great innovation. I find the connection very stable (better than MSN!). I can keep contact and do video conferencing with my overseas friends. Portability is the major reason which could make it successful. For me, as an intermediate foreign language learner will be happy to read the teaching material whenever and wherever I want.
I have never read a cell phone novel, so I have no comment.
cell phone novels are not very widely promoted in Hong Kong.
I think it takes a lot of effort and consideration in implementing education with cell phone novels. Primary and Secondary schools in HK prohibit their students to use mobile phones in the school. Using cell phone novels as educational tool might cause many problems.
teachers may use cell novel to be an example of bad language. content of mobile phone novel: Japan-mostly about love story; China-have some meaningful cell novels rewrite from previous literature, cell phone novel is not popular in HK, many ppl have never heard of it,
Because only small parts of the whole novel is delivered every time, this might create a new reading style which does not require concentration for long time., SMS novel is a new platform for people expressing themselves. Its high availability and low cost in "publishing" might attract new writers of younger generation, Because of the highly accessibility of mobile phones and SMS, people might be encouraged for writing more, The new style of reading might encourage people who do not usually like reading; on the other hand, it might discourage readers who prefer continuity in novels,

nature, has taken advantage of new technology to develop in new dimensions. A cardinal theme of this book, that new technology demands new forms and styles of language, is borne out, and can indeed then also be extended thus: new technologies engender new sub-genres of literature.

Will the mobile or cell phone be with us for a long time to come or will it be a nine-day wonder? As with many products of the rapidly developing information technology environment, it is hard to predict the future. What is certain is that the cell phone, if it stays with us, will continue to take on new forms. This is partly due to the ever-changing nature of technology and also, according to Baktin, partly due to the ever-changing nature of the novel.

After exploring in this chapter how computer mediated communication practices engender new forms of literacy and literature, I shall in the next chapter take a closer look at computer mediated communication practices and their implications for the school curriculum.

REFERENCES

Bodomo, A., and C. Lee (2002). Changing forms of language and literacy: technobabble and mobile phone communication. Literacy and Numeracy Studies: An International Journal in Education and Training of Adults , 12(1), 23–44.

Holquist (1981). *The Dialogic Imagination: Four Essays by M. M. Bakhtin*. University of Texas at Austin.

Japan Echo Inc. (2004). *Novels Delivered to Your Phone*. Retrieved April 20, 2008, from Trends in Japan Web site: http://web-japan.org/trends/lifestyle/lif040310.html

Katayama, L. (2007). *Big Books Hit Japan's Tiny Phones*. Retrieved April 20, 2008, from Wired web site: http://www.wired.com/culture/lifestyle/news/2007/01/72329

Marx, W. D. (2007). Koizora: *Empathy and anonymous creation*. Retrieved April 20, 2008, from Clast web site: http://clast.diamondagency.jp/en/?p=88.

Onishi, N. (2008). *Thumbs Race as Japan's Best Sellers Go Cellular*. Retrieved April 20, 2008, from The New York Times web site: http://www.nytimes.com/2008/01/20/world/asia/20japan.html?_r=2&th&emc=th&oref=slogin&oref=slogin

Sina. (2007). *Koizora*. Retrieved April 20, 2008, from Sina Entertainment Web site: http://ent.sina.com.cn/m/f/koizora/

The Associated Press. (2005/2008). *Next hot cell phone trend: Reading?* Retrieved April 20, 2008, from MSNBC website: http://www.msnbc.msn.com/id/7232995/

Wilson, F. (2007). *Japan falls in love with mobile phone novels*. Retrieved April 20, 2008, from World Business website: http://www.worldbusinesslive.com/article/645160/japan-falls-love-mobile-phone-novels

Zhou Zhou Tiffani. (2008). *A Closer Look at Mobile Phone Novels: Features and Implications*. Term Paper for the course, Language and Information Technology. HKU

ENDNOTES

1 I thank Zhou Zhou Tiffani for collecting and gathering facts and analysis about cell phone novels for this section of the book as part of a term paper (Zhou Zhou Tiffani 2008) for my course in Language and Information Technology

2 I thank Lucille Hu, who as TA for my course, helped me administer the questionnaire and write a report of the statistics obtained from the questionnaire

3 I thank Julia Kuehn for drawing my attention to this discussion within the field of novel studies.

Chapter 9
New Languages, New Literacies and the School Curriculum

INTRODUCTION

We have in previous chapters undertaken a quite detailed analysis of the features of language and literacy practices within the medium of communication technologies, including email, ICQ, MSN, mobile phone speech, and mobile phone texting. This has given us the opportunity to examine carefully new or peculiar linguistic structures, and new ways of communicating and expressing oneself within these new technological environments.

Of course, it is not everybody who is comfortable with these new ways of using whatever language it is by experimenting with these new structures and strategies of encoding such as shortening words and explicitly disregarding standard grammatical encoding. In particular, parents and teachers have been quite critical about

DOI: 10.4018/978-1-60566-868-0.ch009

these new languages and new literacies produced within the context of new communication technologies by their children and students.

To address these concerns, I shall in the rest of the book, beginning with this chapter, take a closer look at the educational implications of these forms and ways of communicating in the age of IT. These concerns will be addressed by means of a case study based on a funded project whose aim was to do a focused and careful analysis of email, ICQ and mobile phone practices among young people in Hong Kong with a view to examining if there are any consequences of these language and literacy practices on the way they learn and use language in the classroom. The bulk of this chapter is a case study based on a survey report (Bodomo 2006) about the role of ICT in the curriculum, with particular reference to the issue of whether or not CMC language adversely affects standards of correct grammar and language use. This chapter has a simple organization. The next section is the bulk of the chapter since it constitutes the case study on which the chapter is based while the part that comes after the case study concludes the chapter with a summary of the main findings and implications for future curriculum planning.

CASE: NEW OF LANGUAGE AND THE SCHOOL CURRICULUM

Introduction: Rationale for the Study

There has been a general concern about the falling standards of language among students in Hong Kong. However, language use among the youth in some new channels of communication, such as texts from email, ICQ, and mobile short messages/ SMS, reveals a good amount of creativity, eloquence, dexterity, systematicity, and multimodality. The issue is, if standards are really falling, how should we handle this paradox whereby students manifest falling standards within the formal context of the educational system and yet manifest considerably high levels of proficiency and linguistic sophistication in IT communication modes? This case study aims to:

i. Address the issue of 'language standards' through analyzing texts produced by the youth in new information communications technologies (ICT texts),
ii. Investigate the plausibility of using ICT texts in language education, and
iii. Develop a considerably large corpus of ICT texts for educational purposes.

It is, therefore, expected that a better understanding of the changing paradigms of communication can help language teachers, educationists, and curriculum developers produce competent graduates to better serve society's manpower and human resource development needs.

This report presents findings from three school visits, involving over a hundred local secondary school students and their English teachers. Through surveying and interviewing these students and teachers about their attitudes towards ICT communication, we hope to better understand the relationship between 'Netspeak', the language of the Internet, and formal language education.

The case study is organized as follows: Part two outlines the methodology of the research, including two sets of questionnaire survey and in-depth interviews. Part three reports on the findings from the two sets of questionnaire, in terms of habits of using ICT tools, the ways in which language is used, and how it is related to formal language education. Results from the interviews are also summarized. Part four attempts to compare results from the student and teacher surveys.

Methodology

The Subjects

The age of the target students mainly falls within the range of 15-20. This is not surprising as a 2004 Nielsen//NetRatings report reveals (Mingpao, 2004) that about 86% of ICQ users are under 30 years old. This certainly indicates that the youth constitutes the majority of Internet users in Hong Kong.

The Survey: Some General Remarks

The survey took place from February to October 2004. Some 184 Cantonese-speaking secondary school students and 15 English language teachers were involved. Table 1 outlines the details of these school visits.

Two methods were used to collect data from these students and teachers - investigator-administered questionnaire survey and individual in-depth interviews with students and teachers.

In each of the three school visits, at least two research assistants (RAs) were assigned to one class of students. Since the questions were written in English, one of the RAs had to go through each question with the students to make sure that they fully understood the meaning of each question. Students were also allowed to ask questions in the course of filling out the forms.

During each school visit, we also asked the class teachers to fill out a questionnaire for us and we also left a few copies of the questionnaire behind for distribution to other English teachers in the schools.

After the questionnaire survey, about 3-4 students were selected from each class to be interviewed by the RAs. We successfully interviewed 12 students from these 5 classes we surveyed, details of which are reported in part three.

Table 1. Details of school visits (as of Dec 2004)19219

Date/time	School	No. of students involved	Number of teachers involved
6 Feb 2004, 2:40pm	**Good Hope School**, Clear Water Bay Rd, Hong Kong	26 (F.2)	1
25 Feb 2004, 3:20pm	**SKH St. Simon's Lui Ming Choi Secondary School**, Tuen Mun	78 (F.3, F.4)	7
27 May 2004, 2 pm	**Madam Lau Kam Lung Secondary School of MFBM**, 22 Castle Peak Rd, Lam Tei, Tuen Mun, NT	44 (F.2)	6
21 Oct 2004	**Yan Chai Hospital Lan Chi Pat Memorial Secondary School**	44 (F.1)	1
		184	15

Results

This section reports on the results of the two sets of questionnaire. Findings from the student survey are first presented (Tables 2 and 3), followed by findings from the teachers' responses.

On the Habit of Using ICT (Qs.1, 2, 3)

The first question of the questionnaire aims to identify what kinds of communication tools the respondents used most often. The results are summarized below in Table 4.

The most popular ICT tools among students, according to the respondents, were ICQ, Bulletin Board, email, and voice mobile phone. This piece of information did

Table 2. Age distribution of respondents

Age	N	%
10-12	30	16
13-15	126	69
16-18	25	14
19 or above	25	1
unknown	1	1

Table 3. Gender distribution of respondents

Gender	N	%
M	56	31
F	126	69
Unknown	2	1

provide us with a direction of text data collection – aiming to collect more ICQ data than those from other ICT tools.

In order to confirm the degree of prominence of ICT in young people's everyday life, we asked the respondents whether they preferred electronic communication to traditional modes of communication.

The majority of the respondents said they actually preferred online communication to other forms of communication. For the respondents who actually 'preferred' using ICT tools, we further asked them to justify their choice.

Table 4. How often do you use the following communication tools or media?

	A. Always (at least once a day)	B. Often (almost every day)	C. Quite Often (3-5 times a week)	D. Seldom (less than 3 times a week)	E. Never used it	F. Never heard of it
AOL Instant Messenger (AIM)	2	2	5	9	37	127
Bulletin Board	45	25	19	26	20	46
Email	47	36	46	47	5	3
ICQ	87	26	16	21	33	1
Chatroom (e.g. IRC)	12	20	23	53	47	24
Land-line phone	69	22	18	24	19	28
Letter (hand-written)	16	9	31	71	34	13
Mobile Phone (voice)	58	38	31	35	12	8
Fax	5	2	16	57	92	11
MMS	9	13	12	30	72	45
MSN Messenger	31	21	24	43	43	21
Newsgroup	11	5	10	29	68	60
Online Forum	17	9	18	21	53	64
SMS	23	20	17	37	73	13
MUDs and MOOs	1	4	3	5	40	128
Video Conferencing	9	10	12	23	82	48
Yahoo Messenger	15	21	28	53	61	6

Figure 1. Do you prefer online communication to other means of communication (e.g. face-to-face conversation, letter writing)?

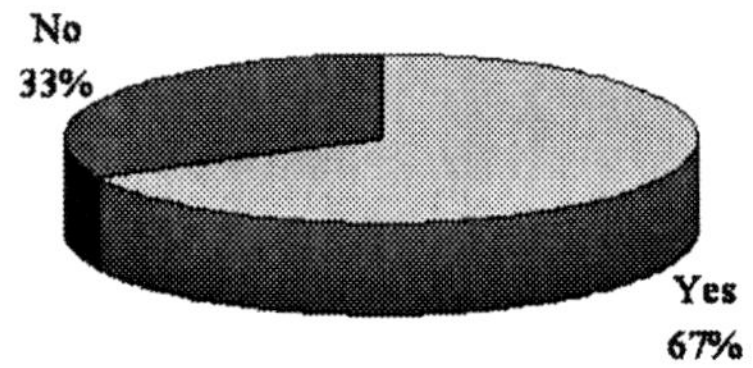

Convenience and Speed were the two major factors influencing the prominence of electronic forms of communication. Some other respondents preferred online communication because it is economically friendly (as compared to mobile phone communication or letter writing). Some said that it is easier to communicate with people overseas through online communication.

The following points may be generalized from the above findings:

i. The respondents use electronic modes of communication (e.g. email, ICQ, mobile phone) more frequently than the 'traditional' ones (e.g. landline phone, letter writing).
ii. The fact that digital modes of communication are more prominent among this rather young group of people also corresponds to the initial observation of this study – young people are the most frequent users of the Internet and thus they constitute the most representative group of subjects for providing data for this study.
iii. A majority of the respondents had a preference for electronic modes of communication. The reasons for this preference were that ICT tools were more convenient and fast. However, one should bear in mind that some students actually found it easier to express themselves through electronic means of communication.

Table 5. Reasons for preference for communicating online

Reason	*N*	*% (out of 184 respondents)*
It is entertaining.	45	37
It is fashionable.	32	24
It is convenient.	81	58
It is fast.	79	58
It allows me to express myself freely.	44	29

Table 6. What language do you usually use in the following forms of communication?

	A. Chinese / Cantonese Only	B. English Only	C. Chinese / Cantonese + English	D. Other (specify)	E. N/A
AOL Instant Messenger (AIM)	3	3	20	1	155
Bulletin Board	22	3	87	4	62
Email	27	15	123	1	15
ICQ	12	5	129	3	31
Chatroom (e.g. IRC)	24	3	86	2	65
Land-line phone	56	3	72	2	45
Letter (hand-written)	55	4	90	3	29
Mobile Phone (voice)	70	4	83	4	16
Fax	18	2	54	6	101
MMS	13	2	40	7	119
MSN Messenger	20	7	84	3	66
Newsgroup	17	3	46	3	112
Online Forum	19	4	53	3	101
SMS	21	7	75	3	75
MUDs and MOOs	4	2	17	3	155
Video Conferencing	19	2	38	2	121
Yahoo Messenger	29	5	91	2	54

On 'Internet Language' in General (Q. 4-5, 10)

Another set of questions aim to elicit opinions with regard to using language in online communication in general. First of all, we asked the respondents what language(s) they usually used in various modes of communication, including both traditional ones (e.g. letter writing and telephone calls) and new ICTs (email and online chat).

More Chinese-English code-mixing is found in the three most familiar modes of electronic communication – ICQ, bulletin board, and email. While there is a growing amount of public discourse complaining about the high degree of mixed codes in electronic communication, there is a very high tendency of 'code-mixing' phenomenon in *both* traditional and electronic modes of communication. Indeed, the phenomenon of code-mixing has already existed in the Hong Kong society for a very long time (e.g. Li 1998, 2000). Besides, rarely do students write 'all' English in any of the specified modes of communication. In this case, I doubt to what extent one should blame ICT altogether for the deteriorating standards of language. The

Figure 2. Do you pay attention to the correctness of your grammar and style when you write online messages?

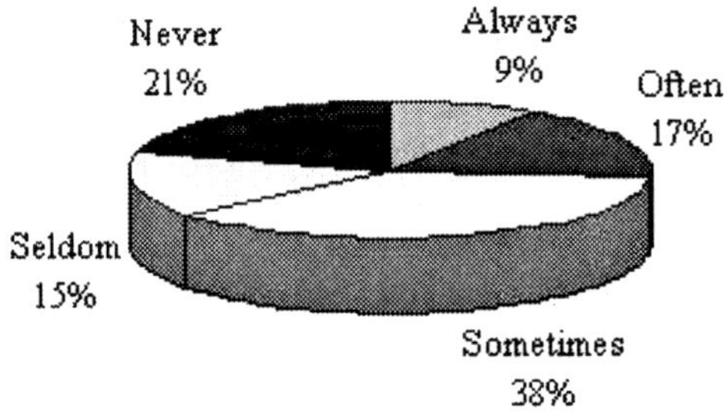

issue here is – Do students care about grammatical correctness when communicating online? Figure 2 shows that only 9% of the respondents *always* checked the grammar of their online messages. But about half of the respondents said they *often* or *sometimes* paid attention to grammar in online communication. This may imply that students are indeed aware of the issue of 'correctness'. They do not always do so because they probably do not feel that there is such a need in some communication contexts, e.g. when chatting with close friends or any informal situations or when an exchange is synchronous and the messages have to be transmitted within a very short time in order to sustain the 'conversation'.

To further investigate whether there is an emerging pattern of features of online communication, the respondents were asked to write out the expression "***Thank you. See you tomorrow. Good Night***" using the form of language they would use on ICQ. We collected 151 expressions from students, which are listed in Appendix 3. Patterns are identified and summarized as follows:

(i) General Features

A large number of shortened forms are used, for example, 'thx' instead of 'thank you', 'c' instead of 'see', 'u' instead of 'you' and all these shortened forms including abbreviations, clipped forms and especially homophonous or near-homophonous words indicate a high degree of creativity. Major strategies of shortenings are summarized in Table 7:

(ii) Detailed Analysis

The following table summarizes the frequency of shortenings used for different parts of the given expressions:

The rule of capitalization is not strictly followed in these expressions given by

Table 7. Types of shortenings used in the reported expressions (Q.10)

Different types of shortenings	Examples
Acronym of phrase/sentence	*G.N.*(good night), *ttyl* (talk to you later).
Compression of word/phrase	*Thx*/*Thz*/*Ths* (thanks), *gd nite*/*Gd nite* (good nite), *tmr*/*tomoro*/*tomo*/*tml*/*tomolo* (tomorrow).
Homophony in number	*8* (bye), *88* (bye bye), *9* (night), *2* (to).
Homophony in letter	*CU*/*cu*/*CY* (see you), *U* (you).
Combination of processes: compression of word & homophony in number	*gd 9* (good night).

Table 8. Tendency of using shortenings for 'Thank You. See you tomorrow. Good Night.'

	thank you	see	you	tomorrow	good	night
Frequency of shortened forms for these expressions	127	79	99	51	58	25
Percentage(%)	84.1	52.3	65.6	33.8	38.4	16.6

the respondents. Very often, small letters are used at the beginning of a sentence.

There is indeed a very high tendency of using English in the expressions. One possible explanation to this is that phrases like 'Thank you' or 'See you' are already very familiar and conventionalized and thus the issue of language barrier is not significant. Some of the respondents used codes other than English. Within the 151 messages, 134 messages were written in English and 3 of them were in Chinese. Fourteen of them were of mixed codes. Some of them even contained numbers like '88', '9' and '2morrow'. The result is shown in Table 10.

Other than the verbal messages, some respondents expressed 'emotions' as well. 14 emoticons were found in the data. The result is shown in Table 11 below.

For the expression 'Thank you', only 5 (3.3%) of the respondents used the full form while 127 (84.1%) used other shortened forms. The most popular form for

Table 9. Omitting rule of capitalization

	Omitted	Not omitted	Not applicable
Number of Respondents	54	92	5
Percentage(%)	35.8	60.9	3.3

(One of the messages is discarded because it is written in Chinese.)

Table 10. Language(s) used

	English	Chinese	Mixed
Number	134	3	14
Percentage(%)	88.7	2.0	9.3

Table 11. Occurrence of emoticons

	Yes	No
Number of Respondents	14	137
Percentage(%)	9.3	90.7

Thank You was 'thx', which was reported by 93 respondents (61.6%), whereas 6.6% of the respondents used 'thanks' and 5.3% used 'thz' respectively. Other shortened forms used were 'thanzzz', 'thnx', 'ths', 'thxx' and 'thankyo' (Table 12).

For the word 'see', 44 (29.1%) of the respondents used the full form. (Table 13).

For the word 'you', only 17 (11.3%) of the respondents used the full form while 99 (65.6%) used other shortened forms. (Table 14) The most popular form 'u' was used by 94 respondents (62.3%).

For the word 'tomorrow', 47 (31.1%) of the respondents used the full form while 51 (33.8%) used other shortened forms. Unlike the case of 'see' and 'you', the shortened forms for the word 'tomorrow' are very inconsistent. The most popular form 'tmr' was used by only 24 respondents (15.9%) while the forms 'tom' and

Table 12. Expressions for 'thank you'

	Thank you	Thx	Thanks	Thz	Thank	Others	Missing
Frequency of Occurrence	5	93	10	8	4	20	11
Percentage(%)	3.3	61.6	6.6	5.3	2.6	13.2	7.3

Table 13. Expressions for 'see'

	see	c	Others	Missing
Frequency of Occurrence	44	79	13	15
Percentage(%)	29.1	52.3	8.6	10.0

Table 14. Expressions for 'you'

	you	u	ya	Missing
Frequency of Occurrence	17	94	5	35
Percentage(%)	11.3	62.3	3.3	23.2

Table 15. Expressions for 'tomorrow'

	tomorrow	tmr	tom	tomolo	tomoro	Chinese equivalence	Others	Missing
Frequency of Occurrence	47	24	5	4	3	12	15	41
Percentage(%)	31.1	15.9	3.3	2.6	2.0	7.9	10.0	27.2

'tomolo' came the second (2.6%) and the third (2.0%). Other shortened forms used were 'tomoro', '2 more ro', '2morrow', 'tml' and 'tm' (Table 15).

For the word 'good', only 39 (25.8%) of the respondents used the full form while 58 (38.4%) used other shortened forms. The most popular form 'gd' was reported by 54 respondents (35.8%).

For the word 'night', 69 (45.7%) of the respondents used the full form while 25 (16.6%) used other shortened forms. The most popular form used was 'nite' (8.6%). Other shortened forms used were 'N', '9', 'nt', 'nig', and 'nitez'. It has to be noted that 47 respondents (31.1%) omitted the meaning 'good night' in the message (Table 17). Instead, they used expressions with the meaning 'goodbye' and expressions like '88' and '886' were found in these messages.

On 'Internet Language' in School and Education (6-9)

Questions 6 to 9 of the questionnaire constitute the core part of the survey. These questions were used to investigate the possibility of incorporating ICT texts into the formal classroom.

We first asked the respondent whether he/she had used 'Internet language' in their

Table 16. Expressions for 'good'

	Good	gd	G	Missing
Frequency of Occurrence	39	54	4	54
Percentage(%)	25.8	35.8	2.6	35.8

Table 17. Expressions for 'night'

	night	nite	N	9	Others	Missing
Frequency of Occurrence	69	13	6	6	10	47
Percentage(%)	45.7	8.6	4.0	4.0	6.6	31.1

assignments. The majority (73%) said they never used such language in their school writings. There was, however, some 26% of the respondents who did unconsciously incorporate 'online language' (supposedly different from the 'norm' of language education in HK) into their written assignments in school.

Would the teacher consider these elements as 'errors'? Quite surprisingly, only about half of the respondents reported that their teachers considered their 'online language' features as 'errors'. Of course, there are many other factors affecting the result of this question, for instance, the quality of teacher feedback and the form of assessment.

One of the major questions we asked was whether students would accept that they were allowed to use 'online language' in their school assignments.

Although only 30% of the respondents hoped that they could use online language in their written assignments, one should not overlook the fact that 29% of the students were 'neutral'. This group of students did not completely reject the idea, but was probably hesitant to accept the plausibility of being allowed to use 'Internet language' in their homework.

We also asked the respondents whether they thought their language skills could be improved through online communication. About 34% of the respondents reported that reading and writing online messages could improve their language skills and 46% of them did not think so, whereas 20% of them remained neutral. This means that, among those who actually answered 'Yes' or 'No', almost half of the respondents agreed that they could improve their language skills through communicating

Figure 3. Have you ever used 'Internet language' in your written assignments in school?

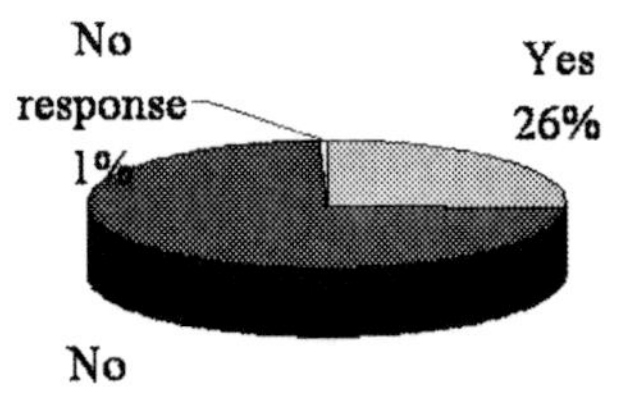

Figure 4. Did your teacher consider those parts of your writing incorrect?

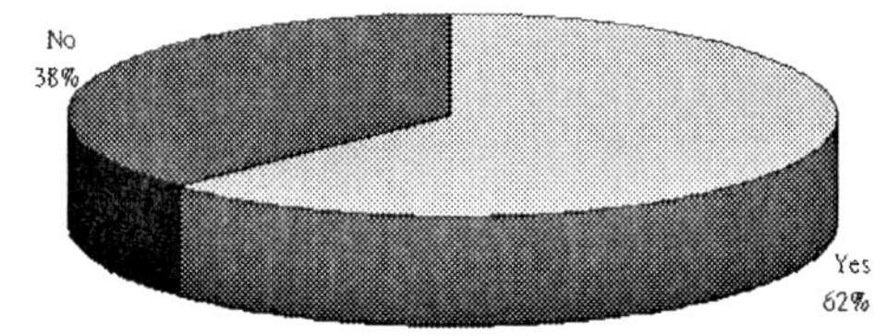

online. This may suggest that implementing ICT texts in the formal curriculum could be an effective teaching and learning strategy.

Individual Interviews

We further conducted in-depth face-to-face interviews with 12 students from the three schools. These interview sessions were semi-structured and mainly dealt with the following topics:

i. The interviewee's general habits of online communication – e.g. How often they do online communication; what language(s) the interviewee used in online communication;
ii. The pros and cons of electronic communication;
iii. Whether the interviewee thought that online language had any influences on one's language standards; and

Figure 5. Do you wish that you were allowed to use Internet language in your writing in school?

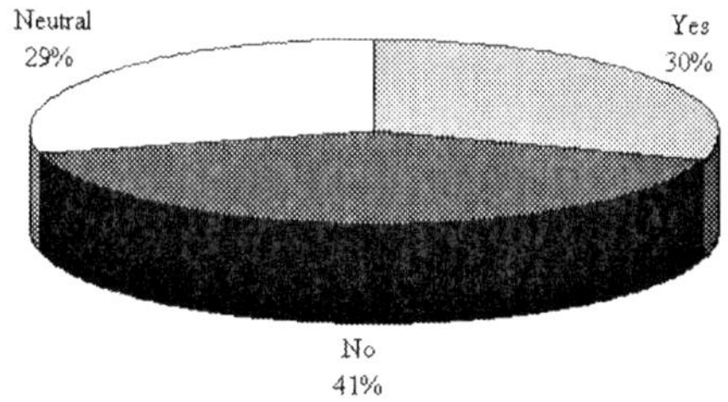

Figure 6. Do you think reading and writing online messages can improve your language skills?

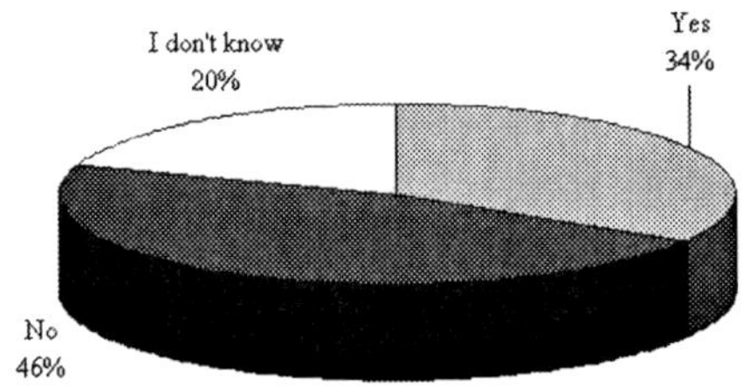

iv. The interviewee's views regarding the use of internet language in the traditional classroom context.

Of the 12 respondents, 11 were daily users of ICQ. The other one used it every week.

Investigator: Which of the two languages, Chinese and English, do you usually use in electronic communication? Why?

A: I usually use both Chinese and English. Since I didn't know Chinese typing before, I used more English. But now I'm better at Chinese typing, so I am using more Chinese. I find it more convenient to express myself in Chinese than English.

B: I like to use English letters to express Cantonese because it is easy for people to understand.

C: Usually Cantonese, with some English because it is more expressive

D: English – it's faster (in terms of typing)

E: Cantonese – it is more convenient – it is my own language

F: Chinese – because that is my native spoken language

G: I usually use both Chinese and English but I would use more Chinese because I speak Cantonese most of the time – easier to express myself in Chinese

H: Both Chinese and English but relatively more Chinese because I don't know how to express everything in English.

I: Use both Chinese and English but more English because I am not very good at Chinese typing.

J: I use both languages, esp. when I want to be playful (e.g. adding Cantonese final particles).

K: Both languages are used. I use Cantonese because I can be more expressive using my mother tongue. But English is also used because I can type faster in English.

L: Cantonese and English – but I prefer English because I am more used to using English in school.

Investigator: What are the pros and cons of using email, ICQ and SMS as means of communication?

Investigator: Do you think that online language has negative influences on one's language standards? Give reasons and state what these possible impacts are.

A: One may unconsciously use ungrammatical sentences in writing since people seldom care about the grammatical accuracy of online messages.

B: Many people like to 'translate' Chinese into English when using ICQ – this may worsen English writing.

C: Yes – short forms may affect compositions – esp. in terms of tenses/grammar/ vocabulary.

D: Yes – one may forget the use of correct grammar.

E: No, because we don't use this kind of language in school.

F: Yes – only if people replace formal language with online language in school writings

G: Yes – one may unconsciously use online language in school, e.g. in compositions.

Table 18. What are the pros and cons of using email, ICQ and SMS as means of communication?

	Pros	Cons
Email	- E-cards, save money, - convenient - More space allowed - Fast - Avoid face threats - Free - Can send one email to many people at the same time - Can be sent to people in different parts of the world - Easier than letter writing	- No face-to-face interaction, can't express emotions - Delayed reply - Difficult to use - Too much typing involved - Inconvenient (e.g. need to get one's email address before sending a message) - Insecure – accounts may be hacked - bulky
ICQ	- fast file transfers, - convenient - easier to contact people at the same time (better than using phones) - allows abbreviations - avoid face threats - synchronous - can 'speak' one's mind easily, more straight-forward	- Lack face-to-face interaction, can't express emotions - Limited space - Doesn't have any concrete functions – just for chit-chatting - Shortenings may cause misunderstanding - Chat partners have to be online at the same time - Too much typing - Wrong grammar - Virus problem - May affect my English
SMS	- can be used anytime anywhere (with a mobile phone) - for leisure - fast - convenient - avoid face threats	- costly - delayed reply - slow in character inputting

H: Yes, one may use online language in compositions and may therefore cause misunderstanding

I: Yes, and if one gets used to this, it is hard to switch back to formal writing.

J: Possibly – but it all depends on the context of communication.

K: Yes, because we are supposed to write 'pure' English in school assignments.

Investigator: Have you ever used Internet language in your written assignments? If so, why did you do so?

A: Yes – when I'm in a hurry to finish up an assignment.

B: Yes & No – I only use short forms in note-taking because it's easier and faster. But I won't write this way in school because it is not allowed.

C: I don't do so personally– but for other people, I think they are at the risk of forgetting the standards if they use Internet language too often.

D: No – other people use it because they are simply lazy.

E: No – it is important to maintain accuracy in English writing – using online language in school writing is a bad habit.

F: Yes – when I am in a hurry – e.g. u for you and ppl for people.

G: Yes – I have used this kind of language in school assignments because I have formed a habit of using Internet language and therefore used it in my assignments unconsciously (in both Chinese and English assignments).

H: No – using Internet language is a bad habit because not everybody can understand the abbreviations.

I: No – bad habit may be formed.

J: No, I don't because I am aware of the formality of my writing. Besides, people use Internet language in their writing simply because they are careless. I think it is important to use the appropriate form of language in the appropriate context.

K: Yes, I have used Internet language in my school assignments. I am too used to writing online so I unconsciously used it in my school work. But I am aware of the disadvantages of doing so.

L: Yes, I have used 'u' for 'you' in my English composition.

It is noted from the interview sessions that:

i. Expressivity of the messages is a key issue, no matter which language is being used. For students, the aim of online communication is to let people understand what they mean. The issue of grammatical accuracy comes the second.
ii. Students have a lot of assumptions about what is required in school writing, which often emphasizes formality, purity, accuracy of language.
iii. Only very few students have used the presumably 'bad' form of Internet language in school. Most of them are well aware that they should write the 'standard' form of language in school and that ICQ language is discouraged.

TEACHERS' RESPONSES

Although not a large number of teachers were involved in this study, the findings, in general, already provide us with some information for generalizing teachers' attitudes towards using ICT texts in language teaching.

A total of 15 secondary school English teachers were surveyed, among whom 12 were native Cantonese speakers whereas the other two were native English speakers. All but one of them had at least two years teaching experience. The age range of the participating teachers was between 21 and 55.

Most of the participating teachers were over 25 years old, which may mean that they are not frequent users of the Internet or online communication tools, given the idea that "digital natives" are mostly younger than 25 years.

In terms of ICT tools, a majority of the respondents were most familiar with voice mobile phone communication and email, and very few of them always used other means of online communication like ICQ or chat programmes.

As far as language is concerned, most of the respondents said they would use only English when writing online messages, in general.

Unlike the findings from the student survey, where most students would use a mixture of English and Chinese, English teachers tend to be 'purists'. This probably has to do with their profession, in which code-mixing is often discouraged.

Quite interestingly, the results from another question revealed that all teachers under survey thought that students can benefit to a certain extent from online communication.

It is worth noting that 11 out of the 14 respondents thought that through online communication, "Students can communicate with their peers outside the classroom without any language constraints." So why is it that while Schools are striving to teach students to communicate effectively, teachers are still thinking that their students can "communicate" "without any language constraints" *outside* the classroom? At the same time, there was the concern that "Students may use Internet language in their writing unconsciously."

However, when asked whether these teachers had actually come across signs of Internet language in students' writing. Only 29% of the teachers said they had *often* (once every 2 or 3 assignment) seen such language in students' writing. The rest said they *seldom* saw such signs.

Why then would teachers worry about this at all, if ICT allows students to communicate more effectively? This is probably because being language teachers themselves, who often conform to a set of prescribed rules of language, they had to consider any forms of 'unconventional' language as 'mistakes'.

We further asked them whether they think students should use Internet language outside the classroom.

Most respondents answered 'neutral'. Further investigation is needed in order understand what exactly this 'neutral' implies. Apparently, some teachers did not want to take a position on this.

When asked whether they found it possible to adopt examples of 'Internet language' as language teaching materials, half of them answered 'unlikely' or even

Table 19. How much do you know about the following communication tools?

	A. Very familiar and have used it	B. Quite familiar and have used it	C. Only heard of it but never used it	D. Never heard of it
AOL Instant Messenger (AIM)	1	1	5	6
Bulletin Board	0	2	10	2
Email	12	2	0	0
ICQ	4	4	6	0
Chatroom (e.g. IRC)	0	3	10	0
Land-line phone	7	1	1	5
Letter (hand-written)	10	4	0	0
Mobile Phone (voice)	13	1	0	0
Fax	6	8	0	0
MMS	3	4	1	5
MSN Messenger	2	2	8	2
Newsgroup	1	3	8	2
Online Forum	1	4	8	1
SMS	5	2	3	4
MUDs and MOOs	1		2	10
Video Conferencing	1	3	10	
Yahoo Messenger	3		10	1

'absolutely impossible'. However, among the rest of the respondents, three of them still thought that it was 'very likely', 'likely' or 'quite likely' to use ICT language as language teaching materials, while the other one remained 'neutral'.

SUMMARY AND CONCLUSION OF THE CASE STUDY

Summary of Findings

Although this case study did not involve a large number of students and teachers, a number of issues can already be observed:

i. The majority of the respondents preferred online communication to other forms of communication.
ii. Bulletin Board systems, email, and ICQ are the three most popular types of electronic communication tools among students – They are also the top three

Table 20. How often do you use the mode(s) of communication that you have chosen in Q.1?

	A. Always (at least once a day)	B. Often (almost every day)	C. Quite Often (3-5 times a week)	D. Seldom (less than 3 times a week)	E. Never	F. Never heard of it
AOL Instant Messenger (AIM)	0	1	0	1	5	6
Bulletin Board	0	1	0	0	11	1
Email	9	0	3	1	0	0
ICQ	1	3	0	4	5	0
Chatroom (e.g. IRC)	0	1	1	2	9	0
Land-line phone	5	2	0	1	3	2
Letter (hand-written)	0	1	2	11	0	0
Mobile Phone (voice)	9	3	1	1	0	0
Fax	0	1	3	10	0	0
MMS	0	2	1	1	5	4
MSN Messenger	0	3		2	7	1
Newsgroup	0	1	1	2	8	1
Online Forum	0	1	1	3	7	1
SMS	1	3	1	1	4	2
MUDs and MOOs	0	0	0	1	3	5
Video Conferencing	0	0	1	2	9	1
Yahoo Messenger	1	1	0	1	9	1

Figure 7. What language do you usually use in online communication, in general?

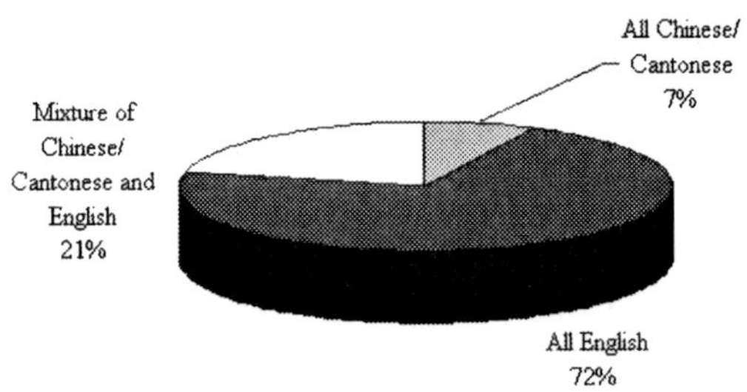

Table 21. In what ways can Hong Kong students benefit from online communication?

Opinion	*N*
Students can develop their linguistic creativity by inventing their own means of communication e.g. abbreviations, emoticons and code-mixed language.	7
Students can expose themselves to different languages and culture via electronic modes of communication.	6
Students can communicate with their peers outside the classroom without any language constraints.	11
Students can express themselves more easily.	9
It is a means to show intimacy and friendliness.	9
No benefits at all	0

ICT tools on which Chinese-English language mixing can be found.

iii. Teachers do not altogether discourage the use of ICT language.
iv. To some teachers, ICQ may still be the cause of the falling standards of language but a majority of them admit that students can be more expressive by using this kind of language outside the classroom context.
v. ICQ and other Internet communication tools may not be the immediate causes of the "falling" standards of language – (i) both students and teachers rarely used or came across online language in formal assignments, and (ii) many of them did not think Internet language could become part of the formal education. Tools like ICQ are, therefore, simply for pleasure and convenience of communication among friends. Students are well aware of the difference.

Here I attempt to make a comparison between students' and teachers' perception and experience of using ICT tools and texts:

Possible Implications and Further Issues

CMC language is the kind of genre pursued and desired by the new generation since it suits better the online communication environment which emphasizes immediacy,

Table 22. What is/are the disadvantage(s) of using 'Internet language'

Opinion	*N*
Internet language leads to the falling standards of language.	7
Internet language hinders formal language education	7
Students may use Internet language in their writing unconsciously.	12

Figure 8. Should students use Internet language outside the classroom?

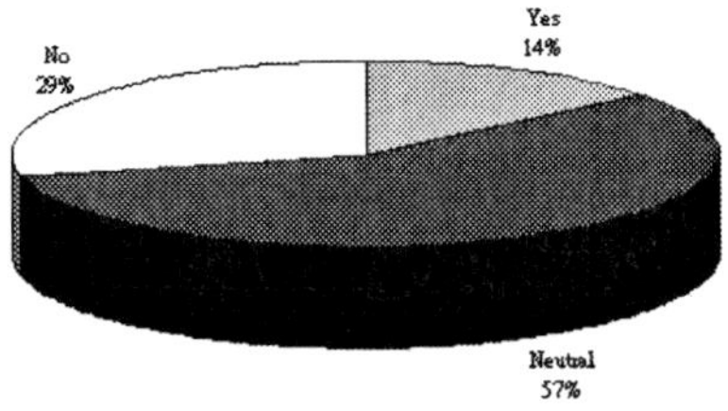

simplicity, and idiosyncrasy. What I mean by 'immediacy' and 'simplicity' here is that online communication can let two or more conversational partners interact without any time and spatial constraints and hence people are able to build up immediate virtual communities by using simple and 'shared' forms of language. In the online communication environment, one can even freely construct a brand-new language according to one's preferences and needs without worrying about the violation of the rules of formal and standard languages (cf. Crystal [2001]'s 'Netspeak').

Apparently, the emergence of CMC language is an irreversible process, thanks to rapid technological development. Instead of suppressing and marginalizing the growth of this new genre of language, one should remain open-minded and recognize the novelty of the language of IT. One possible way we can rely on this is to incorporate CMC language into our formal educational curriculum. A systematic introduction of the nature, functions, and features of CMC language may enhance students' awareness of the appropriate use of this language and thus they will not be confused about when it can be used and when it can not. In other words, we should try to persuade educational conservatives to do away with their ingrained bias towards CMC language.

Furthermore, if we look at ICQ and other kinds of CMC language more deeply and thoroughly, we can observe that students nowadays indeed possess an exceptionally high level of skills in the manipulation of this language, but they themselves may not even notice it. For example, students are able to use the letter homophone *C* to represent the word '*see*', *U* for '*you*' and the number homophone *2* for '*to*' and *9* for '*night*' etc.. At first glance, these may look minor or insignificant, but in fact, they are strong clues which reveal the capability amongst the students in manipulating sounds to express themselves in the most direct and *economical* ways. Therefore, '*see you*' becomes '*cu*' and '*good night*' changes into '*gd 9*'. All these also uncover the creative power of students in handling language. Criticizing and barring students from using ICQ and other forms of CMC language is more or less like strangling the creativity and innovativeness of the next new generation.

Table 23. Comparing students' and teachers' perception and experience of using ICT tools

	Students	*Teachers*
General Habit	Students are the 'insiders' of ICT - frequent users of ICT tools and their associated linguistic practices.	Teachers are the 'outsiders' of ICT (?) – they are not frequent users of ICT tools (esp. ICQ).
Chinese-English code-mixing	Chinese-English mixing is very prominent in ICT communication among students.	Teachers generally discourage code-mixing in writing.
Expressivity of ICT language	Both students and teachers agreed that one can be more expressive when using non-standard forms of language in online communication (outside the traditional classroom context).	
Possible impact of ICT language on the 'standard'	Both students and teachers were aware that the familiarity with the non-standard Internet language may lead to the falling standards of language.	

CONCLUSION

In this chapter I have raised questions about the way the youth create and use language on the computer and allied technologies and the implications these have for standards of language, ways of learning, and the general curriculum as a whole. I here summarize our major findings and draw some conclusions.

Based on a survey of about 200 (184 students and 15 teachers) people in the educational sector, I have come up with the following findings to address the questions we have raised about whether or not and how we should incorporate youth language and literacy practices in the school curriculum.

The first finding is that the majority of the respondents preferred online communication to other forms of communication. Students are now even spending more time online than reading books and watching TV. This finding has an important implication for the way we deliver our teaching and encourage students to learn. We would obviously have to incorporate online platforms in our course delivery, assignments and assessment if we want to catch the attention of our target students.

A second finding is that Bulletin Board systems, email, and ICQ are the three most popular types of electronic communication tools among students in Hong Kong – they are also the top three ICT tools on which Chinese-English language mixing can be found. However this finding cannot stand the test of time as new types of online communications are springing up by the day. Currently, MSN messaging has replaced ICQ as the most common synchronous tool among Hong Kong students (though the ICQ version in mainland China, called QQ, is still very popular there), while blogging has replaced Bulletin Boards as the main asynchronous tool for communication among the youth in many parts of the world. It is very common for students to request to have their own blogs on particular courses. In a recent course the author taught, he encouraged all course participants to start their own student pages on WebCT but one particular student insisted on creating his own blog. One implication here for course delivery and administration is that teachers have to be prepared to be flexible and allow many choices if they want to take on board the interests of most or all the students.

Thirdly, an interesting and relieving finding is that teachers do not altogether discourage the use of ICT language. This is relieving because when my research assistants and I began the project that is the basis of the case study in this chapter, many teachers we spoke to were rather negative about the language used online by students and were vey wary that it might indeed affect formal language standards. They were therefore rather skeptical or even negative altogether regarding the incorporation of ICT and especially youth linguistic practices on ICT into the curriculum.

Fourthly, in relation to the above finding, while ICQ and other forms of online communication may still be the cause of the falling standards of language, a majority of the teachers interviewed on our project admit that students can be more expressive by using this kind of language outside the classroom context.

Finally, despite all the finger-pointing, ICQ and other Internet communication tools may not be the immediate cause of the "falling" standards of language – (i) both students and teachers rarely used or came across online language in formal assignments and (ii) many of them did not think Internet language, as shown in the appendix at the end of this chapter, could become part of their formal education. Tools like ICQ are, therefore, simply for pleasure and convenience of communication among friends. Students are well aware of the difference, and most of them do not incorporate this way of speaking and writing into formal communication environments. However, one should be wary of this happening in the future more extensively. The linguistic and pedagogical implications for this then are that teachers must introduce and focus particularly on teaching the appropriate registers – using the right kind of language in the right kind of context.

In the next two chapters we will see how we can design and then evaluate learning environments based on computer mediated communication to enhance education, in general, and natural language learning, in particular.

REFERENCES

Bodomo, A. (2006). *Project report: New forms of language and literacy in the Age of IT*http://www.hku.hk/linguist/research/bodomo/QEF/

Crystal, D. (2001). *Language and the Internet.* UK: Cambridge University Press.

港人玩ICQ亞洲第一多 [Hong Kong has the highest number ofICQ users in Asia]. (2004, May 6.). *Mingpao* <明報>.

Li, D. C. S. (1998). The plight of the purist. In M. C. Pennington(Ed.), *Language in Hong Kong at Century's end* (pp. 161-190). Hong Kong: HKU Press.

Li, D. C. S. (2000). Cantonese-English code-switching research in Hong Kong: A Y2K review. World Englishes , 19(3), 305–322. doi:10.1111/1467-971X.00181

ENDNOTE

1 The number in brackets indicates the number of respondents who gave the same expression.

APPENDIX: Q.10. EXPRESSING 'THANK YOU. SEE YOU TOMORROW. GOOD NIGHT' VIA ICQ.

Answered: 90

Unanswered: 14

唔該…睇日見啦^^" 88~

Thx. 聽日 see u Good 9

Thx. 聽日見. Good night

ByeBye ~ 明見la…….

Thx!! c u聽日!! 886.

多謝~聽日見~早頭~bye!!!

Thx_cu tmr_早D抖la*

Thx, cu, 886 (2)[1]

Thx. 88.

Thx, 886

88. cu

thx. See u 88

Thanks, see u tomorrow. 88

Thank, C U tomorrow 8

Thx Cu tomorrow 88

Thx. see u torrom. 88

Thank. C U tomorrow. 88

Thx, later talk to u, 886

THX. CY. 886

Thx Cu next day 886

Thank. C you tomoro. 886

thanks~ see u ~ 886

Thank u, 86

Thx. see u tomorrow. Good 9.

Thx. cu tm. gd 9 (2)

Thx~! c u 2 more ro.

Thx, see you tomorrow, Good night. (6)

Thx. C U tomorrow. Good night (3)

Thx. C U tomorrow. Gd night (3)

Thxx. See u tomorrow. G.N.

Thz, see you later, Good Night

Thx. See u. tomo. Gn.

Thz. C u tml. Good night

Thx. see u tml. Good Night.

Thx, see you, GN

Thx, c u tmr, Goodnight

thx. cu tmr. gd night.(4)

Thz. C u tmr. Gd night. (2)

Thz! C u tmr…gd nite.

Thx, See u tomorrow Good night.

Thx. See u. Good Night

Thx ~ CU tomolo, gd n

Thx. c u tomolo. gd night.

thx. c u tomlo. night night.

Thz. C u tomolo. Gd nite.

Thx. C U tomr. Good night

Thx. See u tom. Gd night

Hey. See you tomorrow. Good night

Thanks, see you Tomorrow. Good Night

Thx u. c u tmw. Gd nite.

Thanks, See u tomorrow. BB

Thx. see u tmr, good night

Thx….C u tomoro. gd night

Thank. see u tomorrow. Good night

Thank X. See U Tom. Good N.

thx…see u tomorrow ~ Good night ^^

Thx u.

Thank you, cu tomorrow. Good Night.

Thx. cu la tomoro. gd night.

Thx. See you tmr. Goodnight.

Thnx. See u tmr. gd nite.

Thankyou, nice chat with you, ant wiedersehen!!! =..=

Thanks*! c u!! ttyl*, nitez

Thanzzz. c ya tomolo. gd nite.

Thanks* c u tomorrow…gd nite*..= =zzz

Thank you. See u tomorrow. Gd night.

Thank you! c u tomorrow. Gd nite.

Thanks, cu tmr, gd night

Thx, c ya tmr, gd nite. (2)

Thz, c u tomo. gd nt.

Thx, cu tmr. Gd Nt.

Thx. c u tomor. Gd nig.

Danks. C ya tmr. Gd nite

thx…c ya tmr, nite nite…

Chapter 10
Educational Technologies (WebCT)
Creating Constructivist and Interactive Learning Communities

INTRODUCTION

In this chapter, I will discuss how to take advantage of youth interests in these new technology that have been discussed at length in previous chapters by showing how we can turn them into learning and pedagogical technologies, leading to a model for language learning using these media – the Conversational Learning Community. Two main concepts, constructivism (Bodomo 2005b, 2007) and interactivity (Bodomo 2006, 2008, will be highlighted as important concepts in the area of using learning technologies to create good pedagical environments for teaching issues of linguistics and literacy, and for that matter any other subject. The first study is on constructivism while the second part deals with interactivity.

The empirical basis for this chapter is a series of studies undertaken in the context of a teaching development project at the University of Hong Kong led by me, the author[1]. I will be particularly concerned with how we can achieve interactivity in the learning environment I term Conversational Learning Community (CLC). CLC is based on a constructivist theory I term conversational learning theory (CLT). Interactivity is by far the most important issue when it comes to evolving web-based and distance learning environments. This is because, in most cases, learner and teacher are not in a physical face-to-face environment. It is therefore crucial to

DOI: 10.4018/978-1-60566-868-0.ch010

evolve learning materials and learning processes that can capture the attention of the learner, making interaction between instructors and learners as natural as possible to the point of mimicking or even enhancing physical face-to-face learning environments.

After a background information sketch on the actual study, I explain the key issue of constructivism and do a survey of constructivist learning theories before delving into a detailed experimental study of the bulletin board as a constructivist learning environment.

Within the field of education, and in particular educational technology, dominant paradigms of instruction include active, interactive, constructivist and student-centered approaches as opposed to passive and teacher-centered approaches to learning and teaching. There is hardly much argument as to the relevance of such constructivist approaches in the contemporary world. It is no more of a question of whether we have enough information but what critical and analytical skills we need to sift out relevant information from the huge barrage of information churned out of the pervasive Information Communications Technologies (ICTs) of our time and construct meaningful knowledge to enhance human life. Within such constructivist learning environments, digital ICTs are deployed to enhance the best possible methods. This chapter aims to provide an overview of the meaning and theories of constructivism, and to further illustrate the concept with a case study of using online bulletin board in a university undergraduate course.

WHAT IS CONSTRUCTIVISM

To construct, literally, means to build or to create something by combining different parts. In the field of education, the idea of constructing knowledge and meaning is highlighted. This philosophy of constructing knowledge and meaning is often called constructivism. The central idea of constructivism is that learners construct their own knowledge of the world. Learning is, therefore, a process of creating meaning by the learners themselves, and the instructor simply serves as a facilitator in this process.

An Overview of Constructivist Learning Theories

Theories of learning within education and related fields such as psychology and cognitive science have proliferated over the years. New pedagogical methods based on these theories are turning away from passive methods of teaching which require no action on the part of the student beyond listening and taking notes to interactive delivery methods which enable the student to control and manipulate the instruction environment. These active approaches to instruction may be situated within the framework of what may be called constructivist theories of learning. The following subsections outline various views and theories of constructivist learning, including Piaget's cognitive constructivism, Dewey's theory of experiential education, Bruner's theory of active learning, and Vygotsky's social constructivism.

Piaget

The Swiss psychologist Jean Piaget is considered to be one of the earliest proponents of constructivism. Piaget's theory of cognitive development in the field of epistemology has greatly influenced today's theories of learning and child development. According to Piaget (1955, 1973), children's abilities to acquire knowledge are attributed to the fact that they are born with the ability to adapt to the environment. This adaptation is achieved through two processes: assimilation and accommodation. Piaget's views provide the foundation for many instructional models. Pedagogically, Piaget's theory implies that children learn 'naturally' through their interaction with the environment they live in, but not through information given by the teacher.

Dewey

John Dewey's educational theory (Dewey 1913, 1956, 1963) also focuses on the fact that knowledge is created by the learners themselves, but unlike Piaget's theory which mainly stresses cognitive development, Dewey stresses the importance of society in the process of acquiring knowledge. As stated in Dewey (1956), learning is an active process that involves reaching out of the mind and organic assimilation starting from within. It is the learner and not the subject-matter which determines both quality and quantity of learning. For Dewey, the learner plays an extremely integral role in the learning process, where knowledge is constructed by the learners themselves through their experience of the real world.

Bruner

The theoretical framework of Bruner is similar to that of Dewey. Learning, for Bruner (1966, 1983, 1986, 1990), is an active process in which learners construct new ideas or concepts based upon their current/past knowledge. The learner selects and transforms information, constructs hypotheses, and makes decisions, relying on a cognitive structure to do so. As far as instruction is concerned, the instructor should try and encourage students to discover principles by themselves. The instructor and student should engage in an active dialog (i.e. Socratic learning). The task of the instructor is to translate information to be learned into a format appropriate to the learner's current state of understanding. The curriculum should be organized in a spiral manner so that the student continually builds upon what they have already learned.

Vygotsky

Another theory of much relevance to constructivist learning approaches to learning is the social development theory as conceptualized by the Russian psychologist Lev Vygotsky. In his work (e.g. Vygotsky 1934, 1978), Vygotsky further expands on Piaget's view and emphasizes that social interaction plays a fundamental role in the development of cognition. Another aspect of Vygotsky's theory is the idea that the potential for cognitive development is limited to a certain time span which he calls the 'zone of proximal development' (ZPD). Furthermore, full development during the ZPD depends upon full social interaction. The range of skills that can be developed with adult guidance or peer collaboration exceeds what can be attained alone.

Summary of Main Tenets of Constructivism

Based on the various views presented above, the main tenets of constructivism can be summarized as follows:

- The learner plays an active role in the learning process. (Bruner)
- Learners build their own knowledge through experience, but not through 'given' information. (Piaget, Bruner)
- Instructors should only serve as facilitators and encourage students to discover new knowledge by themselves. (Bruner)
- Learning is a social activity that takes place in an environment that stresses the role of the cultural context. (Dewey, Vygotsky)

These tenets are considered influential to today's theories of constructivist learning, in particular in the area of computer-based learning (Duffy and Jonassen 1992; Bryson and Scardamalia 1996; Scardamalia and Bereiter 1999; Bodomo 2001, Bodomo 2005b, 2007; Bodomo, Luke and Anttila 2003). According to Blurton (1999:9), '[M]odern constructivist education theory emphasizes critical thinking, problem solving, "authentic" learning experiences, social negotiation of knowledge, and collaboration - pedagogical methods that change the role of the teacher from disseminator of information to learning facilitator'. It further describes constructivist methods of learning as involving 'self-paced, self-directed problem-based... learning processes'. In the 21st century and the age of information technology, the issues and questions then are how best we can make use of technologies to achieve such an active and learner-centered environment. What techniques can a teacher develop and deploy at the beginning of a new class in order to turn it into a constructivist learning environment? What are the features of a student-centered, interactive and constructivist class? This article seeks to address these and other research issues by documenting and analyzing a case study presented in the following section.

CASE STUDY: ONLINE BULLETIN BOARD AS CONSTRUCTIVIST LEARNING ENVIRONMENT

This case study analyzes experiences gained from managing and guiding a class of 18 second and third year students of the three-year BA programme in Linguistics at the University of Hong Kong. The subject matter of the course on which most of the observations are based deals with the relationship between Language and Information Technology. Over a semester, students, teaching assistant, and teacher had weekly two-hour face-to-face classroom meetings. This was largely complemented by web-based interactive course activities built on a WebCT platform. To a large extent then, the course was based on a mixed-mode type of learning.

The following are some concrete ways and activities through which I aimed at creating a sense of community among participants in this learning situation.

In-Class Debate

Topic/Motion: 'The Internet is Changing the Ways We Use Our Language.'

Effect: Students were divided into two groups of 'for' and 'against' the motion. Each group had a leader who organized his/her group to work together to come up with evidence in support of their positions. They soon began to express their views among themselves quite freely.

Preliminary Online Assignment: Information Mining and Homepage Creation (Figure 1)

Instructions: Search over the Internet for 5 language- or linguistics-related websites and present a survey of URL findings on your WebCT student homepages

Effect: Creating a sense of community through sharing research findings with classmates by using the 'student homepages' tool. In addition to sharing their findings, many students actually made available information about themselves which

Figure 1. A student presents her web-based research findings on her WebCT homepage

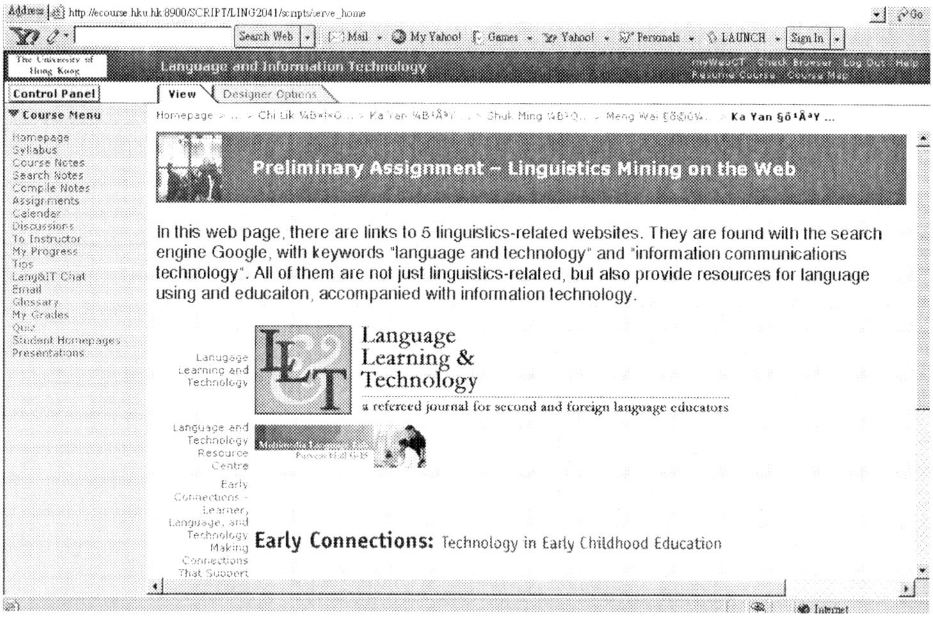

enabled their fellows to get to know their study interests and specializations. This helped strengthen the community of partnership in learning.

A large number of learner activities were recorded for this course. Many of these activities were centered on the various WebCT modules. These may be grouped into two as follows:

WebCT Tools Available for Students

- Course Notes
- Discussions (bulletin board)
- Glossary
- Quiz
- Chat
- Calendar
- Email
- Student Homepages
- My Grades / My Progress
- Assignment Dropbox

Through the Course Notes module, students could download and discuss the latest versions of the course materials. The Glossary served as a kind of specific online dictionary for explaining the technical terms in the notes. Online exercises could be done through the Quiz module. The Calendar module served as a kind of course-book journal for noting deadlines. The Email module is available for off-public communication with the course instructor and the teaching assistant. The Student Homepages module, as mentioned earlier, is each student's little corner in the community for documenting useful information. The My Grades and My Progress modules help students to track their own progress in the community. The Assignment Dropbox module is a kind of community post office for communicating with course organizers.

Tools for Instructor to Track Student Activities on WebCT

- Student Tracking
- Log-on time (first access, last access)
- Log-on frequency (hits)
- Number of bulletin messages read/posted
- Page tracking (time spent of each page of course contents)

There are also tools for instructors and their assistants to monitor and enhance student activities towards the establishment of the desired learning community. These include log-on time, indicating when students log-on for the first time and when their latest access occurs. Log-on frequency helps the instructors to monitor student activities and identify those that need more prompting if need be. Numbers of bulletin board messages read and posted help to calibrate the level of community participation among the learners. Page tracking helps to identify how much time learners allot to interacting with various course materials.

The Bulletin Board Discussion

Each of the above modules deserves a detailed description and explanation of its role in facilitating a learning community and helping learners to construct knowledge for themselves through their own initiative. In this chapter, however, I will concentrate on the bulletin board discussion, one of the most salient modules, that provides telling evidence for the student initiative, community building, and knowledge creation. The bulletin board serves as a community notice board or newsletter through which members interact and exchange vital bits of information. Figure 2

Figure 2. WebCT Opening page and bulletin board of the course LING2041 language and IT

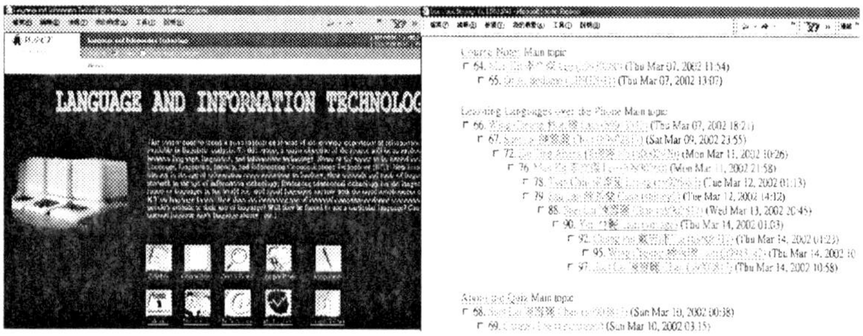

documents a sample of online learner interaction. Further analyses of bulletin board activities are indicated in the next section.

Gradual Transformation from Passive to Active: Student Initiatives in Bulletin Board Discussions

There were altogether 136 messages posted on the bulletin board from the beginning of the course (Jan 2002) to the beginning of April, of which 91 were posted by students. Students were slow and passive at the beginning but as they began to have a sense of community they started doing many things on their own initiative, ending with very comprehensive discussions. These discussions quickly turned into material that informed students about the topics treated. These are illustrated with the following chart (Figure 3):

Three important observations about this graph, indicative of a process of transformation into a student-centered learning community, include the following:

For much of the time, students posted more messages than the instructor or teaching assistant. This point would illustrate the fact that students, rather than teacher and TA, are the core participants in online learning.

As can be observed, not much student interaction took place at the beginning of the course. This observation implies that the instructor/TA ought to intervene to encourage active participation. Mid-way through the course, the amount of discussion increased gradually, indicative of a stage of maturation in the community transformation process.

Figure 3. Discussion postings: Gradual transformation

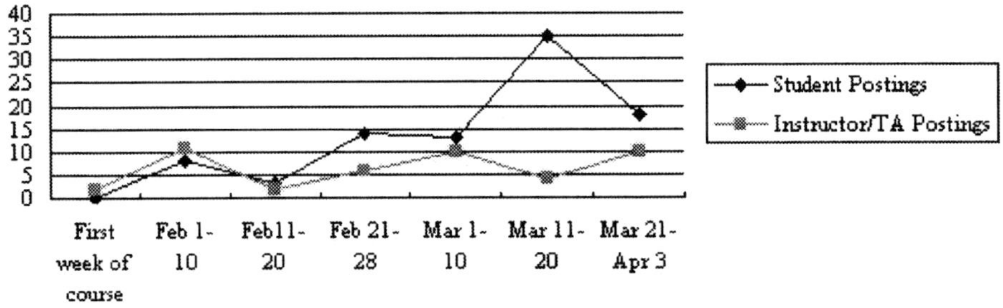

Significantly, there was an explosion in the volume of student interaction in the period of March 11-20 (i.e. a month and a half from the beginning of the course). Students began to take the initiative to raise discussion topics or other course-related issues. A peak was reached when two major discussion topics were proposed, in which around 20 follow-up postings were involved. Concrete examples of these student-initiated topics are shown in figures 4 and 5.

Figure 4 is a message posting where a student shares a webpage link with her fellow classmates, and more importantly, raises an open-ended question at the end of her message. The student in Figure 5 bases her discussion on what she has learnt from the lectures, and further addresses the issue of using educational technology in secondary schools. These are clear indications of students' own initiatives towards the second half of the course (i.e. around early to mid-March, according to Figure 3), where knowledge is constructed through sharing, questioning, and interactive online debates with very little instructor intervention.

Figure 4. Student-initiated topic (i)

Subject: Learning Languages over the Phone
Posted by xxx
Thu Mar 07, 2002 18:21
Please see below: http://www.logoi.com/phone.html This is a very interesting idea. However, how could one ensure if the 'teachers' on the phone are qualified?

Figure 5. Student-initiated topic (ii)

- **Subject: Problems when using educational technology**
  Posted by xxx
  Mon. Mar 11, 2002 22:00
  Apart from the three points in the lecture notes, I would like to propose another problem language professions may face when using educational technology:

  There may be a mismatch between the investment and the use of education technology. I am thingking of the IT-enhanced education, and then hardware/software are installed in secondary schools. But such plan may not take into account the adequacy of IT training for the teachers and most importantly, the schools' curriculum development of varius subjects may not be able to catch up with such vast investment. To what extent the secondary school teachers (actually this problem is not just addressed to language professions) should use IT in their teaching is a problem.

Implications and Future Trends for Web-Based Course Design and Management

Our experiences from the observation, description, and analysis of learner activities within our class may be generalizable onto, and thus have some implications for, the enterprise of web-based course design and management from a constructivist perspective.

These implications include (i) an identification of the stages in the process of community transformation, (ii) the need to intervene at appropriate times, and (iii) the evolution of effective strategies to enhance knowledge building.

For those course guiders and managers who aim at active, interactive and constructivist learning communities where students should take control of their own learning environment, careful attention must be paid to the process of community transformation that has been sketched in this chapter. Web-based course instructors should identify and be conscious of the various stages, which, according to us, as illustrated Figure 3, include (i) an initial period of low community activities where each student seems to grope around on their own, (ii) a period of gradual rise in activities, (iii) a peak, in which there is an explosion of student activities, and (iv) periods of intermittent lulls.

The exact role of the instructor in a constructivist learning environment is a commonly discussed topic. A purely laissez-faire approach would be one extreme, leading to a lazy, irresponsible and do-nothing teacher, while an overly interventionist approach would be the other extreme, and is symptomatic of a teacher-centered learning environment. There has to be a balance. The experience presented indicates that teachers must only intervene at the most appropriate time, and this will imply an understanding of the process of transformation. The appropriate presence of the teacher can be termed as interaction-oriented intervention.

Not only must the teacher understand the process of transformation, and know when to intervene appropriately, he or she must also evolve effective methods to encourage students to construct their own knowledge from interaction with the course materials. The case study has presented concrete ways of encouraging student participation and knowledge creation. This includes information mining in the form of individual or group web search, in-class group debates, and encouragement of online bulletin board discussions.

A final issue introduced in this article is the issue of initiative and innovation in a constructivist course design and management. It has been noted that it is often difficult to gauge if a particular class is showing signs of active interaction and student-centeredness, leading to an active group of knowledge constructors. What are the signs of arrival at the threshold of student-centeredness and knowledge construction? This article has proposed that one of the indicators of identifying a community of knowledge constructors is to examine the initiatives and processes of innovation occurring in the community. This has been illustrated with some student postings.

In conclusion, as a future trend, to succeed in the complex enterprise of knowledge construction in a constructivist learning environment, web-based course designers and managers must identify the processes and stages of community building, practice interaction-oriented intervention and develop appropriate measures to promote student initiative and innovation.

A STUDY ON INTERACTIVITY

The foregoing section of the chapter has presented a case study on constructivism as a learning and education theory. In this section of the chapter I present a second study of the chapter dealing more explicitly with how we can make our learning environment more interactive. This will lead to the formulation of a learning theory and model titled Conversational Learning Theory and Community.

Study Background

At the beginning of the 21st century, we are faced with an age of rapid technological development in information and communication. Issues of educational reform have never been more urgent than now. The main challenge is how we can design our educational system, in general, and our methods of instruction, in particular, to produce graduates who are better prepared to take up jobs in a knowledge-based environment characterized by a pervasive use of information communications technology (ICT). The Hong Kong-based South China Morning Post (SCMP) newspaper editorial of March 9th 2000 formulates the challenge quite appropriately as follows:

If Hong Kong is to remain a vibrant city, it needs a workforce peopled by creative thinkers and problem solvers. That cannot be supplied by pupils who sit obediently taking notes as a teacher intones instructions.

The main issue raised here is that passive or receptive methods of instruction must give way to more active and interactive methods of instruction. Freire (1970) describes passive methods of teaching or what are termed digestive and nutritionist pedagogies in the following way. A pedagogy based on a digestive concept of knowledge is suggested by controlled reading, by classes which consist of only lectures, by the use of memorized dialogues in language learning, by bibliographical notes which indicate not only which chapter but which lines and words are to be read and the methods of evaluating the students' progress in learning.

Interactive learning, on the other hand, promotes a more active approach in the knowledge dissemination and acquisition processes. Blurton (1999: 9) describes interactive or constructivist methods of learning as involving "...self-paced, self-directed problem-based...learning processes".

While it can be said that it is the rapid changes in information communications that has created such an educational challenge, interestingly enough, this situation of rapid changes in technology does actually present teachers, course designers, and university administrators with opportunities to successfully produce a literate workforce for our society.

ICTs, especially modern digital ones, include various types of computers; digital cameras; local area networking; the internet and the World Wide Web; CD-ROMs and DVDs; and applications such as word processors, spreadsheets, tutorials, simulations, electronic mail(email), digital libraries, computer-mediated conferencing, videoconferencing, and virtual reality (Blurton 1999). Four main features of these modern digital ICTs make them stand out as very useful educational tools. These are integration of multimedia, flexibility of use, connectivity, and interactivity (Blurton 1999).

The main focus of this paper is an examination of just one of these features: interactivity (but see, for instance, Kwok and Bodomo (2000) in which some of the other issues are also discussed). While interactivity has been a subject of considerable attention in the search for newer and more active methods of teaching and learning (Brogan 1999, Parker 1999), there still remains a lot to be discussed as to how it can be enhanced in learning situations involving a mixture of web-based course administration and face-to-face classroom instruction. It is quite clear that the introduction of ICTs into distance learning curricula is crucial in enhancing interactivity, given the situation where teacher and student are separated by distance. It is shown here, based on a course that is designed for both distance learners and traditional face to face classroom students where there is unity of time and unity of venue, that the use of the web, one of the new digital ICTs enumerated above, along with other accessories and software that together give us what is termed web-based teaching in a course, plays a crucial role in enhancing interactivity. The paper is organized as follows. Section two defines interactivity and shows the important role it plays in constructive/active learning theories. In section three, I describe the main features of my course and show how interaction was achieved. The fourth section of the paper points to certain challenges that should be overcome to create more opportunities for enhancing interactivity in web-based teaching.

Interactivity and its Role in Constructive Learning Theories

What is Interactivity?

Thompson (1996), which is the Oxford Compact English Dictionary, is a first useful place one should turn to in an attempt to understand the term interactivity. Certain words in this lexical field including the verb interact and the adjective interactive should help us to better conceptualize the term interactivity. To interact according to the OCED is to act reciprocally; act on each other, while the adjective, interactive, according to the OCED means 1. To be reciprocally active; acting upon or influencing each other. 2. (of a computer or other electronic device) allowing a two-way flow of information between it and a user. From these related words, we can see quite clearly that interactivity involves active communication between two or more individuals or between an individual or groups of individuals and one or more electronic devices.

Beyond these general definitions of interactivity, there are more specialized studies and approaches to the term in works that include Daniel and Marquis (1983), Moore (1992), Wagner (1994), Markwood and Johnstone (1994), Barnard (1995), Parker (1999), Brogan (1999) and Bodomo 2006, 2008). The key concepts that run through most of these studies include 'active learning', 'two-way communication', 'critical conversation', and transactional distance learning (Moore 1991, 1993), etc. All these contrast sharply with what would take place in traditional passive/lecture type instruction as described in Freire (1970) and the SCMP (2000) referred to much earlier in the paper. Parker (1999:14) indicates that 'interaction can be defined as active learning and can be as simple as pushing the "play" button on the VCR.' An interesting aspect of understanding the nature of interactivity is an enumeration, within the literature, of different types of interactivity.

Moore (1992) offers three types while Markwood and Johnstone (1994) provides four types of interactivity. In Moore's typology we have learner-content, learner-instructor, and learner-learner interactivity. Learner-content interactivity is illustrated by a student reading a book or a printed study guide (Parker 1999). The interactivity or otherwise of the content is very much a function of how the material is structured and accessed. This point is crucial in deciding how best to place course notes on the web. Instructor-learner interaction is the core of the teaching process. The success of the course design will depend largely on whether the conversation between teacher and learner is such that the learner can increase self-direction and construct new knowledge or not. Learner-learner interaction involves students working together to discuss, debate and attempt to solve problems that arise in their study of the course materials. Moore (1992) provides us with a very useful framework to discuss how interactivity was achieved in our teaching. Indeed, his notion of transactional distance theory (Moore 1991, 1992, 1993) has contributed immensely in defining relations between participants, not only in a distance learning situation, but also in traditional face to face classroom learning situations.

Markwood and Johnstone (1994:94) describe interaction as the "silent, critical, creative conversation within the learner's mind that is spurred and supported by the learning environment." The study outlines four different types of interaction that trigger what it calls critical conversation. The first is interaction with media where individual students scrutinize textbooks, videotapes or any other course material. In our case, this involves a major textbook supplemented by a number of other book sections and course notes. The second is interaction with resources. Here individual students or groups may collaborate with tools such as those used by professionals, including word processors, electronic libraries, laboratories and studios. The third type of interaction according to Markwood and Johnstone (1994) involves interaction with experts. This would mean students conversing with instructor or other experts in real time. This aspect of interaction is yet to be much explored in my course. The last type of interaction is one of interaction through electronic exchange, with students electronically or digitally sharing the results of newly formed knowledge over a period of time (Markwood and Johnstone 1994).

Moore (1992) and Markwood and Johnstone (1994) provide a solid foundation on which to build our idea of interaction and draw up a typology of interaction within the larger framework of what I call a Conversational Learning Community (CLC). In conceptualizing a CLC, one may see the pedagogical process as taking place in an interactive conversational learning community. In this community, we have instructor(s), learners, course materials, and links to remote experts and resources.

This constellation then gives us the following typology of interactivity:

Instructor-learner interaction is either via physical face-to-face interaction (at lectures, tutorials, demonstrations, and consultations), or via digital ICTs (email enquiries, bulletin board enquiries and clarifications, and very rarely chat rooms), or a mixture of both.

Learner-learner interaction is within or without an ICT medium, where students are involved in communication with each other in the classroom, in the corridors, on web-based bulletin boards, in chat rooms, and by emails.

Learner-resource interaction involves learners actively communicating with textbooks, hard-copy hand outs, lecture notes, and with ICT-based current and remote resources such as online lecture notes and outlines, CD-ROMs, glossaries, calendar of activities, progress reports, quizzes, and links to experts and more resources.

For us then instructional interactivity may be defined as active communication in a conversational learning community between instructor(s), learners, course materials, and links to remote experts and resources.

This section of the chapter has attempted to give a close examination of interactivity, ending with a quite specific interpretation of the term. Interactivity has been shown to be the single cementing factor that binds together all the elements of what I have termed a Conversational Learning Community. Before going on to show how web-based teaching strategies were attempts to implement this idea of interactivity within a conversational learning community, I will briefly examine the place of interactivity within current learning and pedagogical theories.

The Role of Interaction in Constructive/Active Learning Theories

Theories of learning within education and allied fields such as psychology and cognitive science have proliferated over the years. New pedagogical methods based on these theories are turning away from passive methods of teaching which require no action on the part of the student beyond listening and taking notes to interactive delivery methods which enable the student to control and manipulate the instruction environment. These active and interactive approaches to instruction may be situated within the framework of what may be called constructivist theories of learning.

According to Blurton (1999:9), "[M]odern constructivist education theory emphasizes critical thinking, problem solving, "authentic" learning experiences, social negotiation of knowledge, and collaboration - pedagogical methods that change the role of the teacher from disseminator of information to learning facilitator…". Works like Piaget (1973) and Strauss (1994) illustrate such new pedagogical theories.

A whole website has been devoted to an exposition of the major theories of learning by Greg Kearsley (1994-2000). This website titled, Explorations in Learning & Instruction: The Theory into Practice Database has outlined about fifty of such theories. These include the following:

ACT* (J. Anderson), Adult Learning Theory (P. Cross), Algo-Heuristic Theory (L. Landa), Andragogy (M. Knowles), Anchored Instruction (J. Bransford & the CTGV), Aptitude-Treatment Interaction (L. Cronbach & R. Snow), Cognitive Dissonance Theory (L. Festinger), Cognitive Flexibility Theory (R. Spiro), Component Display Theory (M.D. Merrill), Conditions of Learning (R. Gagne), Connectionism (E. Thorndike), Constructivist Theory (J. Bruner), Contiguity Theory (E. Guthrie), Conversation Theory (G. Pask), Criterion Referenced Instruction (R. Mager), Double Loop Learning (C. Argyris), Drive Reduction Theory (C. Hull), Dual Coding Theory (A. Paivio), Elaboration Theory (C. Reigeluth), Experiential Learning (C. Rogers), Functional Context Theory (T. Sticht), Genetic Epistemology (J. Piaget), Gestalt Theory (M. Wertheimer), GOMS (Card, Moran & Newell), GPS (A. Newell & H. Simon), Information Pickup Theory (J.J. Gibson), Information Processing Theory (G.A. Miller), Lateral Thinking (E. DeBono), Levels of Processing (Craik & Lockhart), Mathematical Learning Theory (R.C. Atkinson), Mathematical Problem Solving (A. Schoenfeld), Minimalism (J. M. Carroll), Modes of Learning (D. Rumelhart & D. Norman), Multiple Intelligences (H. Gardner), Operant Conditioning (B.F. Skinner), Originality (I. Maltzman), Phenomenonography (F. Marton & N. Entwistle), Repair Theory (K. VanLehn), Script Theory (R. Schank), Sign Theory (E. Tolman), Situated Learning (J. Lave), Soar (A. Newell et al.), Social Development (L. Vygotsky), Social Learning Theory (A. Bandura), Stimulus Sampling Theory (W. Estes), Structural Learning Theory (J. Scandura), Structure of Intellect (J. Guilford), Subsumption Theory (D. Ausubel), Symbol Systems (G. Salomon), Triarchic Theory (R. Sternberg)

To these theories, we must add the important notion of transactional distance theory (Moore 1991, 1992, 1993), a theory that has had a major paradigm shift in distance education.

So what is the role of interaction in these theories of learning? Do all these theories emphasize interaction, or are there some of them which by nature of their very conceptualization are more amenable to the features of interactivity that I have outlined above? I will briefly mention four of these theories which I consider to be the most relevant. They are the constructivist theory of Bruner, the conversation theory of Pask and Vygotsky's Social development theory, and of course, Moore's Transactional Distance Theory (Moore 1993).

An exposition of the constructivist theory is contained in works such as Bruner (1966, 1983, 1986, and 1990). According to Kearsley (1994 - 2000), a major theme in the theoretical framework of Bruner is that learning is an active process in which learners construct new ideas or concepts based upon their current/past knowledge. The learner selects and transforms information, constructs hypotheses, and makes decisions, relying on a cognitive structure to do so. As far as instruction is concerned, the instructor should try and encourage students to discover principles by themselves. The instructor and student should engage in an active dialog (i.e. Socratic learning). The task of the instructor is to translate information to be learned into a format appropriate to the learner's current state of understanding. Curriculum should be organized in a spiral manner so that the student continually builds upon what they have already learned.

The role of interaction is fairly prominent in such a theoretical conceptualization. Once again interactivist terms like 'active process', 'active dialogue' come to the fore.

The next theory that is of immediate relevance to an interactive approach to teaching is the conversation theory as contained in Pask (1975). The fundamental idea of the theory is that learning occurs through conversations about a subject matter which serve to make knowledge explicit. Conversations can be conducted at a number of different levels: natural language (general discussion), object languages (for discussing the subject matter), and metalanguages (for talking about learning/ language). In order to facilitate learning, Pask argued that subject matter should be represented in the form of entailment structures that show what is to be learned. Entailment structures exist in a variety of different levels depending upon the extent of relationships displayed (e.g., super/subordinate concepts, analogies). The critical method of learning according to conversation theory is "teachback" in which one person teaches another what they have learned. Pask identified two different types of learning strategies: serialists who progress through an entailment structure in a sequential fashion and holists who look for higher order relations (Kearsley 1994-2000) (http://www.gwu.edu/~tip/pask.html).

The third theory of much relevance to interactive approaches to learning is the social development theory as conceptualized by Vygotsky (1962, 1978). The major theme of Vygotsky's theoretical framework is that social interaction plays a fundamental role in the development of cognition. Another aspect of Vygotsky's theory is the idea that the potential for cognitive development is limited to a certain time span which he calls the "zone of proximal development" (ZPD). Furthermore, full development during the ZPD depends upon full social interaction. The range of skill that can be developed with adult guidance or peer collaboration exceeds what can be attained alone (Kearsley 1994-2000) (http://www.gwu.edu/~tip/vygotsky.html).

The fourth theory is Moore's notion of transactional distance theory (1991, 1992, 1993), which is very much relevant to distance education and attempts to explain the relations between participants in a distance learning situation. Transactional distance is defined to include the psychological and communicative space between learners and teachers. Moore (1993) highlights the issue of interaction, when he defined transactional distance within the context of interaction in a course, as a function of dialogue, structure and learner autonomy. Dialogue refers to teacher-student interaction, structure refers to how the program is designed, and according to Moore, as dialogue increases structure decreases, i.e., as the interaction between learner(s) and teacher(s) increases the teaching programs structure of objectives, activities, and assessment decreases to accommodate learners need. In other words, learner autonomy leading to self-direction becomes a major fruit in interactive learning situations.

With terms like 'active dialogue', 'conversations about subject matter', and 'social interaction' resonating across these theories, it is clear that interactivity has a central role to play in these theories of learning, which may all be grouped under the general framework/paradigm of constructivist methods of learning as described above.

Indeed these four theories may be seen as forming a useful foundation for the idea of Conversational Learning Community (Bodomo 2005a&b, 2006, 2007, and 2008) that I evolved as a conceptual framework for designing web-based courses. Terms like 'active dialogue', 'conversations about subject matter' and 'social interaction' do form the core of what we may term a Conversation Learning Theory. The main idea of a conversation learning theory is that enhanced interactivity, whether face-to-face or from a distance, such as online instruction, would lead to an effective reciprocal, two-way, communication within the learning situation. This enhanced communication is the backbone for the efficient exploitation of the resources, experts, and links by both instructor and learners within the learning community (Figure 6).

A Description of the Design of a Web-Based Course

I have now explained a number of issues, including the need to use ICT in education, web-based teaching, and interactivity and its role in constructivist teaching methods. I will in the rest of the chapter provide a description of a specific course within my web-based teaching programme, and how interactivity was achieved in the course design. I begin with the choices available for a course designer.

Choosing a Web-Based Course Tool

In deciding to do web-based teaching or facilitate web-based learning, course designers have, at least, two options. They can choose to develop their own tools or they can choose from the repertoire of many course tools called asynchronous web-based software suites (http://www.outreach.utk.edu/weblearning/) that are already available on the market. Robert Jackson (2000) describes the key features of asynchronous course tools as follows on the Web-based Learning Resources Library. "Key characteristics of major players in asynchronous suites typically include capability for

Figure 6. Conversation Learning Theory licenses Conversation Learning Community

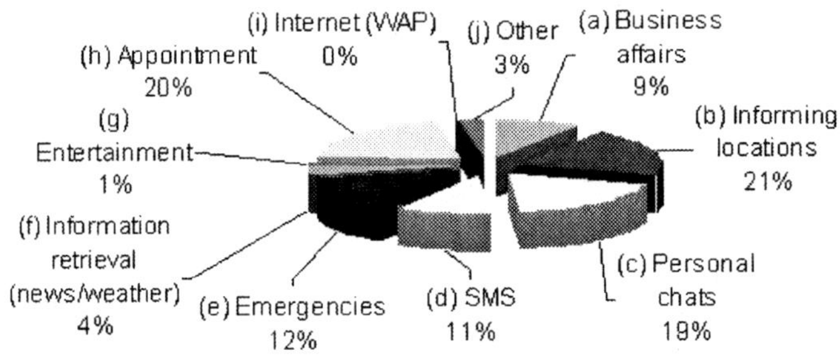

secure student login via standard java browser, centralized database-centered syllabus with links to internal or external web pages, on-line, time-monitored quizzes with randomized dynamically-generated testing, discussion groups, and integrated email. Systems also provide instructor development tools to ease transition from other media to these products" (http://www.outreach.utk.edu/weblearning/).

In the following, I briefly list and comment on a few of them. These include: Blackboard CourseInfo, Lotus LearningSpace, Topclass, and WebCT.

Blackboard CourseInfo claims to enable educators to enhance in-class instruction and/or deliver distance learning by bringing their course materials, class discussions, assignments and quizzes to the Web (http://product.blackboard.net/courseinfo/).

Lotus LearningSpace (http://www.lotus.com/) purports that its version 4.0 will deliver e-learning that's flexible and powerful for all kinds of users -- learners, administrators, planners, course developers, and content providers.

TopClass (http://www.wbtsystems.com/) runs on an Oracle database and so provides a robust and scalable foundation for delivering, managing and measuring online learning in any organization (http://www.wbtsystems.com/products/products.html).

WebCT (http://www.webct.com/) is said to be a low cost, asynchronous delivery and course management system developed by University of British Columbia for higher education faculty. A collection of development tools and custom CGI scripts, it is maturing into an integrated tool suite. WEBCT was purchased by Universal Learning Technology in May of 1999.

Web-Based Course Design via WebCT

Prior to 1998, I used to do web-based teaching by simply putting my course materials on the internet without any course tool. This created a number of problems. First, it was difficult for me, as course designer, to manipulate access to the material in terms of passwords and accounts for the learners. Second, many other interactive features such as bulletin boards, chat rooms, and secured records for student activities were not possible. As such, I hardly was able to design and implement the conceptual notion of conversational learning community that I have referred to above.

In 1998, when I began designing and teaching courses on various aspects of linguistics, I read an announcement from the University's computer centre asking staff members to come to take a course in WebCT. I decided to register for the course and that turned out to be one of the best decisions I have ever made with regards to efforts at teaching development. Since taking that WebCT designer course, I have gone on to computerize all of the four courses that I teach in Linguistics at HKU[2].

I have described the course design of one of the courses in Bodomo (1999), a collection of the first set of WebCT courses at the University of Hong Kong by the Computer Centre. On this score then I am one of the earliest WebCT course developers at HKU.

Further development of WebCT course design has led to the award of a joint Teaching Development Grant (TDG) titled, The Use of Information Technology in the Teaching of Language and Linguistics courses. Further information on the project may be found at the project website at the following address: (http://www.hku.hk/linguist/staff/TDGBodomo.html)

The courses designed under this project include:

- *LING2016 Syntax II: The Theory of Grammar* at http://ecourse.hku.hk:8900/public/B0257/, taught by Dr. A. B. Bodomo
- *LING2006 Syntax I: Describing Grammatical Patterns* at http://ecourse.hku.hk:8900/public/LING2006/, taught by Dr. A. B. Bodomo
- *LING2018 Lexical-Functional Grammar* at http://ecourse.hku.hk:8900/public/LING2018/, taught by Dr. A. B. Bodomo
- *LING2011 Language and Literacy* at http://ecourse.hku.hk:8900/public/LING2011/, taught by Dr. A. B. Bodomo

In the rest of the paper, while giving some excerpts from some of my earlier course designs, I will concentrate on describing just one that I have recently designed to teach a course on the relationship between Language and Literacy.

WebCT Design of a Course on Language and Literacy

The Language and Literacy course is a one semester six credits course for second and third year students of Linguistics and related disciplines.

The course usually begins with an attempt to get students (usually about 20 - 30 in number) to understand the concept 'literacy'. The course materials and lectures and tutorials are designed in such a way that students are supposed to discover for themselves that the concept literacy is NOT limited to just the ability to read and write. Students are supposed to discover for themselves the various linguistic, cognitive, social, and educational issues surrounding the concept. Students are encouraged to gain an understanding of the role of language and literacy in the socio-economic development efforts of many societies through various activities such as discussions, debates, classroom presentation, tutorials and interview of resource persons.

Topics covered in the course often include:

- Definitions and types of literacy
- The relationship between language and literacy
- Writing and other symbolic systems
- Computer Literacy and Language educational technology
- The origins, history, and acquisition of literacy
- Comparative analyses of the language and literacy situations in selected parts of the world, including Hong Kong, Mainland China, and South-east Asia.
- Literacy and socio-economic development

How Interaction was Achieved in my Course

I will now attempt to explain how interaction was achieved in my class. This begins with the creation of a learning community. I have stated in section 2 above that my course design, whether in the form of face-to-face classroom lectures or WebCT course page activities like discussion and presentations, is guided by the conceptual notion of a conversational learning community, comprising instructor(s), learners, current resources and remote experts and resources.

The first task then in my course administration is often to get the group of about 20 - 30 students to communicate and interact with each other and create a sense of community. The first exercise towards this goal is often in the form of internet search. The following excerpt from the course explains the exercise:

LING2011: Reading Assignment/Homework

Literacy Information Mining on the Web

Students should form groups of 2 - 3 people. Each group should search the World Wide Web with key words 'literacy', 'language', (and combinations of these) and choose 10 sites. These sites should be analyzed with a view to finding out what literacy is and what common issues are discussed concerning language and literacy courses. Each group of students should spend five minutes in the next lecture explaining how their understanding of literacy has been affected by these 10 websites.

An Exercise for Building up a Learning Community

This exercise is meant to get students to create physical and electronic networking among them and it often succeeds to a large extent because I have noticed that later groups to be formed in the class often reflect this earlier grouping.

Other exercises include short group discussions on issues in class during the first two weeks.

Once this sense of community is created, the rest of the instructional activities aim to consolidate and strengthen it, developing it into a real conversational learning community. I do this both through my face-to-face classroom activities and my WebCT design activities.

Face to face classroom activities include a mixture of lectures, student presentations and tutorials. Very little ICTs are used here. The main tools employed by me at lectures and tutorials are traditional classroom educational technology such as projector/overhead machines, black- and whiteboard and lecture hand-outs, and assignment feedback sheets. As the semester advances, I have often noticed that student presentations become more and more ICT-mediated, the most typical presentation tool being the PowerPoint.

Surprisingly, greater and more sustained aspects of the interaction between students often take place on the WebCT homepage for this course. In the next subsection, I will describe some of the main features and resources of this tool, showing what kinds of interactivity takes place and how.

Some interactive WebCT features of the course (Figures 7)

Course highlights are shown in Figures 8, 9, 10, 11, 12, 13, 14, 15, 16.

Opportunities and Challenges for the Future

Opportunities

The foregoing has outlined how I have designed my course on WebCT so as to enhance interactivity, a crucial element in a conversational learning community and indeed in any other effective learning situation. A possible question to ask then is how successful I have been. Success, failure and other issues of evaluation are difficult to measure accurately. They may be from the point of view of the instructor or the student. In the following, I shall briefly point to some qualitative features that make me think that, from an instructor's point of view, interactivity has been achieved in thc course. I will also draw from some comments that students made as part of end of semester formal evaluation of my course.

From an instructor's point of view, certain features of communication and academic activity, if they are part and parcel of a course, would serve to indicate

Figure 7. Excerpts from the language and literacy WebCT homepage

that the teaching endeavour is successful. Three of these features include critical thinking, initiative on the part of students, and academic rigour.

Critical Thinking

I noticed that as time went on, not only were students more forthcoming in discussing and interacting with me and with their fellow students, they were also becoming more critical in their thinking. At certain points during the course, students were beginning to question and argue some of the points from me and from their fellow students. Sometimes, I present an issue with regards to the definition and conceptualization of literacy and how it relates to language and then ask students to evaluate these views by applying them to the Hong Kong situation and, indeed, other situations that they know. For instance, the class grappled over a period of time with the relationship between the concepts 'bilingualism' and 'biliteracy'. A lot of discussion ensued on this topic as shown on the bulletin board transcripts (1) (Figure 17).

I consider critical thinking within the conversational learning community as a strong indicator of the success of interactivity in the learning situation. This may be compared with Markwood and Johnstone (1994)'s idea of critical conversation.

Initiative

Another indicator of success with regards to the learning situation is initiative on the part of students. Half way through the course, I noticed with great joy that students did often introduce their own topics of discussion and techniques of information gathering and processing. At one point in time, I had wanted to introduce the concept of biliteracy in the context of our discussion of bilingualism, but was pleasantly

Figure 8. Contents with glossary definitions

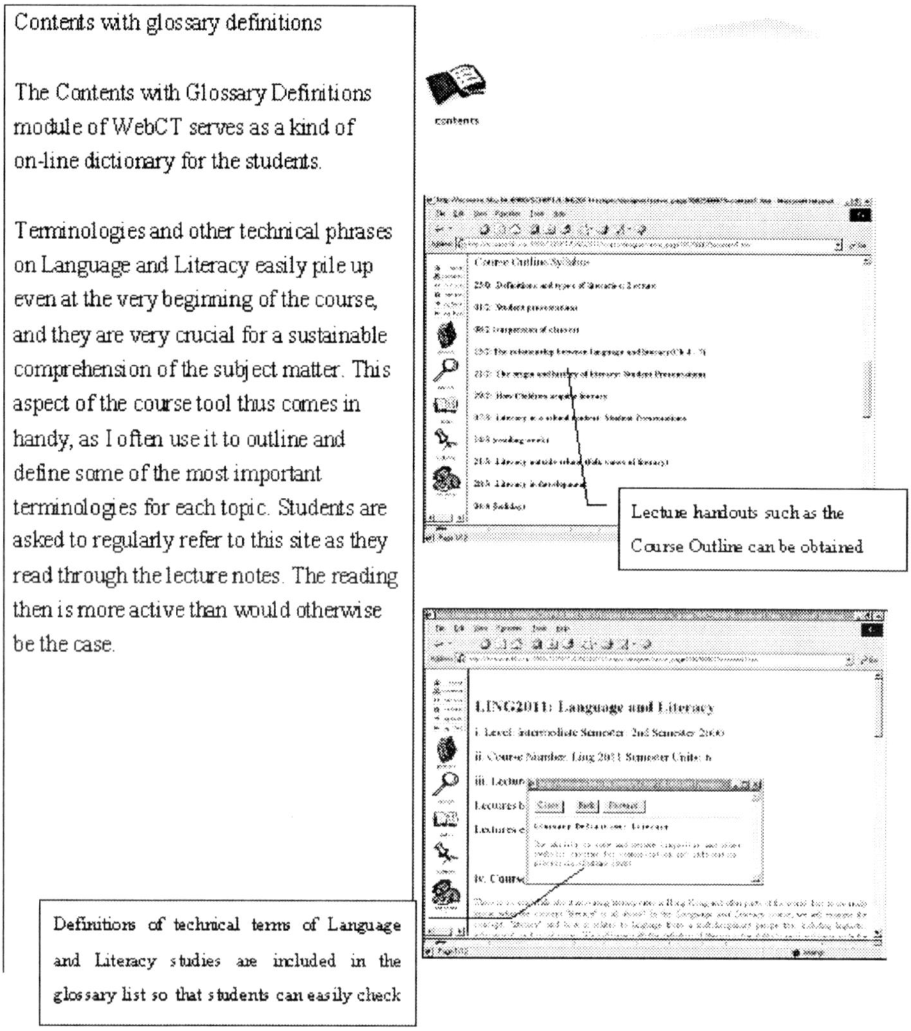

surprised to log onto my WebCT to notice that one student had already started a topic of discussion on it. (2) In terms of methodology, one student surprisingly introduced the notion of annotated URLs on her homepage, something I wanted to introduce to the students. I encouraged students to follow her example. In our knowledge-based economy innovation has become a crucial element of an efficient workforce. Initiative is an essential element of innovation, and the pedagogical process should aim at promoting it. This may be compared to Moore's (1993) ideas of learner autonomy and self-direction, which are products of a successful interactive learning course (2) (figures 18, 19).

Figure 9. Links to useful references

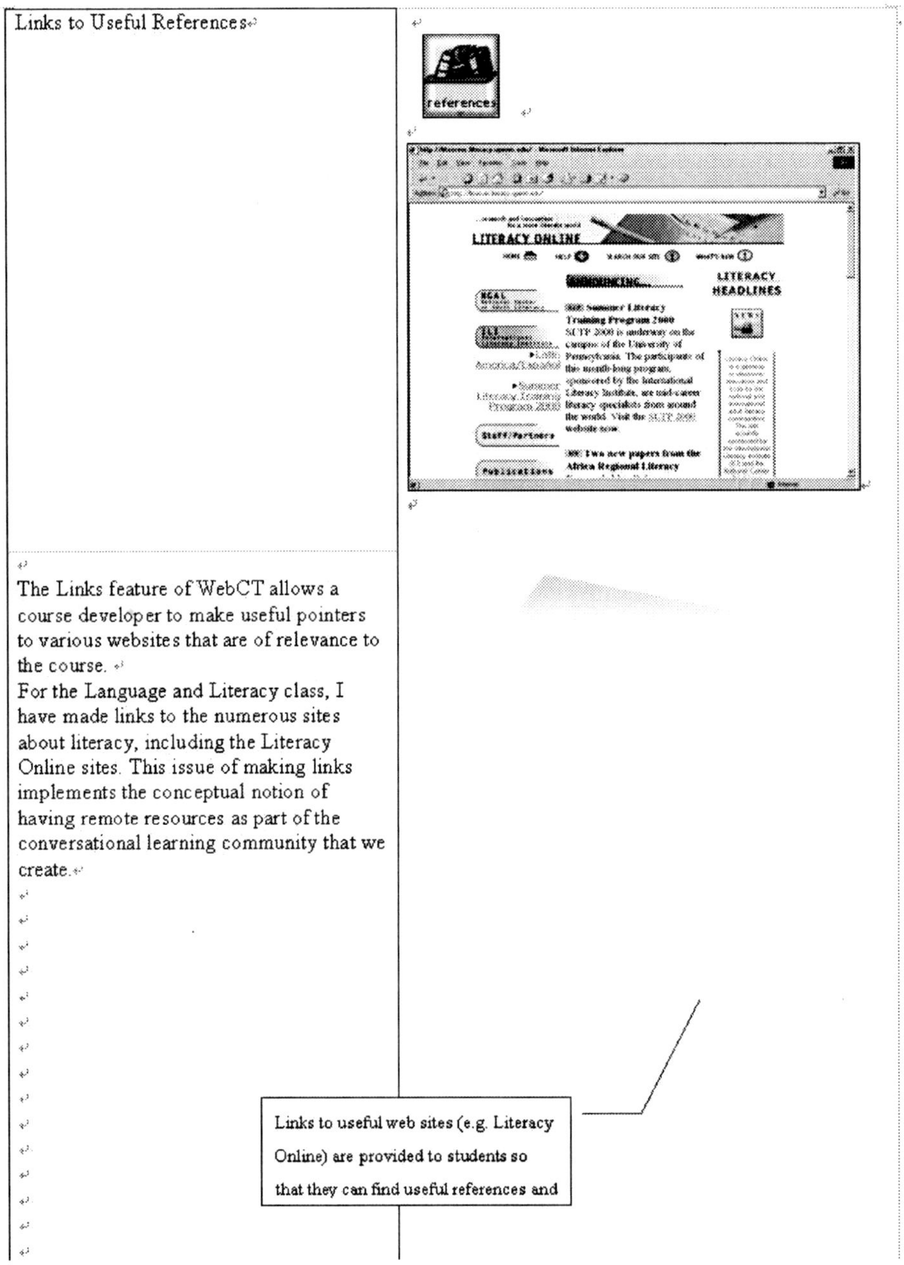

Figure 10. Student access statistics

Student Access Statistics

This feature is a very valuable aspect of WebCT in terms of helping the teacher to track and manage student progress. Each time I post a new course material on the web I demand that students read the material before the next scheduled class. Before the start of the class, I log on to assess how many students have already accessed and, presumably, read the material. This I can gauge by looking at the number of students logging on, and also by what pages they visi
could even have an idea of w
particular students accessed
and their frequency of access. It turned out, however, that sometimes actually more students had accessed the pages than the access statistics indicated. Some students simply asked their friends to download copies of the material for them without they themselves accessing the material from their own accounts! One way to solve this problem, if it is thought of as such, is for the teacher to actively discourage this oblique access to the course material on the web.

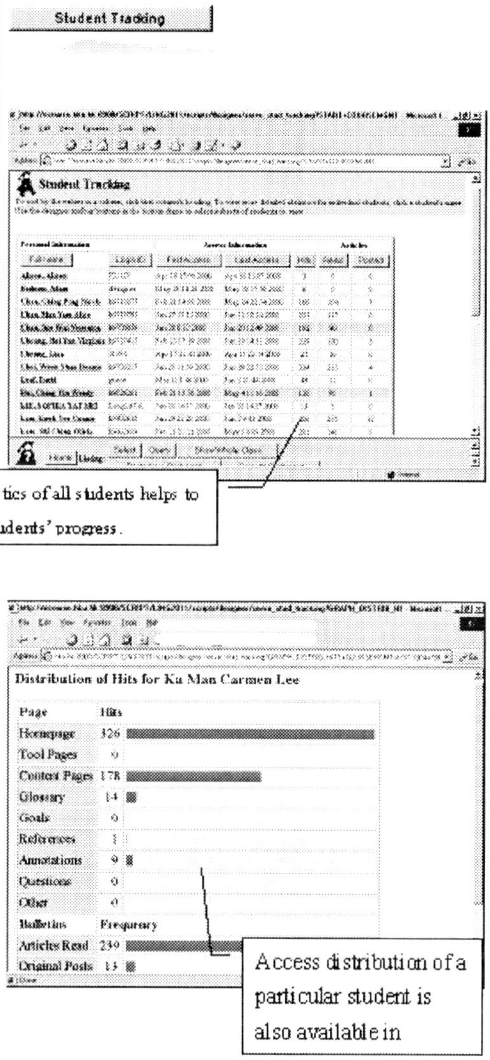

Academic Rigour

A third measure of the fact that my class achieved an enhanced constant interaction was the academic rigour I noticed in the essays that many of the students wrote for me. Students were often generally very knowledgeable about the different shades

Figure 11. Discussion forum

Discussion Forum

WebCT's Bulletin board and Presentation feature together provide a useful discussion forum for participants in my Language and Literacy courses. This is indeed the most useful feature with respect to incorporating interactivity in the course. Through the bulletin board, I can readily send information to the class and to individual students about the course. These include reminders of deadlines for assignments, clarifications about specific points, and pointers to any errata in my course notes. Students, on the other hand, can use this forum to ask me questions on aspects of the course and to post general messages to other students on the course. Groups of students can use the presentation forum to upload and discuss a topic, which they may subsequently write up and present to the whole class.

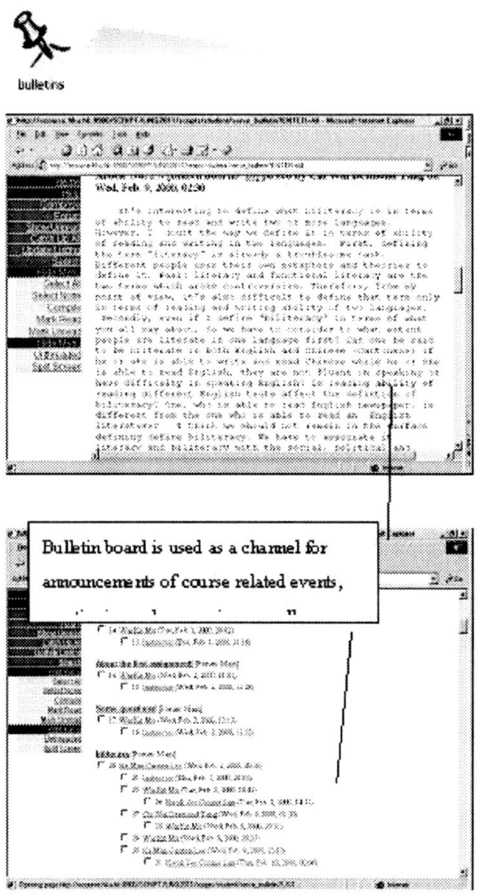

of opinions regarding a particular technical issue. Indeed, some students even began to question some aspects of the textbooks against the realities of the Hong Kong situation that they know best.

In every evaluation situation, it is often best to hear from the horse's own mouth, i.e. those most affected by the situation to be evaluated. The crucial question is how the students perceived this attempt to enhance interactivity in their learning situation by the use of web-based teaching?

Figure 12. A student shares links in her homepage

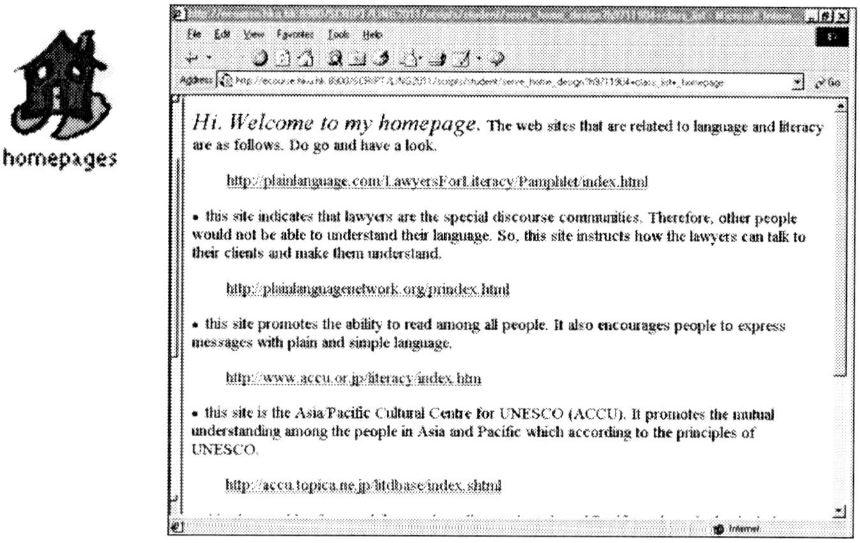

Figure 13. Calendar of important events

Figure 14. Student presentations in lectures are also obtainable on the WebCT

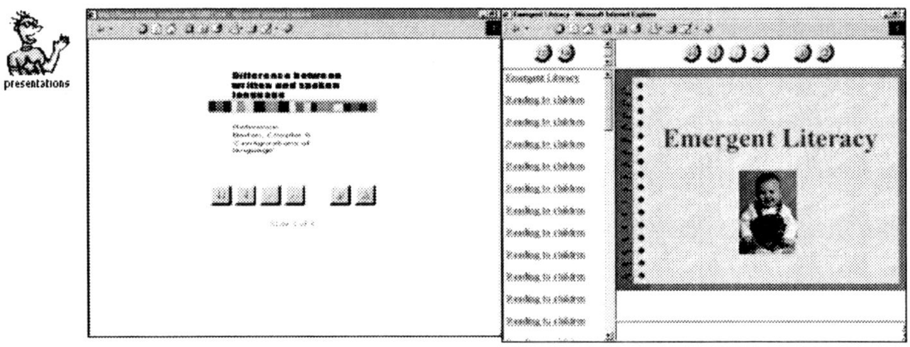

There are several ways that the courses are evaluated. One is to get the students to make open-ended, written comments. The comments point to a positive appraisal of the element of enhanced interactivity in the web-based learning process, as shown by the following anonymous open-ended questions:

Department of Linguistics

Anonymous open-ended comments about the course: LING2011

Figure 15. An interactive revision quiz

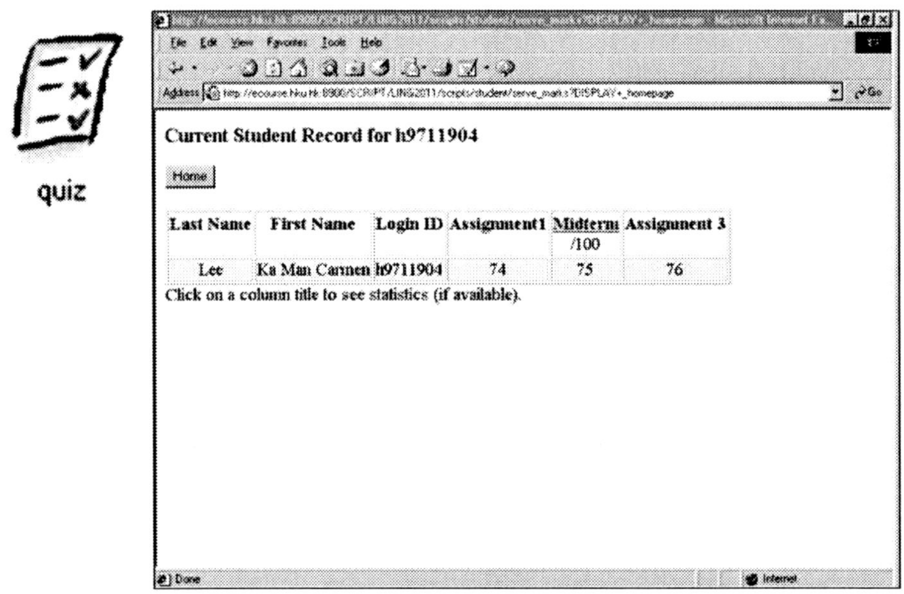

Figure 16. Student record (assignment grades)

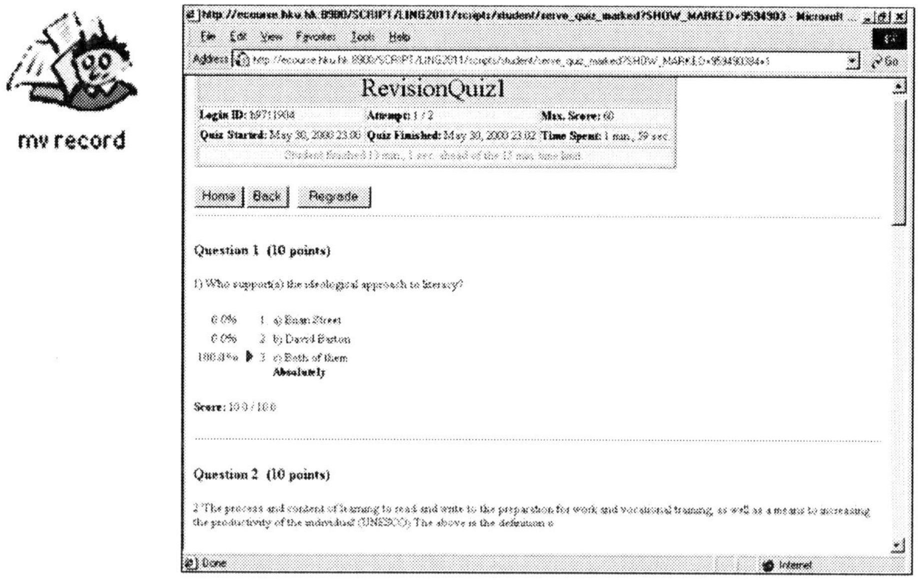

- What aspects of this course did you find most beneficial?
- WebCT: useful tool for interaction.
- The web, i.e. bulletin board is very useful in helping students to learn more about the topics by talking about it.
- I know more about WebCT and become more "computer-literate".
- The use of WebCT
- Handouts and WebCT
- It provides me the opportunity from knowing nothing about literacy to become knowledgeable on the subject.
- The assignments and the tutorials
- Presentations help us understand it too.
- The WebCT is a good place for students & lecturer to discuss different things outside class.
- Knowledge learned

Challenges

In the course of web-based design of my course on language and literacy and indeed other courses, I have experienced a number of issues which, rather than perceiving them as problems and obstacles, I will perceive as challenges to be overcome towards an improvement of web-based teaching.

Figure 17. Bulletin board transcripts (1)

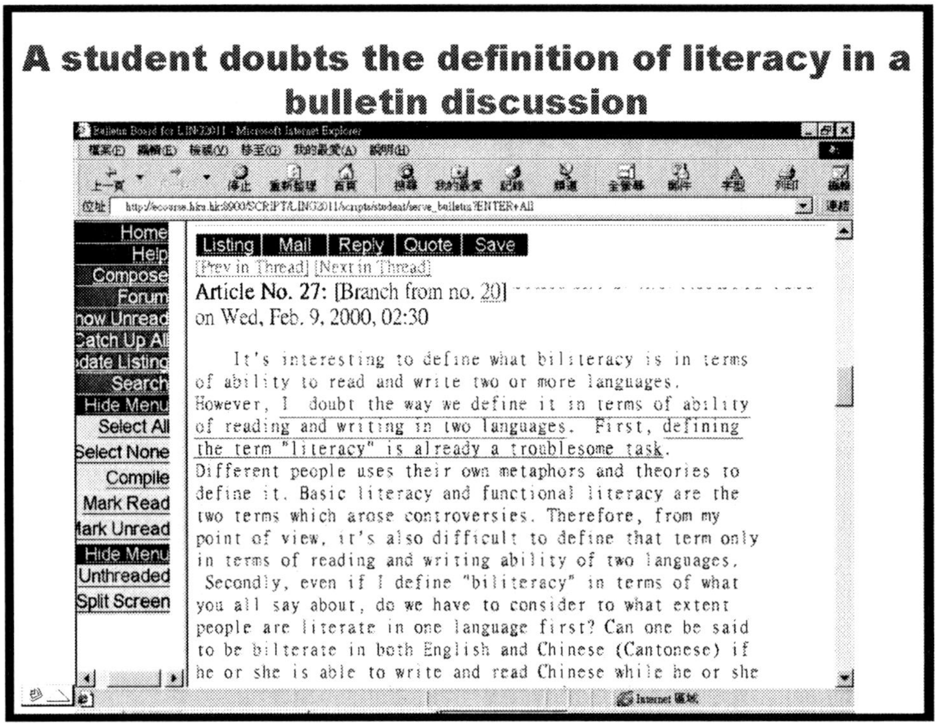

Low Written Interaction at the Beginning

I have described some initial steps I take to get students to form groups and begin interacting with each other. It is however often a bit difficult to get them to start writing and sending messages of discussion on the bulletin board. Indeed some students never post a single message throughout the course, though they may keep reading every bit of discussion going on (4).

I have often made several posts without any responses. In these posts, I ask questions, and exhort students to start making use of the forum. The interesting aspect here is that, it takes just a few students to begin and most come on board. In extreme situations of low participation, I remind them that active participation counts towards the coursework mark (Figure 20).

Multimedia Integration: Graphics and Sound

An aspect of my web-based course design that still awaits attention is the integration of graphics and other kinds of multimedia alongside text (Figure 21).

However, I am yet to add graphics in this course (the above example is an excerpt from the homepage of my course on Syntax). It must however be mentioned that,

Figure 18. Transcripts (2)

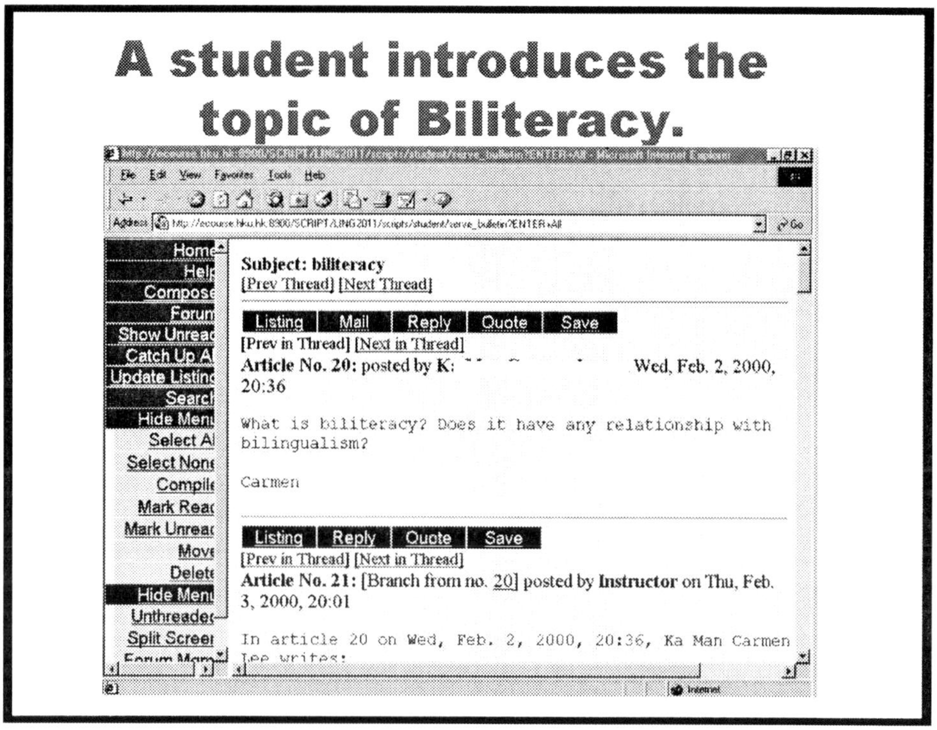

in spite of the fact that graphics are important, recent research has shown that they are not that crucial in processing text information. For instance, a joint study by Stanford professor of Communication, Marion Lewenstein, and others, shows that graphics may not be that important in reading information on the net. According to this study people read 92% of the text on Web pages, but they looked at only 22% of the graphics on the same pages. (http://www.poynter.org/eyetrack2000/index.htm)

This shows that even if one incorporates graphics in web-based design of courses, one should do that sparingly.

Other communication tools that one might consider incorporating into web-teaching are ICQ and web-based mobile phones. I have spoken to some of the students with low written interaction on the bulletin page and they have mentioned that they prefer more synchronous means of communication with their friends, such as ICQ.

Figure 19. Transcript (3)

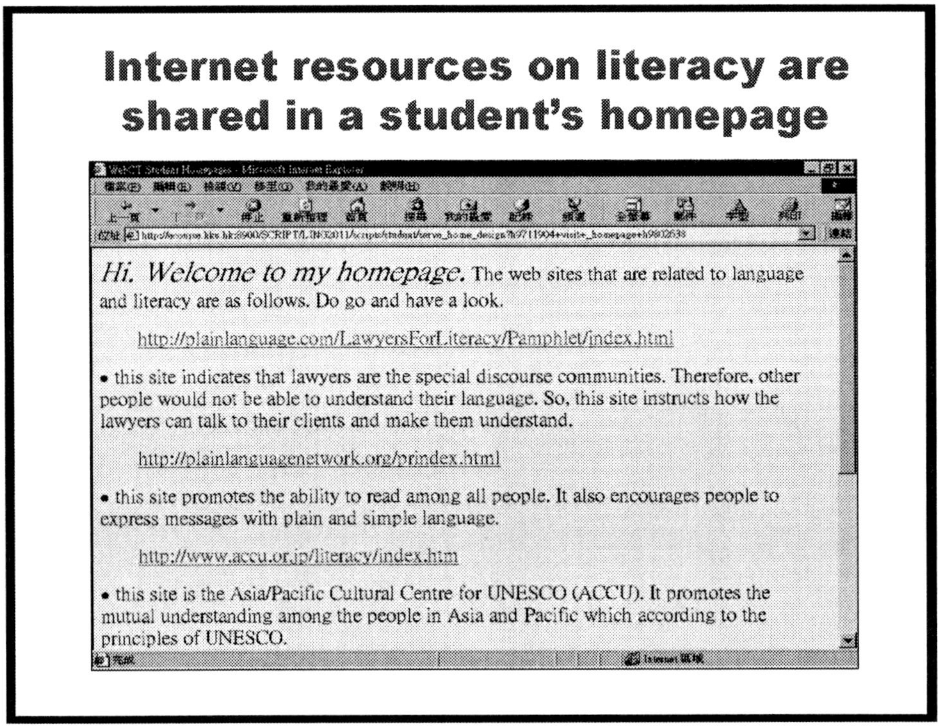

Summary

This study on interactivity has attempted to demonstrate that interactivity is an essential aspect of student-centered course design endeavours, whether in traditional face-to-face classrooms or by distance learning. Society seems to require universities and other learning institutions to produce graduates who are creative thinkers and problem solvers; graduates who are literate enough to function well in a knowledge-based economy where there is a pervasive use of ICTs. To achieve this educational goal, we need to reform our methods of instruction, moving away from more passive methods of teaching to more active and interactive methods. Based on three years of web-based course design and delivery, this paper has proposed some ways of designing more interactive courses.

Basically, teachers ought to construe their learning environment as one of conversation between instructor and learner. Important components in this environment include instructor(s), learners, course materials, and links to remote experts and resources. All these components are glued together by instructional interactivity. Three types of instructional interactivity ought to be recognized. There are instructor-

Figure 20: Low written interaction at the neginning

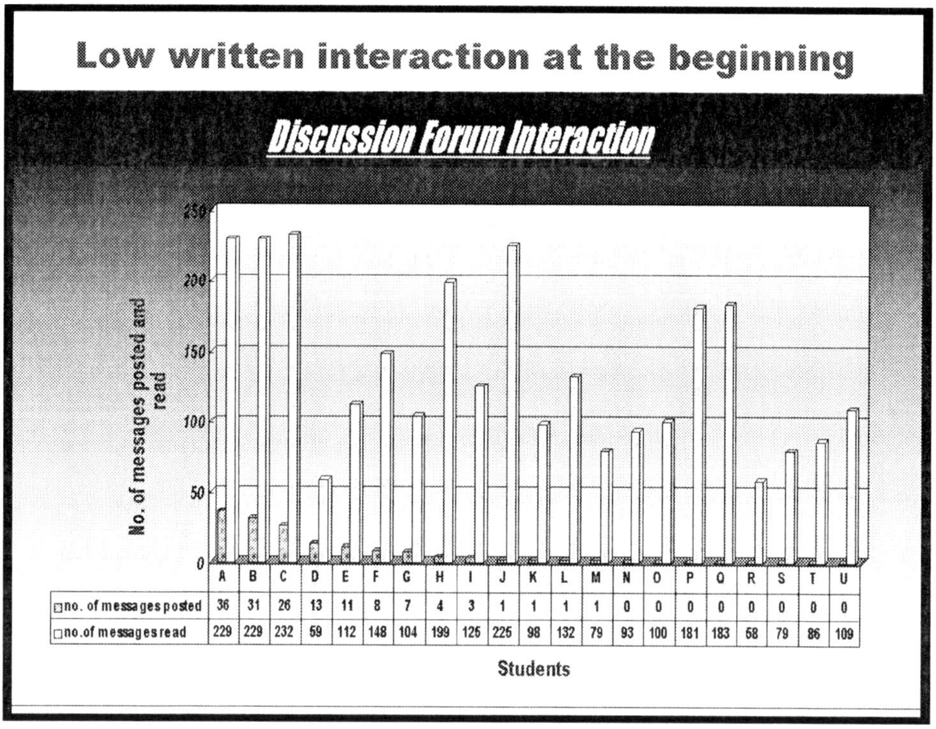

	A	B	C	D	E	F	G	H	I	J	K	L	M	N	O	P	Q	R	S	T	U
no. of messages posted	36	31	26	13	11	8	7	4	3	1	1	1	1	0	0	0	0	0	0	0	0
no.of messages read	229	229	232	59	112	148	104	199	125	225	98	132	79	93	100	181	183	58	79	86	109

learner, learner-learner, and learner-resource interactivity.

While there still remain some challenges, it has been shown that by doing interactive web-based teaching many positive things such as critical thinking, initiative, and academic rigour may be achieved. Moreover, we may conclude that interactivity on the web seems to enhance even traditional classroom and tutorial sessions. Interactive web-based teaching allows teachers to achieve a better management of the course. This issue is relevant for both distance education, such as on cyberspace and the traditional classroom teaching. Interactivity thus has the potential of rendering the gap between traditional face-to-face classroom education and distance education redundant. Distance would no more have to be defined in terms of just space but in terms of the presence and absence of interactivity.

CONCLUSION

In this chapter, I have discussed at length how to take advantage of the communication technologies and techniques associated with them and turn them into educational

Figure 21: Multimedia integration

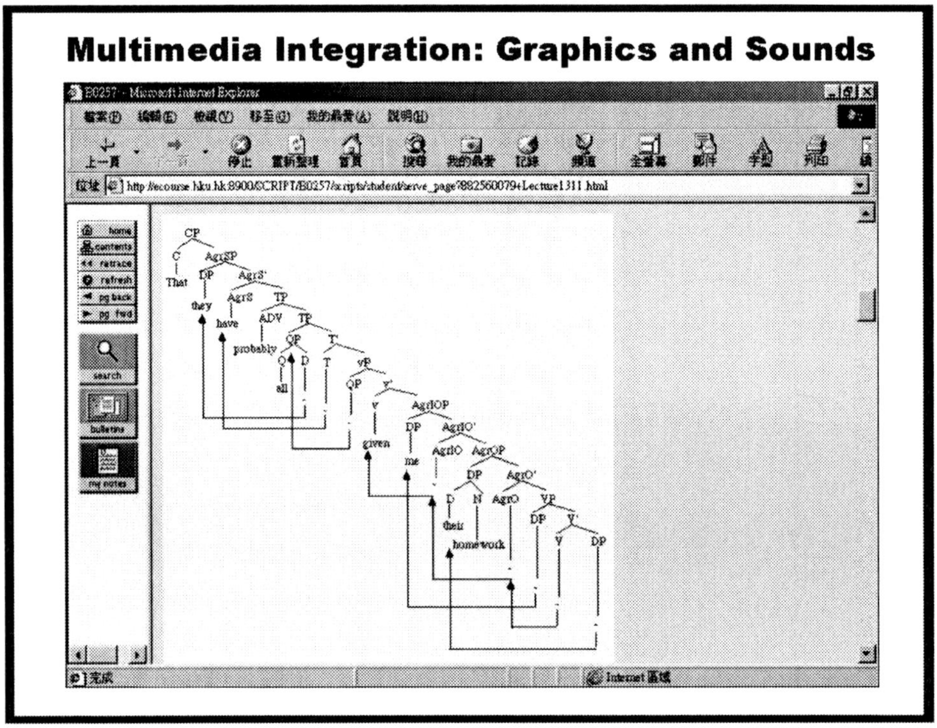

technology for literacy and linguistics. I first introduced and discussed the important theory of constructivism that is well-known in Education and most areas of the Humanities and Social Sciences like Linguistics and Psychology. Constructivism represents a major paradigm shift in terms of learning and teaching. As mentioned earlier, the central idea underlying the theory of constructivism is that learners construct their own knowledge of the world, which signifies that learning is, therefore, a process of creating meaning by the learners themselves, and the instructor simply serves as a facilitator in this process. This is an important paradigm shift not just from the point of view of the learner but also from the point of view of the teacher, who is now seen as not the source of all knowledge but as a facilitator in the learning process whereby the student constructs new knowledge from multiple sources of inspiration.

The second major concept highlighted here is interactivity, the idea that students should be involved in not just passive learning but active learning that requires students to actively interact with their teacher, their fellow learners and even remote learning resources, such as from the many internet resources we have encountered including emails, websites, YouTube, Facebook, and even text messaging and online video games. Interactivity is not a goal in itself but a means to achieving a goal – that of enabling the learner to take charge of the learning situation to actively gather and construct his or her own knowledge base.

All this led us to the creation of our learning theory, the Conversational Learning Theory, which instantiates a learning model called Conversational Learning Community. As indicated earlier, the main idea of a conversation learning theory is that enhanced interactivity, whether face-to-face or from a distance, such as online instruction, would lead to an effective reciprocal, two-way, communication within the learning situation. This enhanced communication is the backbone for the efficient exploitation of the resources, experts, and links by both instructor and learners within the learning community.

From this summary, we may thus conclude that computer-mediated communication tools have an important role to play in creating new ideas and news approaches to learning and teaching in the age of IT. As we have seen through the book, the youth are among the most avid users of CMC tools. One of the best ways then to make learning fun for them is to make use of the technological environment they thrive best ın.

In the next chapter I will continue this theme of learning with new technologies by showing how we can evaluate the learning and teaching process when we create learning environments using these computer-mediated communication tools.

REFERENCES

Barnard, R. (1995). Interactive Learning: A key to successful distance delivery. The American Journal of Multimedia , 12, 45–47.

Blurton, C. (1999). New directions of ICT-use in education. *UNESCO's World Communication and Information Report 1999*. Retrieved from http://www.unesco.org/education/educprog/lwf/dl/edict.pdf

Bodomo, A. B. (1999). WebCT course design for Syntax II: The Theory of Grammar. In *Web-based Course Development Using WebCT at the University of Hong Kong* (pp. 52-54). Computer Centre, HKU.

Bodomo, A. (2001). *Interactivity* in Web-based courses. Paper for the WebCT Asia Pacific Conference, 9th-11th April 2001, Adelaide, Australia.

Bodomo, A. B. (2005a). Interaction in web-based learning. *Encyclopedia of Distance Learning*, edited by C. Howard et al. Idea Group Inc.

Bodomo, A. B. (2005b). Constructing knowledge through online bulletin board discussions. *Encyclopedia of Distance Learning*, edited by C. Howard et al. Idea Group Inc.

Bodomo, A. B. (2006). Interactivity in web-based learning. *International Journal of Web-Based Learning and Teaching Technologies.*, 1, 18–30.

Bodomo, A. B. (2007). Mobile phone texting in Hong Kong. In David Taniar (Ed.), *Encyclopedia of Mobile Computing And Commerce.* Information Science Reference. Idea Group Inc.

Bodomo, A. B. (2008). Instructional interactivity in a web-based learning community. In Lilian Esnault (Ed.), *Web-Based Education and Pedagogical Technologies: Solutions for Learning Applications* (pp. 131-146). IGI Global Books.

Bodomo, A. B., Luke, K. K., & Anttila, A. (2003). Evaluating interactivity in web-based learning. *Global E-Journal of Open, Flexible and Distance Education.* Volume III. URL: http://www.ignou.ac.in/e-journal/ContentIII/Adamsbodomo.htm

Brogan, P. (1999). *Using the web for interactive teaching and learning: An imperative for the new millennium.* A white paper, Macromedia's Interactive Learning Division.

Bruner, J. (1966). *Toward a Theory of Instruction.* Harvard University Press, Cambridge, MA.

Bruner, J. (1983). *Child's Talk: Learning to Use Language.* Norton, New York.

Bruner, J. (1986). *Actual Minds, Possible Worlds.* Harvard University Press, Cambridge, MA.

Bruner, J. (1990). *Acts of Meaning.* Harvard University Press, Cambridge, MA

Bryson, M., and M. Scardamalia (1996). Fostering reflectivity in the argumentative thinking of students with different learning histories. Reading & Writing Quarterly: Overcoming Learning Difficulties. , 12(4), 351–384.

Daniel, J., and C. Marquis (1983). Independence and interaction: Getting the mix right . Teaching at a Distance , 15, 445–460.

Dewey, J. (1913). *Interest and Effort in Education.* Boston: H. Mifflin Co.

Dewey, J. (1956). *The Child and the Curriculum and the School and Society.* Chicago: University of Chicago Press.

Dewey, J. (1963). *Experience and Education.* New York: Collier Books.

Duffy, T., & Jonassen, D. H. (1992). *Constructivism and the Technology of Instruction: A Conversation*. New Jersey: Lawrence Erlbaum Associates.

Freire, P. (1970). The adult literacy process as cultural action for freedom . Harvard Educational Review , 40, 205–212.

Jackson, R. H. (2000). *Web-based Learning Resources Library.* Retrieved from http://www.outreach.utk.edu/weblearning/

Kearsley, G. (1994-2000). *Explorations in Learning & Instruction: Theory in Practice Database* (http://www.gwu.edu/~tip/)(gkearsley@sprynet.com) http://home.sprynet.com/~gkearsley available as at: May 21, 2000.

Kwok, P., & Bodomo, A. B. (2000). *The integration of Information Communications Technology (ICT) into secondary school curricula in Hong Kong and Ghana: Re-conceptualizing educational communication*. Paper presented at the 19th CESE 2000, Bologna, Italy.

Markwood, R., & Johnstone, S. (1994). *New Pathways to a degree: Technology opens the college*. Western Cooperative for Educational Telecommunications, Western Interstate Commission for Higher Education, Boulder, CO.

Moore, M. (1991). Editorial: distance education theory. American Journal of Distance Education , 5(3), 1–6.

Moore, M. (1992). Three types of interaction. American Journal of Distance Education , 3(2), 1–6.

Moore, M. (1993). Theory of transactional distance. In Desmond Keegan (Ed.), *Theoretical Principles of Distance Educatio*n. Routledge, London & New York.

Parker, A. (1999). Interaction in Distance Education: The critical conversation. Educational Technology Review , 13.

Pask, G. (1975). *Conversation, Cognition, and Learning*. Elsevier, New York.

Piaget, J. (1955). *The Construction of Reality in the Child.* London: Routledge & Kegan Paul.

Piaget, J. (1973). *To Understand is to Invent*. New York: Grossman.

Scardamalia, M., & Bereiter, C. (1999). Schools as knowledge-building organizations. In D. P. Keating & C. Hertzman (Eds.), *Developmental Health and the Wealth of Nations: Social, Biological, and Educational Dynamics* (pp. 274-289). New York: Guilford.

South China Morning Post (SCMP). (2000). Newspaper editorial of March 9th 2000, Hong Kong

Strauss, M. J. (1994). A constructivist dialogue. The Journal of Humanistic Education and Development , 32(4), 183–187.

Thompson, D. (Ed.). (1996). *The Oxford Compact English Dictionary*. Oxford University Press, Oxford, England.

Vygotsky, L. (1934). *Thought and Language*. Cambridge, MA: The MIT Press.

Vygotsky, L. (1978). *Mind in Society: the Development of Higher Psychological Process*. Cambridge, MA: Harvard University Press.

Vygotsky, L. S. (1962). *Thought and Language*. Cambridge, MIT Press, MA.

Wagner, J. (1994). Learning from a distance. The International Journal of Multimedia , 19(2), 12–20.

ENDNOTES

1. The pedagogical activities reported in this chapter form part of a completed Teaching Development Grant, titled, *The Use of Information Technology in Teaching Language and Linguistics Courses*, of which the author was Principal Investigator.
2. I thank people at the Computer Centre, University of Hong Kong, including Fanny Chau, Dickson Chau and Alice Lam, for technical support, especially at the early stages of my encounter with WebCT.

Chapter 11
Evaluating Learning Technologies

INTRODUCTION

In this penultimate chapter of the book, I will continue with my discussion of how we can take advantage these youth interests and practices with ICTs for enhancing learning and teaching by actually evolving ways to evaluate these communication and learning environments.

As with most chapters in the book, I focus on a case study as a way to give an in-depth study to the subject matter. In this case, interactivity is the subject matter. Interactivity was discussed at length in the previous chapter, leading the creation of a new learning theory, the Conversational Learning Theory, and a new leaning model, the Conversational Learning Community. Rather than explaining the concept and the learning theory and model again in this chapter, I refer the reader back to

DOI: 10.4018/978-1-60566-868-0.ch011

the previous chapter. Once the basic tenets of these concepts are grasped the reader can now begin to read the discussion about how to increase interactivity in learning environments.

MEASURES TAKEN TO ACHIEVE INTERACTIVITY

Having offered a comprehensive background account of interactivity in chapter 10, ending with a specific idea of what it is within the confines of our notion of conversation learning community, I shall in this section first describe two web-based courses which form the basis for this discussion about interactivity. Following this brief description, I shall then outline measures taken and activities done to achieve interactivity in these two courses.

A Description of Two Web-Based Courses

I have now explained a number of issues, including web-based teaching, and interactivity and its role in constructivist teaching methods. I will in the rest of the chapter provide a description of specific courses within my web-based teaching programme, and how interactivity was achieved in the course design. I begin with the choices available for a course designer.

Web-Based Course Design via WebCT

Web-based teaching on the 'open internet' i.e. without a course tool such as WebCT, Blackboard, etc. can create a number of problems. First, it may be difficult, as course designer, to manipulate access to the material in terms of passwords and accounts for the learners. Second, many other interactive features such as bulletin boards, chat rooms, and secured records for student activities may not be possible. As such, one can hardly design and implement the conceptual notion of conversational learning community that I have referred to above.

In order to solve the above problems, we decided to do course design via WebCT. Further development of WebCT course design has led to the award of a joint Teaching Development Grant (TDG) titled, The Use of Information Technology in the Teaching of Language and Linguistics courses. Further information on the project may be found at the project website at the following address: (http://www.hku.hk/linguist/staff/TDGBodomo.html)

The courses designed under this project include:

- *LING2016 Syntax II: The Theory of Grammar* at
- http://ecourse.hku.hk:8900/public/LING2016/, taught by Dr. A. B. Bodomo
- *LING2006 Syntax I: Describing Grammatical Patterns* at http://ecourse.hku.hk:8900/public/LING2006/, taught by Dr. A. B. Bodomo
- *LING2018 Lexical-Functional Grammar* at
- http://ecourse.hku.hk:8900/public/LING2018/, taught by Dr. A. B. Bodomo
- *LING2011 Language and Literacy* at
- http://ecourse.hku.hk:8900/public/LING2011/, taught by Dr. A. B. Bodomo

In the rest of the chapter, I will concentrate on describing just two of these courses, Language and Literacy and Syntactic Theory.

WebCT Design of the Two Courses

Language and Literacy

The Language and Literacy course is a one semester six credits course for second and third year students of Linguistics and related disciplines.

The course usually begins with an attempt to get students (usually about twenty to thirty in number) to understand the concept 'literacy'. The course materials and lectures and tutorials are designed in such a way that students are supposed to discover for themselves that the concept literacy is NOT limited to just the ability to read and write. Students are supposed to discover for themselves the various linguistic, cognitive, social, and educational issues surrounding the concept. Students are encouraged to gain an understanding of the role of language and literacy in the socio-economic development efforts of many societies through various activities such as discussions, debates, classroom presentation, tutorials and interview of resource persons. Topics covered in the course often include:

- Definitions and types of literacy
- The relationship between language and literacy
- Writing and other symbolic systems
- Computer Literacy and Language educational technology
- The origins, history, and acquisition of literacy
- Comparative analyses of the language and literacy situations in selected parts of the world, including Hong Kong, Mainland China, and South-east Asia
- Literacy and socio-economic development

Syntactic Theory

Like the course on Language and Literacy, the Syntactic Theory course is a one semester six credit course that aims to provide the student with basic notions about formal syntax. Topics often covered include the following:

- Preliminaries: Chomskyan Generative Syntax
- Lexical Categories
- Syntactic Structure
- Empty categories
- Head Movement
- Subjects
- A Movement
- VP shells
- Agreement projections
- Alternatives to Generative Transformational Grammar:
    - Lexical-Functional Grammar (LFG)

How Interaction was Achieved in the Courses

We will now attempt to explain how interaction was achieved in the two courses. This begins with the creation of a learning community. I have stated in section 1 above that course design, whether in the form of face-to-face classroom lectures or WebCT course page activities like discussion and presentations, is guided by the conceptual notion of a conversational learning community, comprising instructor(s), learners, current resources and remote experts and resources.

The first task then in our course administration is often to get the group of about twenty to thirty students in each class to communicate and interact with each other and create a sense of community. The first exercise towards this goal is often in the form of internet search. The following excerpt from one of the courses explains the exercise:

- LING2011: Reading Assignment/Homework
- Literacy Information Mining on the web:
- Students should form groups of 2 - 3 people. Each group should search the World Wide Web with key words 'literacy', 'language', (and combinations of these) and choose 10 sites. These sites should be analyzed with a view to finding out what literacy is and what common issues are discussed concerning language and literacy courses. Each group of students should spend five

minutes in the next lecture explaining how their understanding of literacy has been affected by these 10 websites.

- An exercise for building up a learning community:
- This exercise is meant to get students to create physical and electronic networking among them and it often succeeds to a large extent because I have noticed that later groups to be formed in the class often reflect this earlier grouping. Other exercises include short group discussions on issues in class during the first two weeks.
- Once this sense of community is created, the rest of the instructional activities aim to consolidate and strengthen it, developing it into a real conversational learning community. I do this both through my face-to-face classroom activities and my WebCT design activities.
- Face-to-face classroom activities include a mixture of lectures, student presentations and tutorials. Very little Information Communications Technologies (ICTs) are used here. The main tools employed by us at lectures and tutorials are traditional classroom educational technology such as projector/overhead machines, black- and whiteboard and lecture handouts, and assignment feedback sheets. As the semester advances, I have often noticed that student presentations become more and more ICT-mediated, the most typical presentation tool being the PowerPoint.
- Surprisingly, greater and more sustained aspects of the interaction between students often take place on the WebCT homepage for these courses. In the next subsection, I will describe some of the main features and resources of this tool, showing what kinds of interactivity take place and how.

Some Interactive WebCT Features of the Courses

The following are two excerpts from the Language and Literacy WebCT homepage at http://ecourse.hku.hk:8900/SCRIPT/LING2011/scripts/serve_home (Figures 1 and 2)

Contents with Glossary Definitions

The Contents with Glossary Definitions module of WebCT serves as a kind of online dictionary for the students (see Figure 3).

Lecture handouts such as Course Outline can be obtained from the Contents section.

Terminologies and other technical phrases on Language and Literacy and Syntactic Theory easily pile up even at the very beginning of the course, and they are very crucial for a sustainable comprehension of the subject matter. This aspect of

Figure 1. Excerpts from the language and literacy WebCT homepage

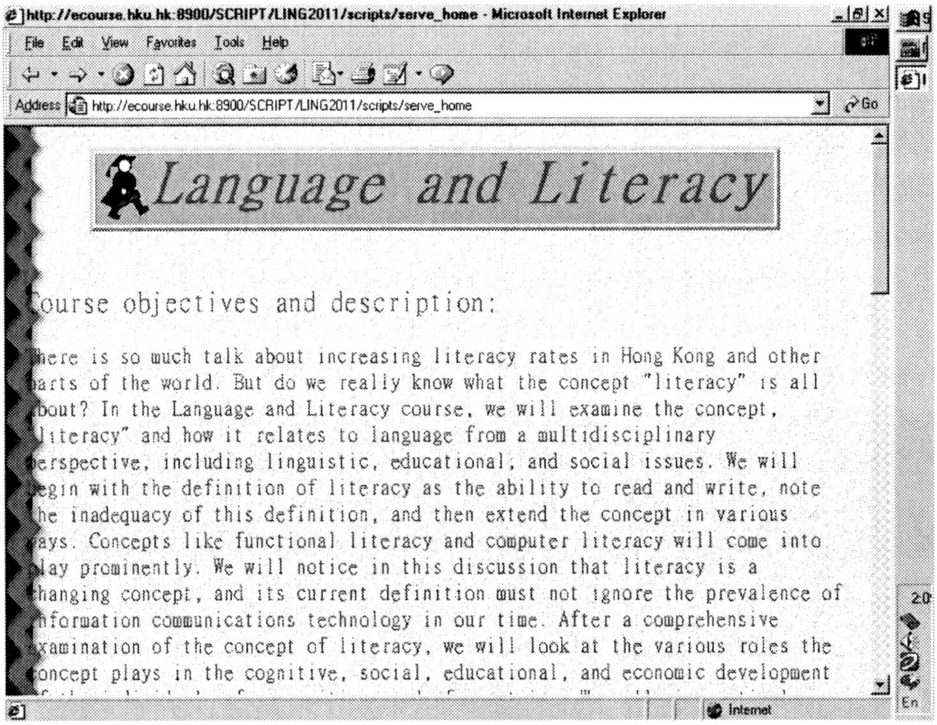

the course tool thus comes in handy, as I often use it to outline and define some of the most important terminologies for each topic. Students are asked to regularly refer to this site as they read through the lecture notes. The reading then is more active than would otherwise be the case (see Figure 4).

Definitions of technical terms of Language and Literacy studies are included in the glossary list so that students can easily check the meanings while browsing the course contents.

Links to Useful References

The Links feature of WebCT allows a course developer to make useful pointers to various websites that are of relevance to the course.

For my classes, I have made links to the numerous sites about literacy, including the Literacy Online sites (Figure 5) and about Syntactic Theory, such as the Linguist's List. This issue of making links implements the conceptual notion of having remote resources as part of the conversational learning community that I create.

Figure 2. Excerpts from the language and literacy WebCT homepage

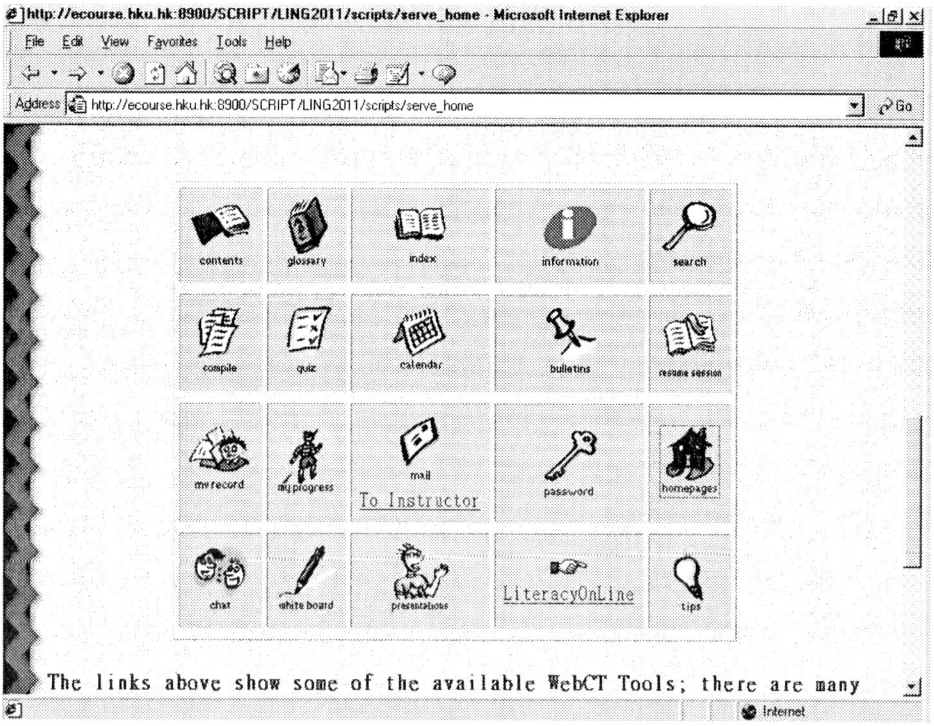

Links to useful websites (e.g. Literacy Online) are provided for students so that they can find useful references and other information related to the course.

Student Access Statistics

This feature is a very valuable aspect of WebCT in terms of helping the teacher to track and manage student progress. Each time we post a new course material on the web we demand that students read the material before the next scheduled class. Before the start of the class, we log-on to assess how many students have already accessed and, presumably, read the material. This we can gauge by looking at the number of students logging on, and also by what pages they visited. Indeed, one could even have an idea of which particular students accessed the material, and their frequency of access. It turned out, however, that sometimes actually more students had accessed the pages than the access statistics indicated. Some students simply asked their friends to download copies of the material for them without they themselves accessing the material from their own accounts! One way to solve this problem, if it is thought of as such, is for the teacher to actively discourage this

Figure 3. Course outline/syllabus

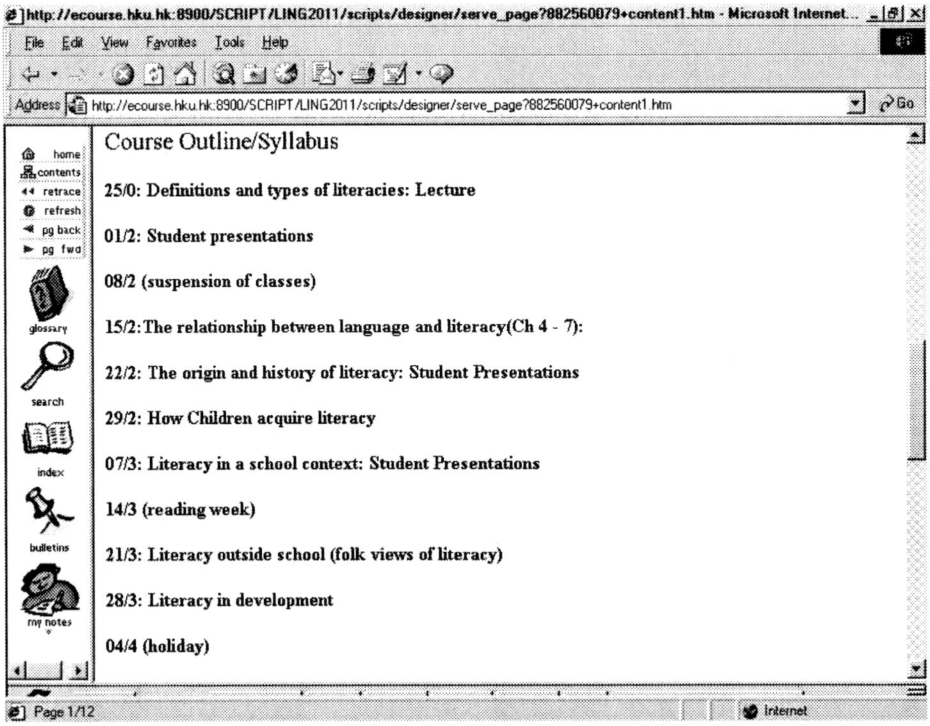

oblique access to the course material on the web (Figure 6).

Accessing statistics of all students helps to keep track of students' progress (Figure 7).

Access distribution of a particular student is also available in WebCT.

Discussion Forum

WebCT's Bulletin board and Presentation feature together provide a useful discussion forum for participants in our courses. This is indeed the most useful feature with respect to incorporating interactivity in a course. Through the bulletin board, we can readily send information to the class and to individual students about the course. These include reminders of deadlines for assignments, clarifications about specific points, and pointers to any errata in our course notes. Students, on the other hand, can use this forum to ask us questions on aspects of the course and to post general messages to other students on the course. Groups of students can use the presentation forum to upload and discuss a topic, which they may subsequently write up and present to the whole class (Figures 8 and 9).

Figure 4. Glossary definition

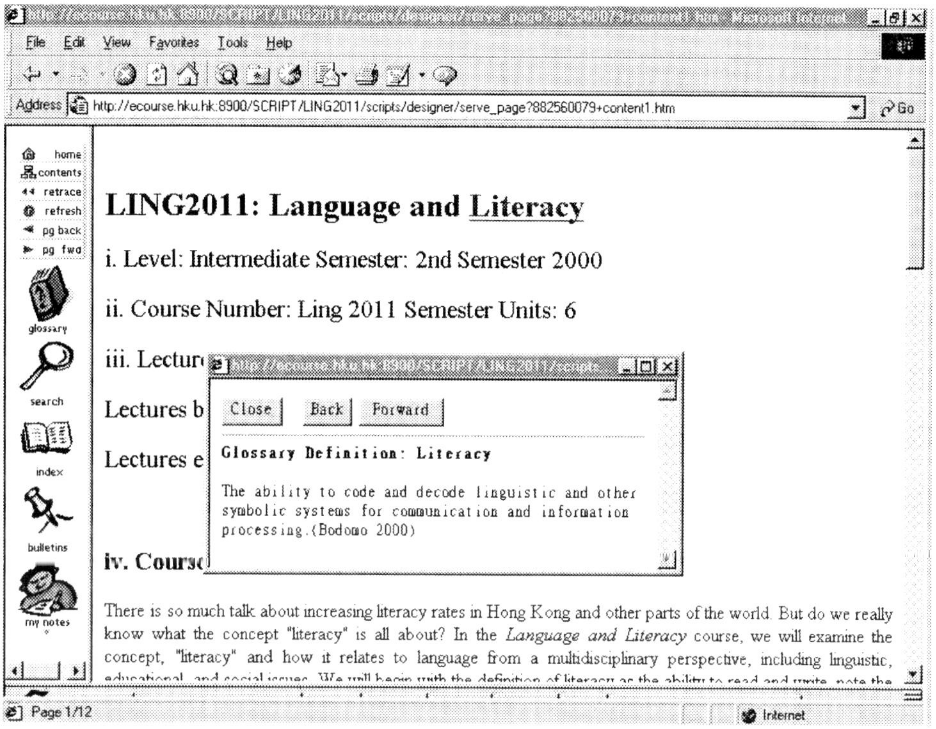

Bulletin board is used as a channel for announcements of course-related events, questioning and answering as well as general discussion about course materials (Figures 10).

Other Features

A student shares links in her homepage. (Figures 11-14)

CRITERIA FOR MEASURING AND EVALUATING INTERACTIVITY

The foregoing has outlined how I have designed my courses on WebCT so as to enhance interactivity, a crucial element in a conversational learning community and indeed in any other effective learning situation. A possible question to ask then is how successful I have been. How can one measure and evaluate interactivity? Success, failure and other issues of evaluation are difficult to measure accurately.

Figure 5. Literacy Online

They may be from the point of view of the instructor or the student. In the following, I shall briefly point to some qualitative features that make me think that, from an instructor's point of view, interactivity has been achieved in the course. I will also draw from some comments that students made as part of end of semester formal evaluation of the courses.

From an instructor's point of view, certain features of communication and academic activity, if they were part and parcel of a course, would serve to indicate that the teaching endeavor is successful. Three of these features include critical thinking, initiative on the part of students, and academic rigor.

Qualitative Aspects

From the above we notice that qualitative criteria would include increased initiative in pursuing certain aspects of the course, measured in terms of number and type of on- and off-web activities done by groups of students without the teacher's prompting, and open-ended student comments during end of course official evaluation, including answers to on-line questionnaire about the course.

Figure 6. Student tracking

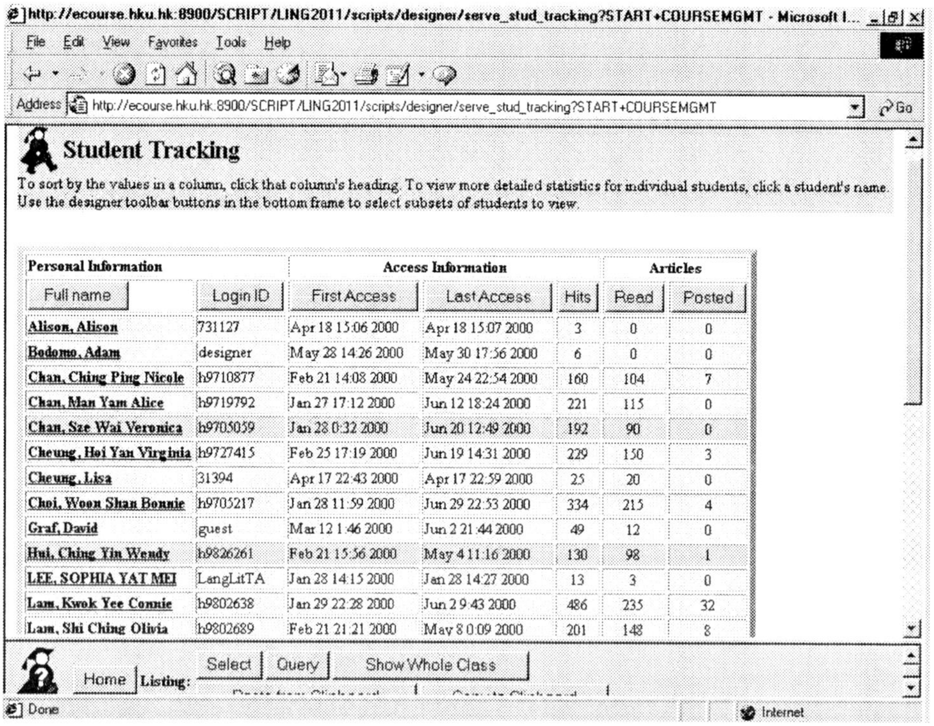

Personal Information		Access Information			Articles	
Full name	Login ID	First Access	Last Access	Hits	Read	Posted
Alison, Alison	731127	Apr 18 15:06 2000	Apr 18 15:07 2000	3	0	0
Bodomo, Adam	designer	May 28 14:26 2000	May 30 17:56 2000	6	0	0
Chan, Ching Ping Nicole	h9710877	Feb 21 14:08 2000	May 24 22:54 2000	160	104	7
Chan, Man Yam Alice	h9719792	Jan 27 17:12 2000	Jun 12 18:24 2000	221	115	0
Chan, Sze Wai Veronica	h9705059	Jan 28 0:32 2000	Jun 20 12:49 2000	192	90	0
Cheung, Hoi Yan Virginia	h9727415	Feb 25 17:19 2000	Jun 19 14:31 2000	229	150	3
Cheung, Lisa	31394	Apr 17 22:43 2000	Apr 17 22:59 2000	25	20	0
Choi, Woon Shan Bonnie	h9705217	Jan 28 11:59 2000	Jun 29 22:53 2000	334	215	4
Graf, David	guest	Mar 12 1:46 2000	Jun 2 21:44 2000	49	12	0
Hui, Ching Yin Wendy	h9826261	Feb 21 15:56 2000	May 4 11:16 2000	130	98	1
LEE, SOPHIA YAT MEI	LangLitTA	Jan 28 14:15 2000	Jan 28 14:27 2000	13	3	0
Lam, Kwok Yee Connie	h9802638	Jan 29 22:28 2000	Jun 2 9:43 2000	486	235	32
Lam, Shi Ching Olivia	h9802689	Feb 21 21:21 2000	May 8 0:09 2000	201	148	8

Critical Thinking

We noticed that as time went on, not only were students more forthcoming in discussing and interacting with teachers and with their fellow students, they were also becoming more critical in their thinking. At certain points during the course, students were beginning to question and argue some of the points from us and from their fellow students. Sometimes, the teacher in the course may for instance present an issue with regards to the definition and conceptualization of literacy and how it relates to language and then ask students to evaluate these views by applying them to the Hong Kong situation and, indeed, other situations that they know. For instance, the class grappled over a period of time with the relationship between the concepts 'bilingualism' and 'biliteracy'. A lot of discussion ensued on this topic as shown on the bulletin board transcripts (see Figure 15).

We consider critical thinking within the conversational learning community as a strong indicator of the success of interactivity in the learning situation. This may be compared with Markwood and Johnstone (1994)'s idea of critical conversation.

Figure 7. Distribution of a particular student

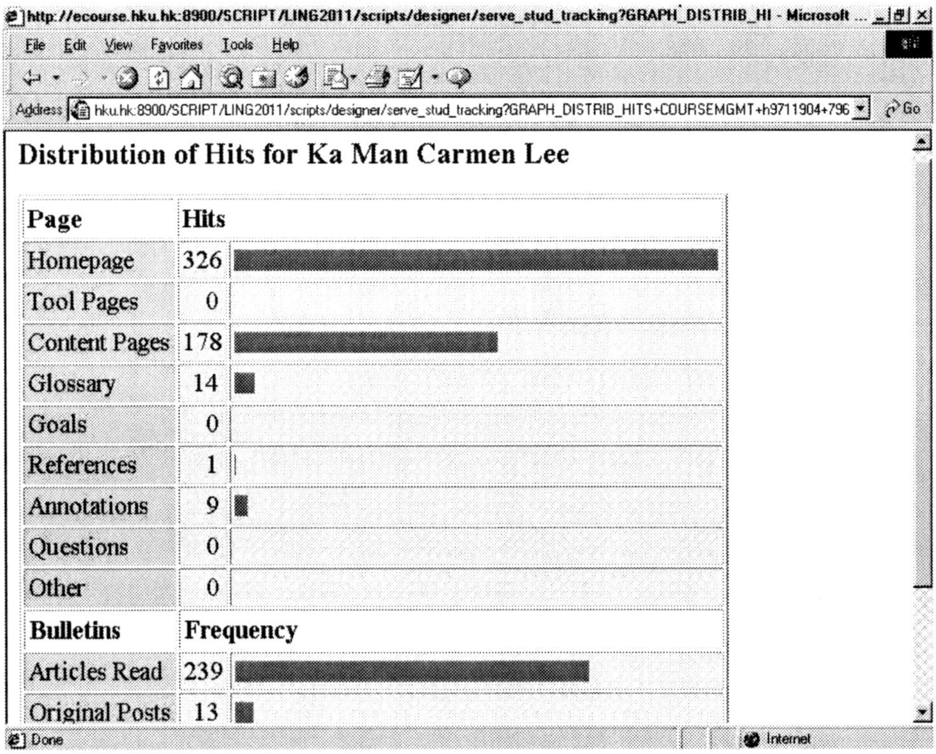

Initiative

Another indicator of success with regards to the learning situation is initiative on the part of students. Half way through the course, we often noticed with great joy that students did often introduce their own topics of discussion and techniques of information gathering and processing. At one point in time, the teacher in the Language and Literacy class had wanted to introduce the concept of biliteracy in the context of our discussion of bilingualism, but was pleasantly surprised to log onto the WebCT to notice that one student had already started a topic of discussion on it. In terms of methodology, one student surprisingly introduced the notion of annotated URLs on her homepage, something the teacher wanted to introduce to the students. The teacher encouraged students to follow her example. In our knowledge-based economy innovation has become a crucial element of an efficient workforce. Initiative is an essential element of innovation, and the pedagogical process should aim at promoting it. This may be compared to Moore's (1993) ideas of learner autonomy and self-direction, which are products of a successful interactive learning course (Figures 16 and 17).

Figure 8. Bulletin board

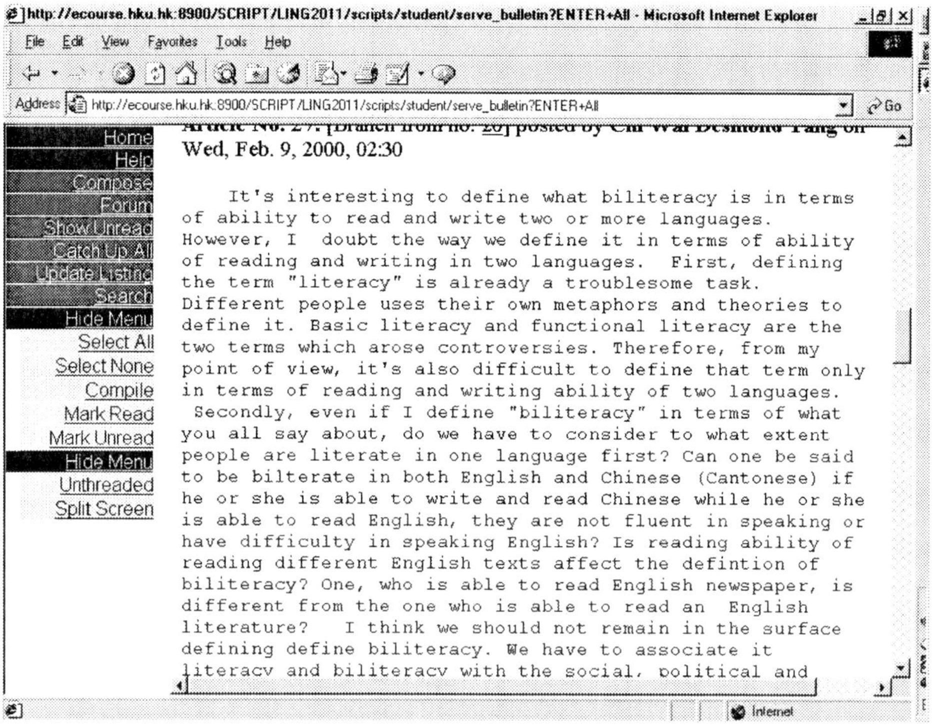

Academic Rigor

A third measure of the fact that our classes achieved an enhanced constant interaction was the academic rigor we noticed in the essays that many of the students wrote for us. Students were often generally very knowledgeable about the different shades of opinions regarding a particular technical issue. Indeed, some students even began to question some aspects of the textbooks against the realities of the Hong Kong situation that they know best.

Student Comments

In every evaluation situation, it is often best to hear from the horse's own mouth, i.e. those most affected by the situation to be evaluated. The crucial question is how the students perceived this attempt to enhance interactivity in their learning situation by the use of web-based teaching?

There are several ways that courses are evaluated. One is to get the students to make open-ended, written comments. The comments point to a positive appraisal of

Figure 9. Bulletin board

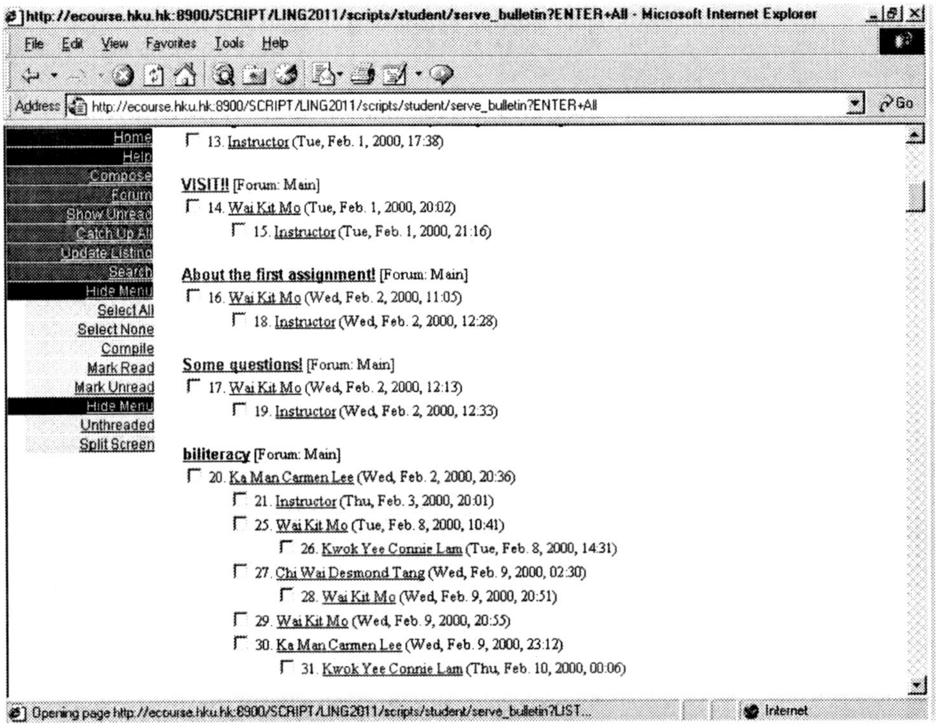

Figure 10. Homepage

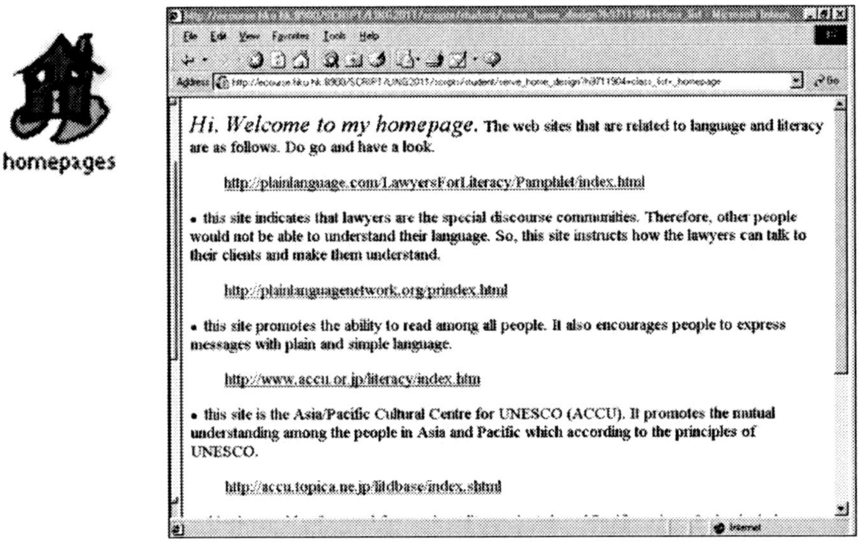

Figure 11. Calendar of course-related events

Figure 14. An interactive revision quiz

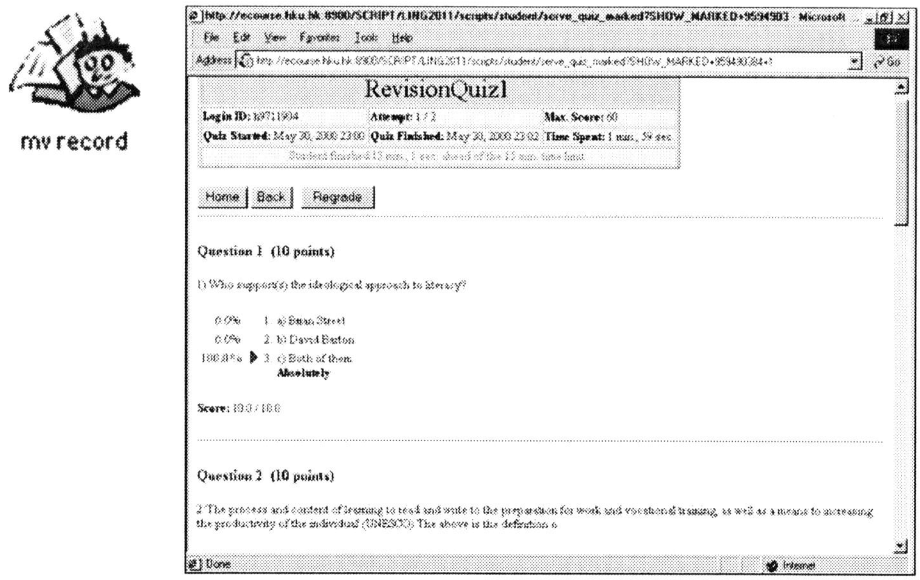

Figure 12. Student presentations

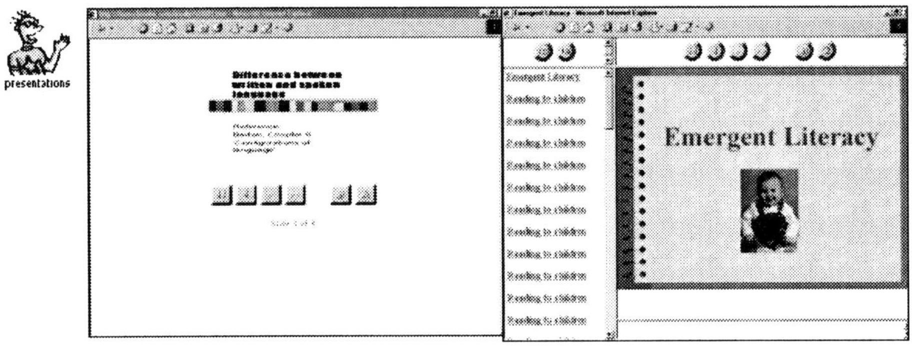

the element of enhanced interactivity in the web-based learning process, as shown by the following anonymous open-ended questions in the two chosen courses:

- Department of Linguistics
- Anonymous Open-Ended Comments about the Course: LING2011
- What aspects of this course did you find most beneficial?
    - WebCT: useful tool for interaction.

Figure 13. Student record (assignment grades)

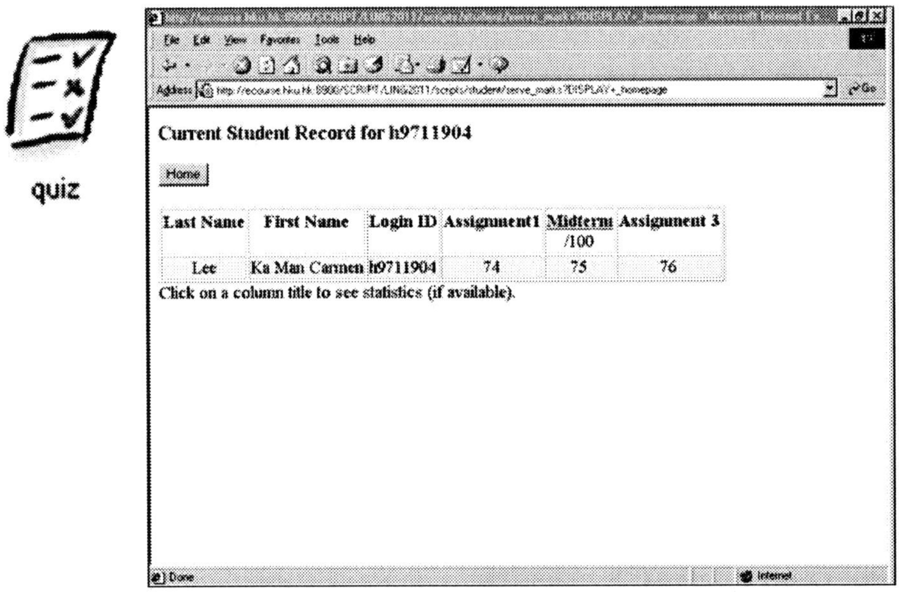

Figure 15. A discussion transcript on the bulletin board

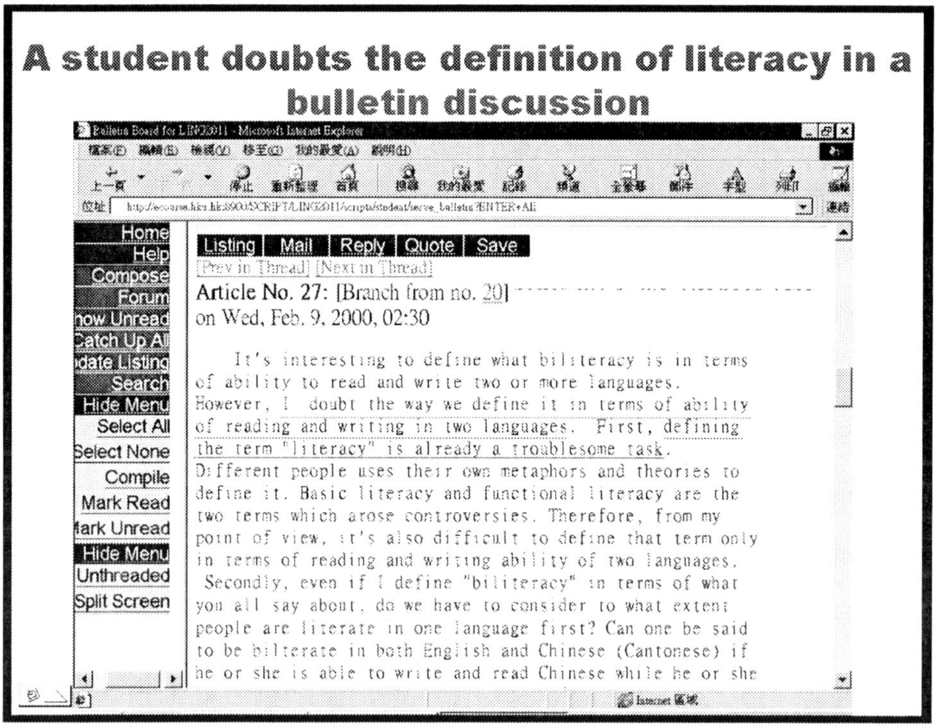

- The web, i.e. bulletin board is very useful in helping students to learn more about the topics by talking about it.
- I know more about WebCT and become more "computer-literate".
- The use of WebCT
- Handouts and WebCT
- It provides me the opportunity from knowing nothing about literacy to become knowledgeable on the subject.
- The assignments and the tutorials
- Presentations help us understand it too.
- The WebCT is a good place for students & lecturer to discuss different things outside class.
- Knowledge learned
    - Anonymous Open-Ended Comments about the Course: LING2016
    - What aspects of this course did you find most beneficial?
- Online quiz in WebCT is useful for revision!
- Web discussion

Figure 16. A student's introduction of the topic of biliteracy

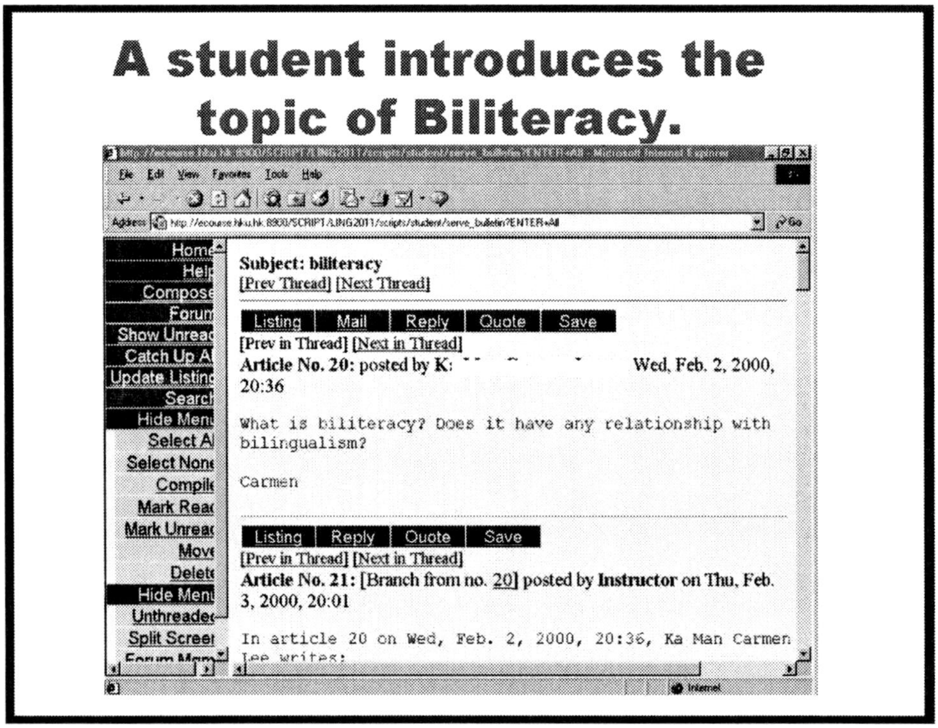

- More thorough understanding of syntax
- Have deeper understanding on syntax
- Lecture notes are clear & useful for revision.
- The lecture notes are well-organized
- Lecture running in an excellent way.
- It teaches me know more about syntax

Quantitative Aspects of the Evaluation

The preceding subsection has outlined a number of factors that should form the basis of evaluating interactivity in our web-based design. I now briefly consider quantitative aspects. I will draw on log-on statistics that point to the amount of interaction on the course home page.

Figure 17. Internet resources on literacy shared in a student's homepage

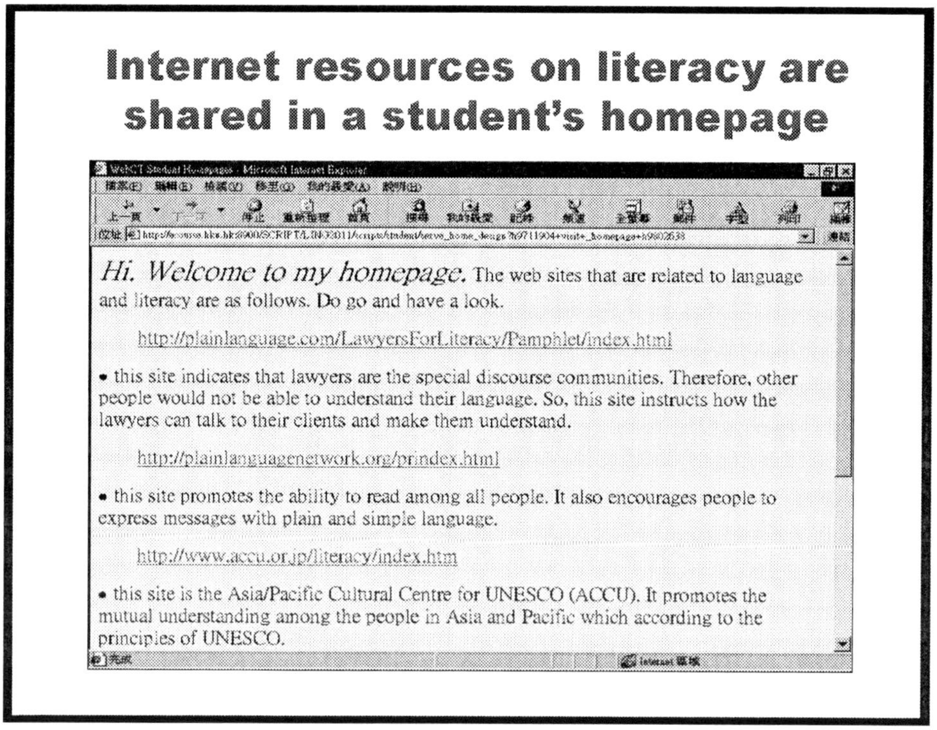

Student Log-On Statistics and Allied Figures

Statistics involving log-on frequencies and the amount of reading, writing and browsing that takes place during the course can serve as indicators to the amount of interaction with the course content. In the following I supply tables and charts showing how many times students log-on to the course website, the number of messages read and written by students and the average number of minutes spent on particular pages containing course notes (see Figures 18, 19, 20, 21, and 22).

Student Tracking

While these may not give the true picture about interactivity they do serve as indicators. For instance, there is not guarantee that students do not just browse through these pages. However, the average number of minutes spent on a page is a strong indicator of some amount of sustained interaction with the page contents.

Figure 18. Student tracking for LING2016

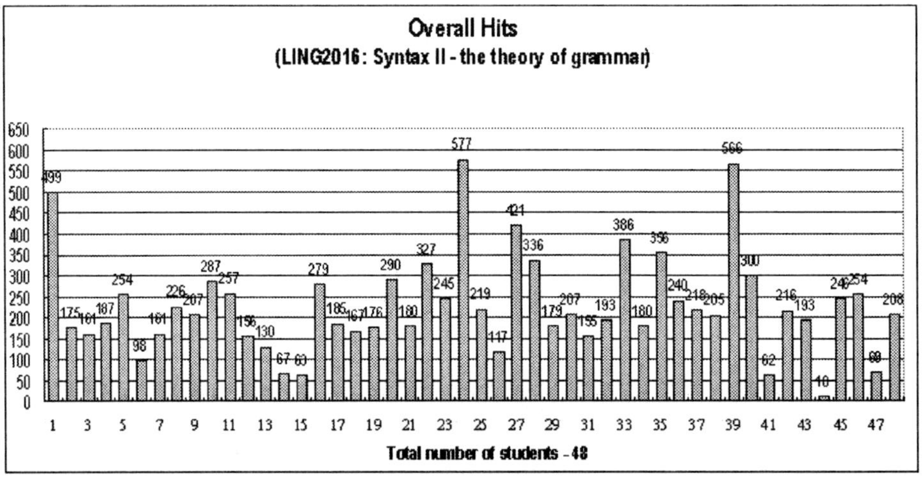

Summary

Interactivity is an essential aspect of student-centered course design endeavors, whether in traditional face-to-face classrooms or by distance learning. Society seems to require universities and other learning institutions to produce graduates

Figure 19. Messages read and posted by the students of LING2016

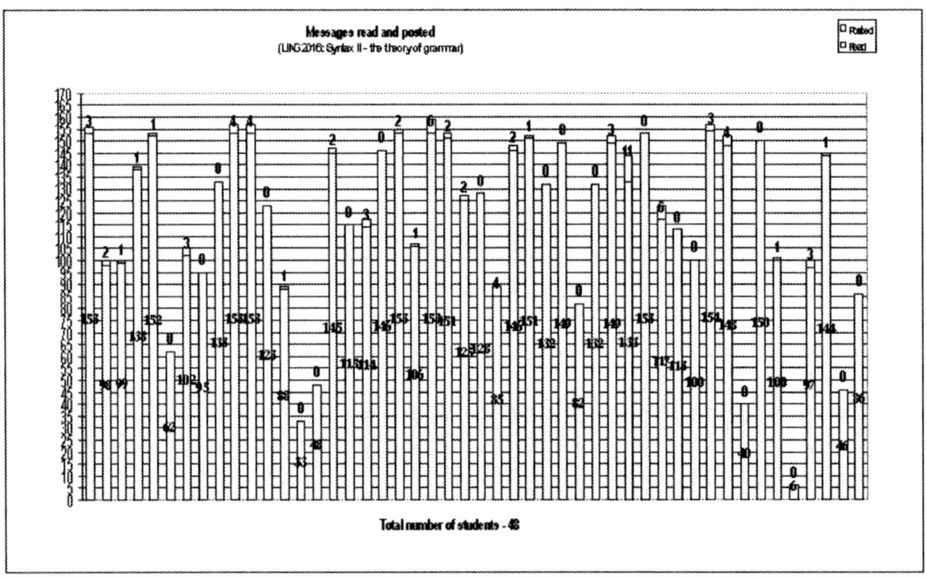

Figure 20. Page hits on the WebCT of LING2016

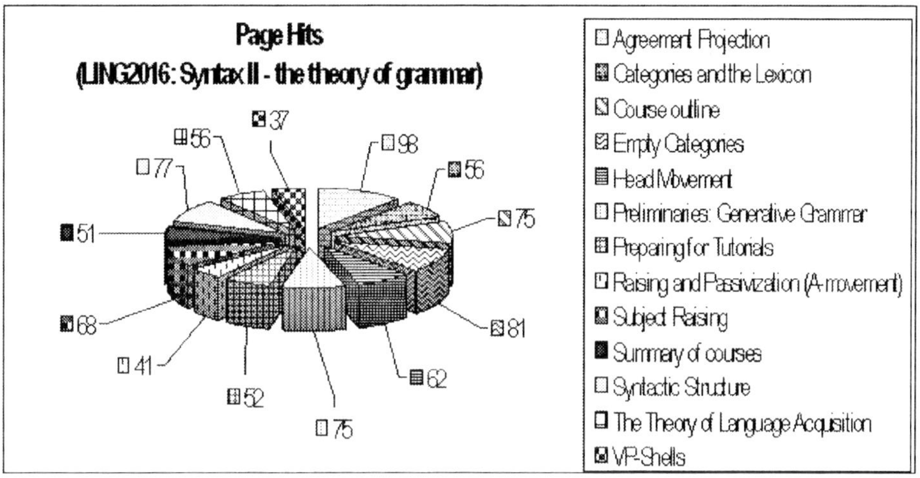

who are creative thinkers and problem solvers; graduates who are literate enough to function well in a knowledge-based economy. To achieve this educational goal, we need to reform our methods of instruction, moving away from more passive methods of teaching to more active and interactive methods.

However, the issue of how to evaluate the desired goal of interactivity in a class one has thought is not often fully addressed, in our opinion. Based on many years of web-based course design and delivery, this chapter has proposed a number of

Figure 21. Total time that students spent on each page

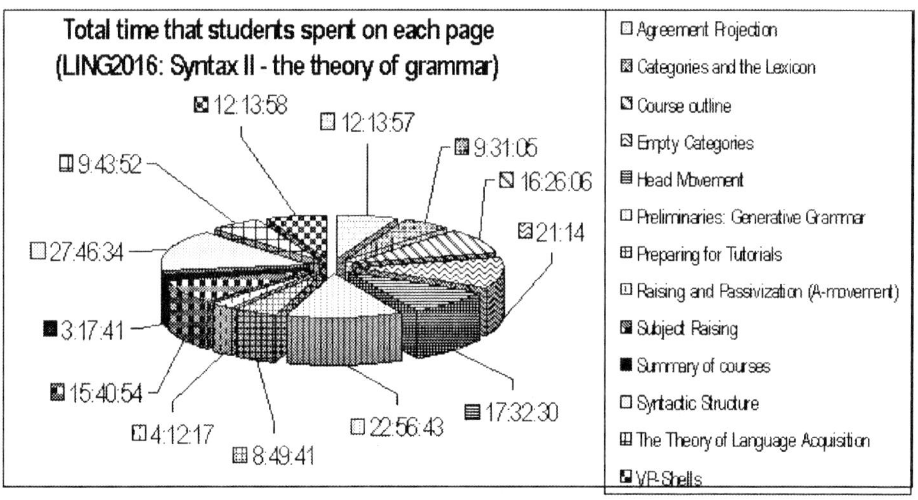

Figure 22. Average time spent per hit

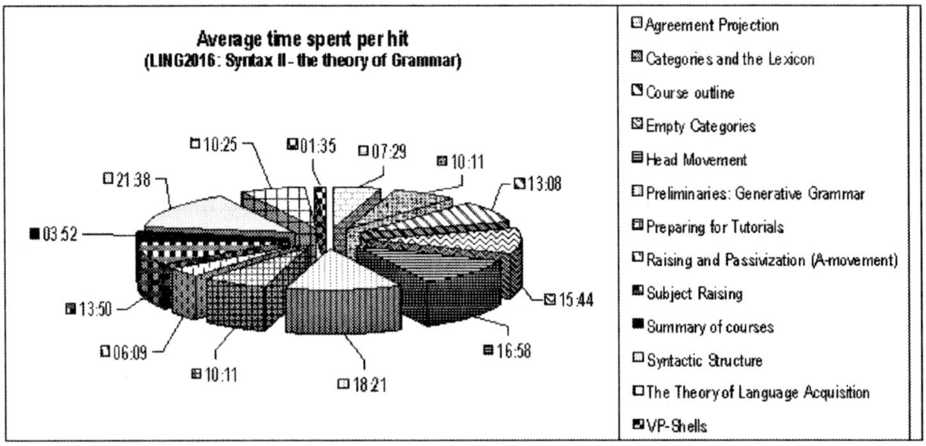

factors and criteria to consider in determining how interactivity can be enhanced and how it can be evaluated. Basically, teachers ought to construe their learning environment as one of conversation between instructor and learner in what I term Conversational Learning Community. Important components in this environment include instructor(s), learners, course materials, and links to remote experts and resources. All these components are glued together by instructional interactivity. Three types of instructional interactivity ought to be recognized. These are instructor-learner, learner-learner, and learner-resource interactivity.

To evaluate interactivity one should consider both qualitative and quantitative factors and criteria. On the qualitative aspects one should pay attention to issues such as critical thinking, initiative, and academic rigor and how far the students seem to be showing signs of these. On the quantitative aspects we may consider various log-on statistics on the course homepage and statistics about student evaluation of the course.

The issue of how to successfully evaluate the extent to which a course design facilitates interaction will remain for a long time a contentious one. Whatever the case with regard to the issue of evaluation, we may conclude that interactivity in computer-mediated communication situations, such as on the web, enhances even traditional classroom and tutorial sessions. Interactive web-based design allows teachers to achieve a better management of their courses, leading to a more effective scholarship of teaching and learning.

CONCLUSION

This chapter has taken up an important issue in education and for that matter in most disciplines dealing with the theory of learning: evaluation. Following from the discussion about creating pedagogical environments using ICT and CMC tools in the last chapter, I have in this chapter outlined how we can evaluate learning and teaching done in the learning environments created using these learning technologies.

We have mainly discussed measures taken to increase interactivity, a means to achieving our desirable goal of constructivist learning. Some of these measures involved encouraging course participants to do team work by mining information from the internet and by actively making use of the discussion forum that is inbuilt into most of these ICT learning environments.

Crucially, a number of criteria have been evolved for measuring how successful a course may be using these learning environments. These include qualitative and quantitative aspects. The qualitative ones involve taking a careful consideration at critical thinking, initiative, academic rigour and student comments. The qualitative ones involve carefully studying student logon statistics to gauge how much they make use of the learning environments created.

In the next and final chapter of the book, which serves both to tie in the major theme running across this book – the use of peculiar linguistic features in computer mediated communicated and their educational implications – and point to future themes and future directions, I look at how language is being used in emerging media of CMC and how these media can indeed serve as pedagogical tools.

REFERENCES

Markwood, R., & Johnstone, S. (1994). *New Pathways to a degree: Technology opens the college*. Western Cooperative for Educational Telecommunications, Western Interstate Commission for Higher Education, Boulder, CO.

Moore, M. (1993). Theory of transactional distance. In Desmond Keegan (Ed.), *Theoretical Principles of Distance Education*. Routledge, London & New York.

Chapter 12
Computer-Mediated Communication
Emerging Media and Themes

INTRODUCTION

Computer-Mediated Communication (CMC), with particular reference to linguistics and Literacy and how the technologies that drive this exciting area of research can be applied for natural language education, is a fast changing discipline. About every month a new technology is introduced by one company or the other that has the potential to impact the discipline in profound ways. So much so that it is hard for a research academic to pretend that at any one moment in time one can capture a representative snap-shot of the discipline that can stand the test of time. New technologies - new media, new issues - new themes are constantly emerging, and it is fitting to emphasize these new technologies and new themes in the final chapter of the book.

DOI: 10.4018/978-1-60566-868-0.ch012

In previous chapters, I focused mostly on what may be termed text-based Computer-Mediated Communication, which may be defined as interaction and transfer of information through the medium of the computer and related digital devices mainly in the written word. We have seen this in extensive discussions and case studies involving email, SMS, and MSN, among others.

One emerging trend, however, is that even within text-based computer mediated communication there was always a tendency to incorporate images and sound, as can be seen in the icons and smileys of Email and SMS, an important aspect of what has come to be known as visual language or visual communication (Horn 1999).

More radically, however, a new theme has emerged from this gradual incorporation of images in text-based CMC: video-based CMC. Video-based Computer-Mediated Communication may be defined as interaction and transfer of information through the medium of the computer and related digital devices mainly in the form of dynamic image streams. This is an even more revolutionary dynamic than the use of still pictures, clip art, images, and icons that has been described in visual language and visual communication studies. Of course, Video CMC still contains the written word, but the written word is mainly meant to just express talk around the main issue, the Video event. Young users of the internet have radically moved away from communication through the written word to communication in the medium of video clips and voice-image interactions through video-based media such as Facebook, YouTube, video games, and Skype. Social networking tools and sites such as YouTube, Facebook, and MySpace are all built with substantial use of video-based CMC.

It is this emerging paradigm shift from text-based to video-based CMC that this last chapter addresses. I do this by focusing on detailed case studies of three of these emerging media. First, I focus on Facebook, which is mid-way between text-based CMC and video-based CMC. Next, I move on to YouTube. YouTube has become the most famous video-based CMC medium and I do a brief discussion of this medium. I then look at Massively Multiplayer Online (MMO) gaming, which is an even more extreme video-based CMC that involves a lot of interaction in the form of competitive video-game contests. After discussing these major media which all incorporate the new theme of video-based CMC, the next part of the chapter is devoted to brief discussions of a miscellany of new video-based CMC media that are just too new to do any extensive surveys on because, even as we read this book, new media are constantly being introduced. Finally, I conclude the chapter – and the book - by summarizing and tying together the emerging themes and media discussed, with a speculation on what the future of linguistics- and literacy-based CMC will look like and how we can expect to harness this paradigm shift from text-based CMC to video-based CMC in the area of education, particularly natural language learning.

FACEBOOK AS A NEW MEDIUM FOR CMC

In the early parts of 2004, a new online social networking community involving college students in the university putting up their personal information on the web and using it as a platform to keep in touch with friends was formed. Facebook, the new CMC medium, would become one of the most popular websites and now attracts more than a 100 million participants. Its popularity has increased so much so that not only the youth but some prominent members of older generations, such as businessmen and politicians, even use it to get in touch with customers, constituents and other target groups.

Facebook then may be defined formally as a computer-mediated communication (CMC) tool, an asynchronous (delay) communication tool which enables communication (one-to-one, one-to-many) and collaboration over a period of time through a "different time-different place" mode (Ashley, 2003). It was invented by a Harvard student called Mark Zuckenberg in February 2004 as a social networking site originally for Harvard students but it quickly gained global currency, especially among the youth around September 2006. By registering an account through email, one can become a member.

There are plenty of functions provided by Facebook: it has photo sharing software, a blog for writing one's diary and getting comments from friends and, finally, a private and public wall which are for friends to leave messages on. As a result, linguistic data for the book is mainly collected from the public wall as they are more visible.

In this part of the chapter, I discuss the impact of Facebook in the area of CMC, based on two case studies, one on a mixed Chinese-English-medium Facebook and the other on a French-medium Facebook, showing how both communities of Facebook users manipulate the Chinese, English, and French languages in this medium for the purpose of communication.

CASE STUDY ONE: THE USAGE OF FACEBOOK AND ITS LINGUISTIC FEATURES AMONG STUDENTS OF THE UNIVERSITY OF HONG KONG

Introduction

This study is based on how a group of HKU students make use of Facebook[1]. In this case study, wall-to-wall messages will be the focus of analysis. Language used in other parts of Facebook is not considered. Firstly, I will look at the ways the students that were interviewed for this study use Facebook, for example, the frequency of logging in, the amount of time spent, the language used, etc. Moreover, the reasons of leaving messages, one of the major functions in Facebook, are outlined and some common topics in Facebook messages among the HKU student community are also listed. Finally, some specific linguistic features of Facebook among this community are discussed in detail. It is argued that expressivity and convenience are mostly treasured in CMC communication.

Facebook in Hong Kong

One point to note right away is that Facebook rather than MySpace, another online community contact group that came at about the same time as Facebook, is more popular in Hong Kong. According to the statistics available at the time of this survey on Facebook (April 2008), there were 16964 students on the HKU network of Facebook, which is quite a big network among students. Though MySpace has a larger number of users globally, it is not as popular as Facebook in Hong Kong.

Formal Features of Facebook

Facebook Profiles

A Facebook profile consists of a number of different sections, including Information, Status, Friends, Friends in Other Networks, Photos, Notes, Groups, and The Wall (which is described in more details in point v.). Figure 1 shows the layout of Facebook.

Photos and Albums

The album of photos is put into users' profile, and other users with the right credentials have the ability to see and comment on it. Figure 2 shows the section where users upload photos.

Groups

Facebook has something called 'groups.' Users can create new ones or join and participate in existing ones. This is also displayed in their profile. Figure 3 shows the section of Groups.

Figure 1. A Facebook Profile layout

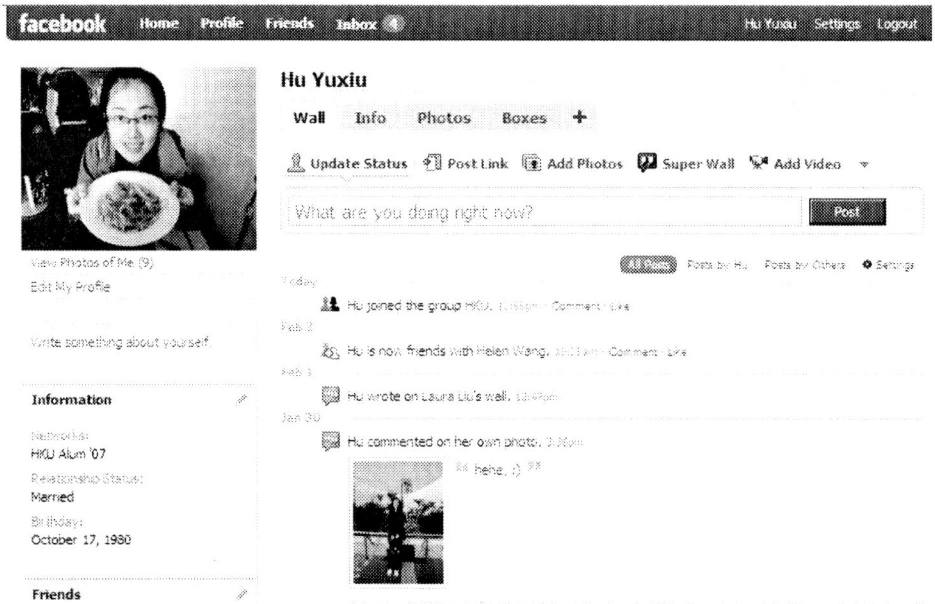

Figure 2. Section for uploading photos

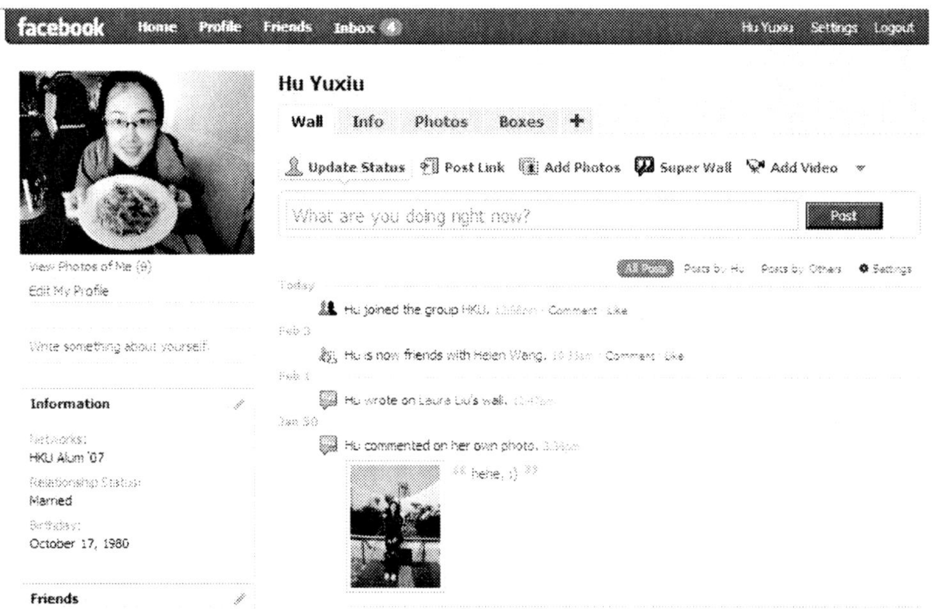

Figure 3. Section of groups

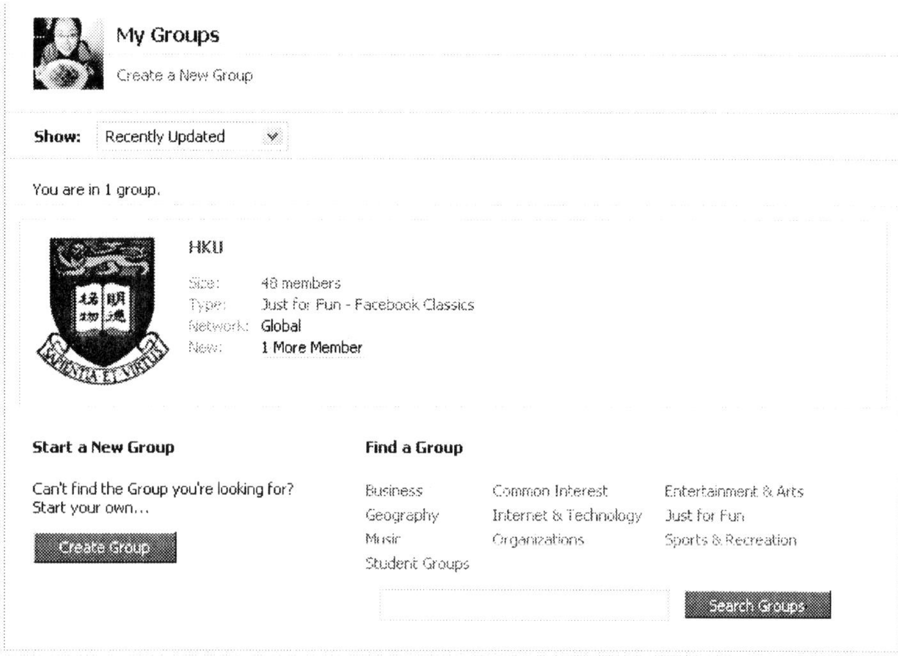

Events

This section makes it possible for users to organize and plan for events or join any events. Figure 4 shows the section of Events.

The Wall

The Wall is a forum for one's friends to post comments or insights about one. Users can always remove comments they don't like from their own Wall. They can restrict who their Wall is visible to, or turn it off entirely, by going to the "Profile" section of the Privacy page. Figure 5 shows the Wall section.

Methodology

The Subjects

Data were collected from six bilingual students of HKU, three each of males and females, all of them are aged between 20 and 25. It must be noted that six participants do not contribute a representative sample of the student population at HKU; how-

Figure 4. The Event section

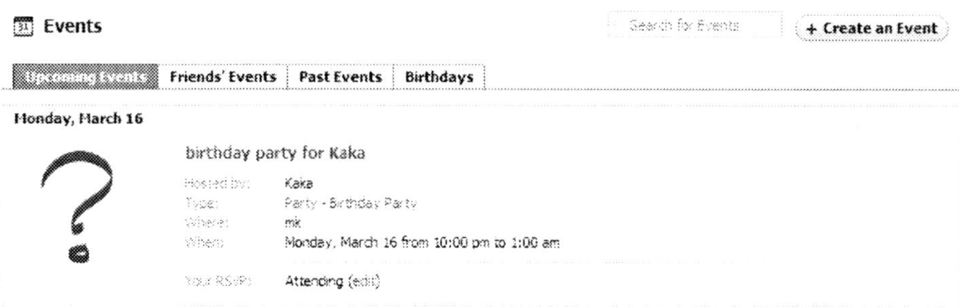

ever the main point of doing this was to get a quick idea about the use of Facebook among these students through a very small scale illustrative case study.

Figure 5. The Wall

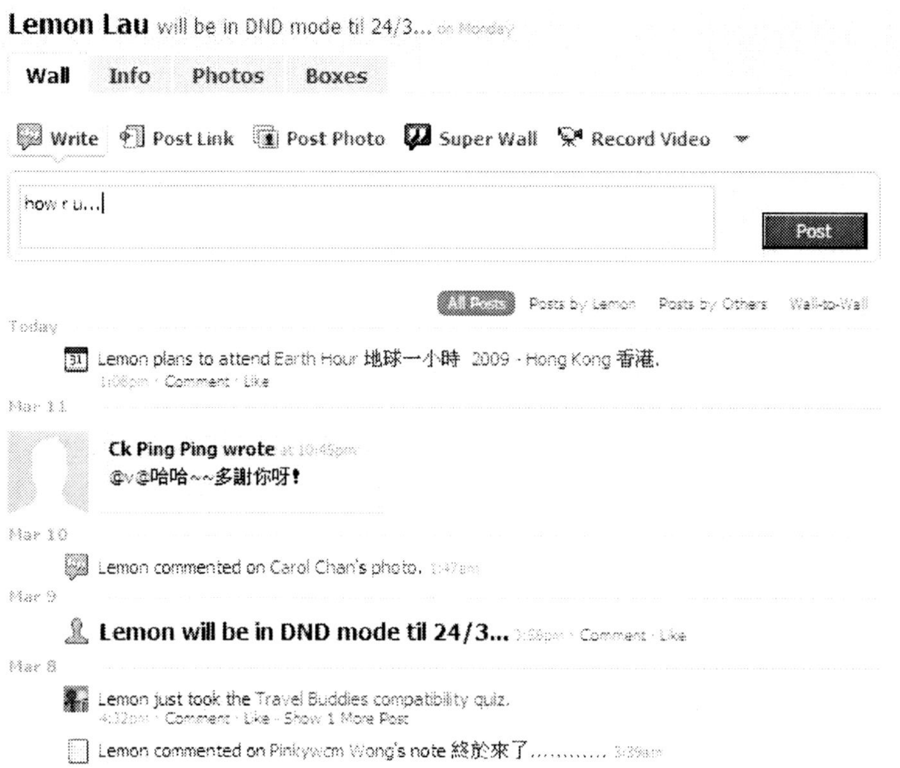

Textual Data

The aim of analyzing textual data is to observe textual features of CMC. In this case, only data from wall-to-wall messages were collected. In the course of textual data collection, we ensured that for a piece of data to qualify for inclusion the following criteria had to be satisfied:

i. Each sample message had to be a product of a one-to-one exchange i.e. only two participants are involved;
ii. Each message had to be written in Chinese/Cantonese, English, or mixture of both;
iii. Authors of the messages had to be native Cantonese speakers, with English as their second language; and
iv. The data providers had to give us permission to use their data.

General Practice of Using Facebook

To better understand the general usage practices involving Facebook, such as why they choose it against competing media, how many times they used it, etc., we elicited responses through questionnaires, and some of the answers are presented below, both descriptively and diagrammatically (see Figures 6 and 7).

It is found that Facebook is much more preferred over MySpace among HKU students.

It is found that half of the respondents use Facebook more than once a day and one-third of them uses it every day. This shows that Facebook is one of the most commonly used CMC systems by HKU students.

Formality of Communication

We also wanted to know to what extent students used Facebook formally or informally, and Figures 8 and 9 provide some answers.

Two-thirds of the students use Facebook for informal communication and only one-third uses it for both formal and informal communication.

Language(s) Used on Facebook

It is found that though some of them use a mixture of English and Chinese, no one uses purely Chinese. English is the language used by two-thirds of the respondents.

Figure 6: Facebook or My Space

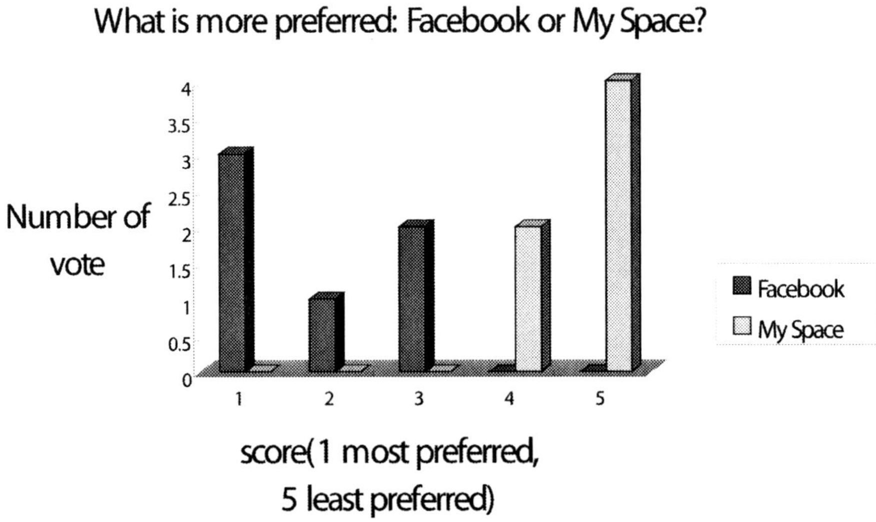

Figure 7: Frequency of using Facebook

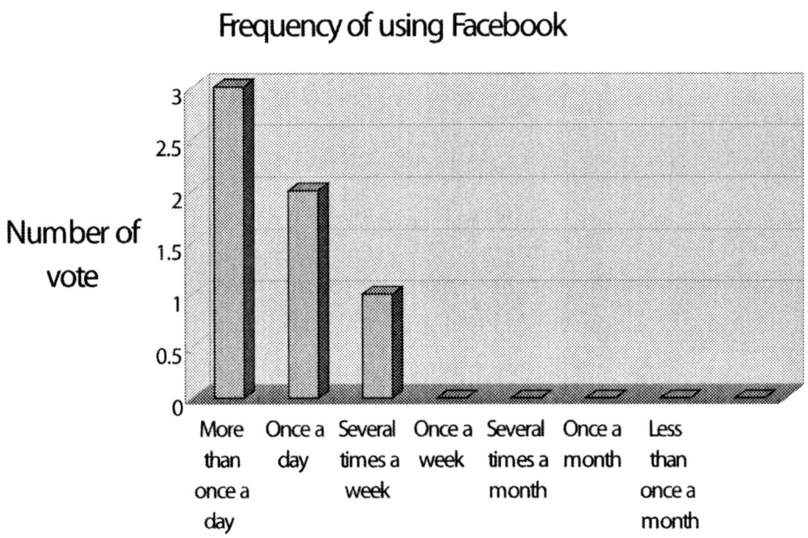

Figure 8: Do you use Facebook for formal or informal communication

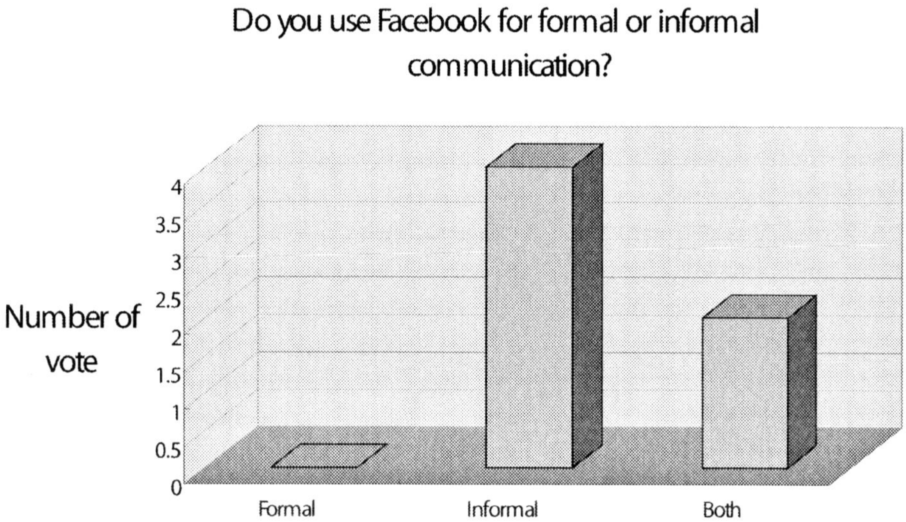

Figure 9: Languages used on Facebook

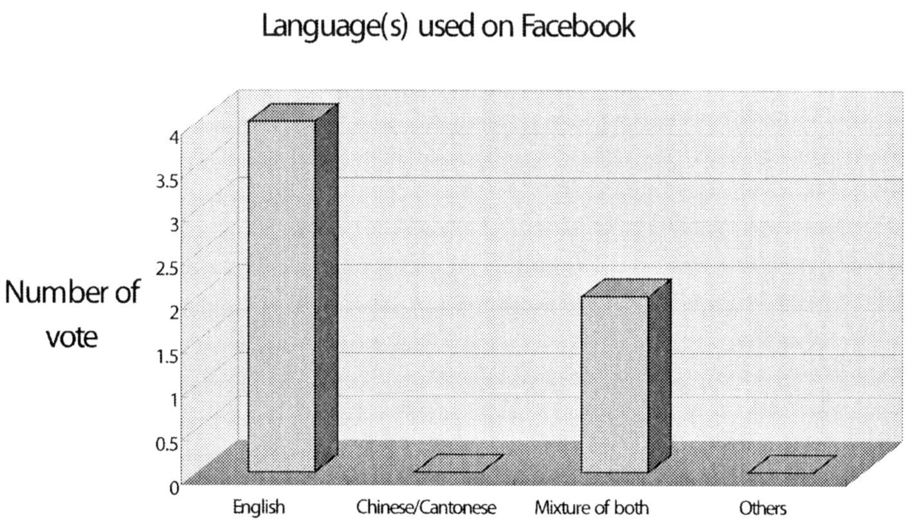

Figure 10: Do you know any Chinese input methods?

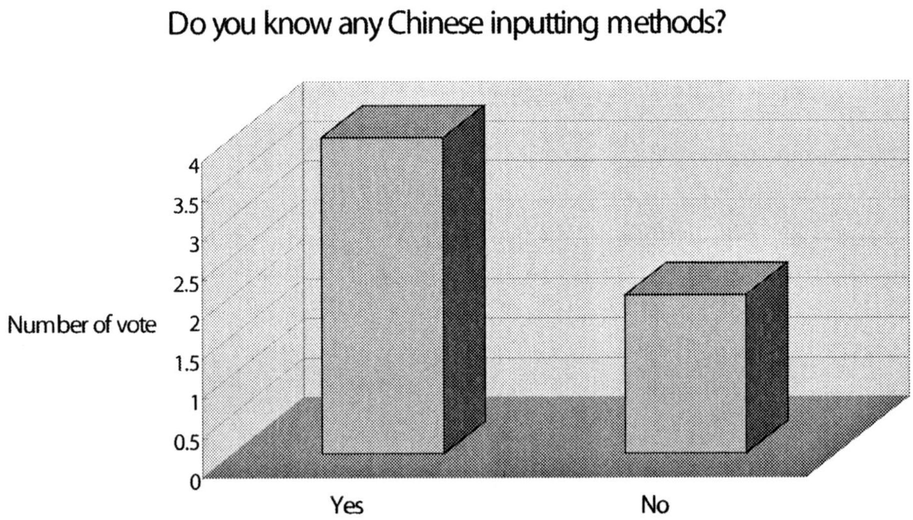

Knowledge in Chinese Input Methods

Knowledge of how to input Chinese characters in their writing is a measure of how much communication through the medium of Chinese would take place. So we sought some answers in our survey (see Figures 10 and 11).

One-third of the respondents do not know any Chinese input methods, which might imply that Chinese input is not very popular.

Half of the respondents pay attention to the grammar of their messages on Facebook and the rest of them either always or often are aware of their grammar.

Purposes of Leaving Messages on Facebook

Keeping in Touch with Friends and Peers

According to the survey, the main purpose of leaving messages on Facebook for all of the respondents is to keep contact with friends. The following examples from the data show this type of messages:

we have a crazy time in the last week......went drinking every night and i got drunk every single night

Figure 11: Frequency of paying attention to the grammar and style of texts

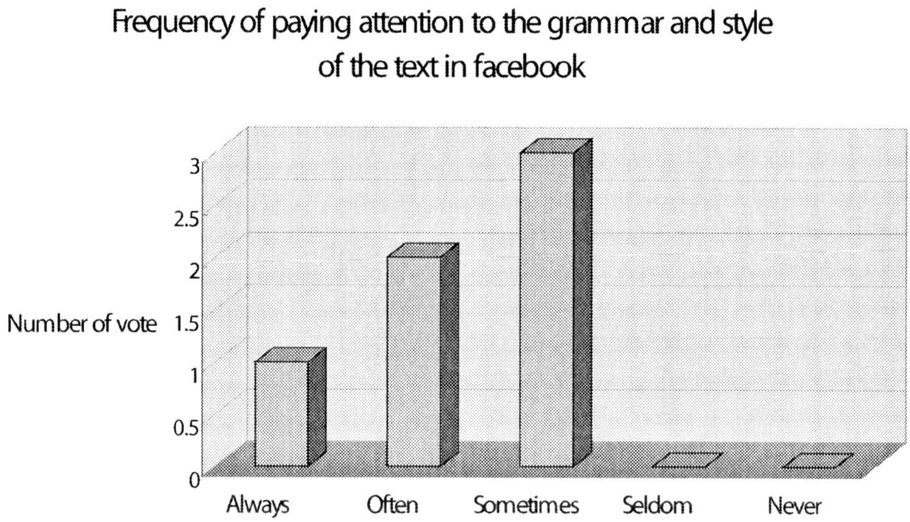

FRANCES!!! every time we bumped to each other i was in a rush...let's find a day to grab a lunch?

Moreover, Facebook is used by HKU students for communicating with people in different places. For example, HKU students keep contact frequently with their peers who are studying overseas or they keep in touch with their foreign friends through Facebook.

weiwei...saw ur photos v hapi wor =)When r u back to HK???

hey Jess, how's everything in UK? miss u very much ar!!! are u still working in the Cafe? any plan for Christmas? Wish u a happy Christmas.

Reconnecting with Old Classmates

The second purpose of leaving messages on Facebook is for reconnecting with old classmates. As there are many networks on Facebook, for example, networks of different institutions, one can easily find one's old classmates through these networks even though one may have lost contact with them. Also, one can search one's friends by just typing their names. These search functions make connecting with old friends much easier.

Arranging Activities

The last reason for leaving messages on Facebook, according to the six people interviewed, is for arranging activities. The following are examples of messages for organizing activities.

Hey, Tom~ would it be ok for u Wednesday lunch? cos i'm not so sure about next week schedule, but definitely have time during lunch on Wednesday.

Nicole our organiser, how is it going???

quickly create an event, otherwise everybody will be fully booked, fai d la fai d la fai d la

Asking for Information and Favours

Though it is not recognized by the respondents as a purpose of leaving messages on Facebook, seeking information and different types of favours are frequently found in their Facebook messages.

did Cathy get the receipt from the bank?"... "could u ask her to make some copies of the receipt for me? thanks~~~~"

Can you tell me Amy's email address?

Common topics in Facebook messages

Food and Eating

Among all the topics on Facebook, food and eating appears to be among the most commonly raised topics.

sure~~~when it comes to food, u know who I will think of? of course you and Nicole la, you guys are experts!!! u decide la

We need to pick a day to hang out and eat! Which day is better for u?

Lectures and Assignments

Another common topic on Facebook among HKU students is about their lectures and assignments. It is not surprising that students talk about their daily routines at the University through Facebook where they can find contact information for most of their peers.

Finally my two days' plans of studies failed! ><

wei...have u started the tutorial assignment yet??

Linguistic Features of Facebook Texts in Hong Kong

In this subsection I outline some of the most salient features of the Facebook texts that we have seen in this case study.

Shortenings

One of the noteworthy linguistic features in CMC is the innovative use of shortenings and abbreviations (Crystal, 2001). To investigate CMC shortenings in the context of Facebook among the community of HKU students, respondents were asked to give five examples of their commonly used shortenings in Facebook messages. The results are summarized in Table 1.

BTW/btw ('by the way'), b4 ('before'), U ('you'), ASAP ('as soon as possible') and bcoz ('because') were found to be the 5 most commonly used shortenings in Facebook.

Some traditional ways of forming abbreviations such as acronymy (e.g. AIDS) and initialism (e.g. AA for Arts Association, CC for computer centre) are retained. On the other hand, several new approaches to shortenings identified in other ICT tools such as ICQ and emails (Lee, 2002) are also found in Facebook with some new examples.

Number Homophony

The method of using numbers to represent the target word which has similar pronunciation has already been recognized in some previous research (Lee, 2002) with examples like the number 2 for 'to'/ 'too', the number 4 for the preposition 'for'. I refer to this here as number homophony. However, most of them were English based homophones. In the data, some numbers are found to be homophones or near-homophones to some Cantonese words, for example, 2 and '易' meaning 'easy' are both pronounced as (ji6); 3 and '衫' meaning 'clothes' have the same pronunciation

Table 1. Examples of Facebook shortenings provided by the respondents

Shortening	Target expression	Number
BTW/btw	by the way	5
b4	before	3
U	you	3
ASAP	as soon as possible	2
Bcoz	because	2
wt	what	1
2gether	together	1
tmr	tomorrow	1
cu	see you	1
me 2	me too	1
sor	sorry	1
fd	friend	1
sth	something	1
nth	nothing	1
dun	don't	1
nvm	nevermind	1
da	the	1
happy bday	happy birthday	1
Gdnite	good night	1

as (saam1); 6 and '綠' meaning 'green' are both pronounced as (luk6). '易', '衫' and '綠' are placed by the above mentioned homophones in the following textual data.

1. "喂, 個 quiz 2 唔 2 架??"
   wai3 go3 quiz ji6 m4 ji6 gaa3
   hey classifier quiz two(easy) NEG two (easy) PART
   'Hey, is the quiz easy?'
2. "你 今 日 件 3 好 6 呀!"
   nei5 gam1jat6 gin6 saam1 hou2 luk6 aa3
   2.SG today classifier three (clothes) very six (green) PART
   'You are so green today'.

Moreover, the number 9 that is previously identified as a near-homophone of 'night' in the example of 99 for 'night night' is now found to have another use. 9 also represents 'line' and is used in the example of on9 representing 'online'. This reflects one of the characteristics of Hong Kong English spoken by many Cantonese-English bilinguals in Hong Kong, which shows an alternation between /n/ and /l/

in their pronunciation of English words.

Letter Homophony

The method of using letters to represent a word that has a similar pronunciation or that is a homophone or near-homophone is referred to here as letter homophony. The most typical example that has always been identified in CMC is U for 'you'. Some new examples of letter homophones are found in this corpus. For example, **t** for 'tea' and **g** for '知' (zi1)meaning "know", **e** for 咦 (ji4), a Chinese exclamation word. The following examples show usage of the above letter homophone.

1. Do you wanna have **t** together after class tmr?
2. **g** ng **g** where I can register for the Mandarin course?
3. **e**…I forgot to bring my camera today ar…

The above examples of number and letter homophony show the influence of Cantonese, the mother tongue of our respondents, on the linguistic features of Facebook language. Structurally, Cantonese is a language with many homophones; as a result one should have a lot of number and letter homophones. However, English is more commonly used in the context of Facebook according to the statistics shown in this study, therefore, not many examples of Cantonese number and letter homophones are found in this corpus.

Character/Symbol Homophony

Other than number and letter homophony, a new type of homophony called symbol homophony is identified in this corpus. This method uses typographic characters and mathematical symbols to replace the target word that has the same or similar pronunciation, such as @ for 'at', + for 'add'. The following are some textual examples in which these new symbol homophones are used.

E.g. 1. "i finish my class **@** 3pm, what about you?"
2. "baby, + oil with your thousand-word essay"

Other methods of shortening like omission of individual letters such as *np* for 'no problem', *msg* 'message', *mb* for 'maybe' and reduction of sound segments, for example, *LING* for 'Linguistics', *Soci* for 'Sociology' are also found.

Although most of the shortened words in Facebook are expressions that are frequently used, there are also some short forms related to jargons in education. It is suggested that abbreviations are traditionally found in long compound words describing technical jargons (e.g. *URL* for 'Uniform Resource Locator') and with the popularity of ICT tools as a communication platform, shortenings are also

Table 2. Other methods of shortening

Shortening	Target expression	Example
LING	Linguistics	i'll be going to europe in the summer ar... becoz of the **LING** fieldtrip
Soci	Sociology	reli wanna cry coz one of my **soci** course has cancelled
TP	teaching practice	how are you doing? busy with the **TP**?

commonly found in daily expressions (Lee, 2002). However, it is not surprising to find the following jargons (see Table 2) about study domains and learning methods on Facebook in this corpus as all of the respondents in this survey are university students, whose main preoccupation is to study.

Code-Mixing

Code-mixing is a linguistic behaviour in which there is alternation between two languages. It is in intra-sentential level in which 'elements from one code become to some extent integrated into another' (Gibbons 1987: 5). In Hong Kong, while Cantonese is the most popular language in daily conversation among the majority of the population, English is always found mixed with it. Cantonese and English code-mixing is therefore known to be a common linguistic behaviour in Hong Kong society (Li, 1998, 2000).

However, as proposed by Lee (2002), a number of codes are available in the context of ICT tools, namely Standard written English, 'attempted' Standard English, Standard written Chinese, Character representation of Cantonese, Coined Cantonese Romanization and Morpheme-for-Morpheme translation. In this case study, only Coined Cantonese Romanization will be handled.

Coined Cantonese Romanization

This refers to the kind of Cantonese romanization coined by Cantonese CMC users. It is not a standard romanization because a Cantonese word, for example, 呀 (aa4) can be coined as *ar*, *aa*, *a*, *ah*, *aar* etc. Cantonese sentence-final particles were found to be most likely to be romanized (Lee, 2002) and the usage of these particles has already received a lot of discussion (e.g. James, 2001, Matthews &Yip, 1994). However, some interesting romanized Cantonese phrases are found in the data of Facebook messages, which have not received much attention before. Table 3 shows some of the examples.

It should be noticed that unlike traditional Cantonese-English code-mixing, English is found to be the dominant language in Facebook. It is already shown

Table 3. Examples of Romanized Cantonese phrases

Romanized expression	Target Cantonese expression	Example from data
aiya	哎呀 (aai1 aa3) 'ah'	**aiya**.... tue n wed will be busy for me as i have a present on thur
fung so	風騷 (fung1 sou1) 'flirty'	Hey FY~ u seems so "**fung so**" in the picture
ging	勁 (ging6) 'very'	that's wut my heart wanna say after working 2 days ... reli **ging** tired
guan sum	關心 (gwaan1 sam1) 'take care of'	I m not teasing you, just "**guan sum**" ja ma...=P
ho jor ho dor	好左好多 (hou2 zo2 hou2 do1) 'much better'	i still have sore throat but ho jor ho dor lar
dim suen ar	點算呀 (dim2 syun3 aa1) 'what should one do'	i have a bad feeling that i wont be hapi working there... **dim suen ar** ...
tat pei	撻皮 (taat3 pei4) 'lazy'	Don't be tat **pei ar**, u only have 3 days left jar...
shun	筍 (seon2) 'with much reward from little effort'	Go to the career talk and see if there is any "**shun**" job for this summer

that the majority of students use English on Facebook. The textual data collected confirms that English is the base language while Cantonese/romanized Cantonese are inserted in the English sentence.

Moreover, it is shown from the textual examples that some of the coined Cantonese expressions are put in quotation marks, for example 'Hey FY~ u seems so "**fung so**" in the picture'. Some previous research suggested that putting the coined romanization in quotation marks implies the awareness of 'non-standardness' of Cantonese spelling (Lee, 2002). The result obtained in the survey about whether HKU students pay attention to the grammar of their texts or not also confirms this. Half of the respondents sometimes pay attention to it; and the rest of them 'at least' often checks the grammar of his/her own writings. However, if the students pay attention to their grammar, why do they still use these 'non-standard' Cantonese expressions? Therefore, the reasons for using romanized Cantonese are worth further investigation.

The technical problem of Chinese input might first seem to be the reason for using romanized Chinese because one-third of our respondents do not know any Chinese input system. However, even when the respondents knew a Chinese input system, they still would prefer coined Cantonese. The response coincides with the opinion of those who know Chinese typing. Therefore, the technical problem should not be considered as a main reason.

One of the reasons for using romanized Cantonese is the specificity of the English language. Many of the respondents think that sometimes English equivalents do not express fully the intended meaning. The use of English would bring about semantic loss (Li, 1998). They believe that romanized Cantonese could help express themselves better. Therefore, expressiveness of their texts is treasured by the Facebook users.

On the other hand, convenience and time saving are important to the Facebook users of HKU students. Though English equivalents might not be able to give the best meaning, Chinese/Cantonese characters are not used either. It is because typing Chinese requires a different inputting system and many of the users find it inconvenient to switch from one inputting system to another. Because convenience and expressiveness are both important to the Facebook users among HKU students, 'non-standard' romanized Cantonese is still used.

Finally, almost all of the respondents said that romanized Cantonese is commonly used by their peers. This makes them communicate with each other in this language without many problems.

Conclusion

From the examination of the specific linguistic features among bilingual Hong Kong students, it is found that code-mixing is very common among their CMC texts. Unlike traditional Cantonese-English code-mixing, more and more new codes are invented, for example, romanized Cantonese. When choosing the language used on Facebook, whether the language is more convenient for the users is an important factor. Besides, time saving is also essential so that many types of shortening are found in CMC texts. Moreover, it is also important for HKU students to express themselves well no matter which language they choose.

CASE STUDY TWO: FACEBOOK: A STUDY OF LANGUAGE USE IN CMC BY FRENCH YOUTH

Introduction

As mentioned earlier, this second case study of Facebook moves from the Chinese-English code-mixed situation to a French language medium by French-speaking youth located in different parts of the world.[2] I will focus on the peculiar ways in which the youth manipulate the French language to suit their particular needs.

As in many parts of the non-English-speaking world, even though Facebook was introduced in the US, which is an English-speaking country, it is now gaining currency in the French-speaking world. In France, a vast majority of the French youth are now using this multi-functional web tool for communication. At the time of doing this survey (April 2008), statistics available showed that Facebook users in France were already in the range of 1.4 million members. In this case study, I will first give general background information about Facebook use among French youth before focusing on the linguistic data collected for this case study, including words collected from these young users of Facebook in France.

As mentioned earlier, there are plenty of functions provided by Facebook - it has photo sharing software, a blog for writing one's diary and getting one's comments from friends and also, private and public "walls" which are for one's friends to write something to one. As a result, linguistic data is mainly collected from the public wall as it is more visible.

In earlier parts of the book, I referred to December (1996)'s definition of CMC is as follows:

Internet-based, computer-mediated communication involves information exchange that takes place on the global, cooperative collection of networks using the TCP/IP protocol suite and the client-server model for data communication. Messages may undergo a range of time and distribution manipulations and encode a variety of media types. The resulting information content exchanged can involve a wide range of symbols people use for communication.

Facebook conforms to this definition since this is a personal site where people can use different media types of functions like video and photo sharing. Also, an important feature is that it is a platform that allows some informal usage of language in communication. According to Liénard (2005), the language used in blogs shares many of the characteristics of SMS language. This is the same in Facebook, in which we can find that a lot of the language uses are characterized by ghetto-style lexicon, simplified spelling and acronymy. It is a thin, expressive and idiosyncratic language, serving as an identity marker. People, especially the young, are far more rebellious in the rules of standard writing, enjoying the freedom of speech with flexibility to express themselves. The data collected on their "walls" will be shown to exhibit these features.

Methodology

The target group involves native French speakers aged from 18 to 27 years old. Some in the group are students in France and some are French-speaking exchange students at the University of Hong Kong. Some 20 questionnaires were distributed through email for those students in France; and for those at HKU. The statistics of the data are shown below:

Age:
18-22: 16
23-27: 4
Gender:
Female: 14
Male: 6
Education Background:
Secondary school student or leaver: 3
Undergraduate student: 12
University Graduate: 5

Phrases and words which are popular and frequently used are collected from the Facebook pages of the interviewees, having their permission to disclose the content for analysis here. In addition, to get more in-depth information, we did an in-depth interview with two Facebook users who participated in our mini survey.

Linguistic Analysis of Facebook

Facebook as a CMC tool meets a level of 'technobabble'. Technobabble itself, by definition, means a form of communication among people in the rapidly advancing computer- and other high-technology industries (Barry, 1991, Bodomo and Lee, 2002). Users of Facebook have developed a number of innovative ways to express themselves. From the collected corpora, new forms of language at the word level are found and can be categorized with reference to their own structure and pattern into four parts, as follows.

The first part concerns shortenings, including acronymy, use of homophones, combination of letters and homophone, and reduction of individual words as a way to shorten expressions. The second and third parts would focus on repetition of letters and code-mixing respectively. The last part is a discussion about the inversed word 'verlan'.

Shortening

Shortening is a prominent and creative feature that is found in CMC in the forms of shortening and abbreviation.

Table 4.

Acronyms of short phrases		
Acronym	**French translation**	**English translation**
1. cc	Coucou	Hello
2. Dmc	Dans mon cul	In my ass
3. Dac	D'accord	All right
4. NTM	Nique ta mere	F*** your mother
5. dr	De rien	Nothing

Acronym

According to SIL International (1997), an acronym is an abbreviation formed from the first letters of a series of words and pronounced as one word. From our Facebook writings, we notice the presence of acronyms of both long and short phrases, which has a total of 100% usage. And from the respondents of the questionnaires, we know that they always use these acronyms in their writings. Some of these are listed in Tables 4 and 5.

Result

Acronyms are made simply by picking the first letter of the words in each phrase, for example, in phrase number 8, "cpg" takes the first initial letters of the long phrase "c'est pas grave" to construct an abbreviation in the expression.

Use of Homophony

Homophones, as seen earlier, are of two main types: letter homophones and number homophones. By definition, homophones are 'words that have the same pronunciation but different spelling or meaning or both…' (Colman, 2001). Therefore, letters and numbers can be used to substitute the original words as they have the same or similar pronunciation. Only 70% of the respondents (14 out of 20) claimed that they have used these forms. The most popular homophones in our corpus are shown below:

Result

In Table 6, the letter homophones and the actual word in French have the same or similar pronunciations. For example, in number 1, c'est /se/ has a similar pronunciation as the letter "c" /se/. Also, in Table 7 (number 3), '7'/sɛt/ has a similar pronunciation as "cette" /sɛt /. As a result, these French writers pick up the letter 'c' only to replace 'c'est' and '7' to represent 'cette' in their expressions.

Table 5.

Acronyms of long phrases		
Acronym	**French translation**	**English translation**
6. Ama	A mon avis	In my opinion
7. Amha	A mon humble avis	In my humble opinion
8. Cpg	C'est pas grave	It doesn't matter
9. Mdr	Mort de rire	Died laughing
10. Mdl	Mort de lol	Laughing out loud
11. Tmlt	Tu me laissses	You let me

Table 6.

Letter homophones	French translation	English translation
1. C	C'est	This
2. G	J'ai	I have

Means of Combination in Words

From the questionnaires collected, two kinds of combination can be traced:

- combination of letter and number homophones
- combination of the capital letter and letter homophones

From the collected data, 85% of the respondents (17 out of 20) use these forms of expressions in their Facebook. These examples are listed on Tables 8 and 9.

As a further illustration, in number 2, Je suis / s it/ is replaced by 'J' / / 's8' /s it/. This combination of letter and number homophones is done firstly by making use of the / /"le e muet" (the silent e), which means that it is optional to pronounce it. So, "j" has the same pronunciation as "je". On the other hand, '8' can replace the syllable in "suis". By adding the consonant, 's' at the initial and combining with '8' in the latter part of the word, 's8' has a similar pronunciation as 'suis'.

Table 7.

Number homophone	French translation	English translation
3. 7	Cette	This (feminine referent)
4. 2	De	Of

Table 8. Combination of letter and number homophone

Combination	French translation	English translation
1. Bil	Bien	Good
2. Js8	Je suis	I am
3. Pti cop1	Petiit copain	Boy friend
4. B8	Bonne nuit	Good night
5. kan	Quand	When
6. ke	Que	What
7. kesk	Qu'est que	What is
8. juska	Jusqu'à	Until
9. jvè	Je vais	I go
10. tkt	T'inquiets	Worry yourself
11. biz	Bisous	Kiss
12. alè	Aller	Go
13. kon	Qu'on	When we
14. cki	C'est qui?	Who is it?
15. koi	Quoi	What
16. kikoo/kikou	Coucou	Hello
17. merki/mici/mci/	Merci	Thanks
18. mwa	Moi	Me

Further still, in number 19, some of the respondents said that they have used the capital letter to combine with the letter homophones in a word like 'pasC'. It is because after joining the letter 'C' /se/ to 'pas'/pa/, obtaining 'pasC', it has a similar pronunciation as the word 'passé' /pase/.

The capital letter is then used as a way to encode the syllable. (Daughmaudytè, 2006). Some of our participants in the study utilize this similarity to replace the whole word.

Reduction of Individual Words

Reduction in individual words is a common operation in our study, with 90% (18 out of 20) of our respondents doing this. The most popular ones are listed in Table 10.

Table 9 Combination of the capital letter and letter homophones

Combination	French translation	English translation
19. pasC	passé	past

Table 10. Reduction in individual words

Reduction	French translation	English translation
1. St	Sont	Are
2. Pr	Pour	For
3. Dsl	Desolé	Sorry
4. Qd	Quand	When
5. Tjr	Toujours	Always
6. Bjr	Bonjour	Hello
7. Bn	Bon	Good
8. Bvo	Bravo	bravo
9. Mrd	Merde	Sh*t
10. Qqn	Quelqu'un	Someone
11. Slt/slu	Salut	Hi
12. Ac/aavc/av	Avec	With
13. Ajd/oj/auj	Aujourd'hui	Today
14. Put	Putain	Whore
15. Abs	Absent	Absent
16. Rep	Répond	Response
17. psk/pcq	Parce que	Because
18. p-t	peut-être	May be
19. ns	Nous	We

Result

The respondents said that they usually pick up one to three letters within the word to shorten so as to represent the whole word. Like number 11, 'slt' and 'slu' for 'salut' and also like number 15, 'abs' from 'absent'.

Discussion

All of the linguistic characteristics above constitute only a partial picture and they cannot cover all the ways of shortening. Indeed, it is so widespread and flexible that there is no fixed set of shortening. Survey remarks from the interviewees indicate that, thanks to the nature of Facebook, an informal tool, Facebook users are able to be more flexible to type anything they want without conforming to the grammar.

It is a matter of setting as well as context. Facebook is so informal that no one will criticize the users' messages and also, the targets are usually their friends. However, on some occasions, like leaving some messages to their teachers or professors, they don't use these kinds of expressions when they write to them in Facebook, but these are rare cases. It is a way to show creativity, like using some numbers to replace the

Table 11. Repetition of letters in words

Repetition	French translation	English translation
1. J'adooooore	J'adore	I like
2. bisouxxxxxxxxx	Bisous	kiss

syllable in word ('bi l' for 'bien'). They admitted that it can show their creativity on one hand and what's more, they want to demonstrate more conformity within their groups of friends. For example, even though some of them do not know these ways of expressions at first, they will learn from their friends' comments to them; then, when they want to respond to them, they pick up the same style to reply to them. It is seen as a means of demonstrating affiliation and belongingness in their friends' network. Sometimes Facebook users are regarded as being not up-to-date or old-fashioned if they use and conform to the standard grammar.

Lastly, due to the economy principles-- time constraints, 'le loi d'economie' (Daughmaudytè, 2006), they want to save the time to type in a message on others' walls.

For example, typing part of the letters in words, like 'pr' from 'pour' and also, those phrases with just typing in the initial letters of each words in phrase, like 'dmc' for 'dans mon cul'. It saves time for reducing the number of the letters in phrases or words. For them, it sounds more economical and faster in processing.

Repetition

Repetition of letters in words is often found, with 65% (13 out of 20) of respondents saying they use these forms of repetitions shown in Table 11.

Result and Further Discussion

It is not very clear that there is a systematic picture of which particular types of vowels or consonants are subject to repetition. While the 'o' in number 1 'J'adooooore' is the nucleus of the syllable and is repeated to indicate intensity of the affection, the "x" repeated in number 2, 'bisouxxxxxxxxxx', is not the nucleus, indeed it is a silent consonant. It means that users really like it a lot. However one way to explain the "x" repetition is that, those repeated 'xxxxxxxx's are just akin to the hissing sounds of prolonged kissing, expressing the great affection for someone. Both of the cases are deployed by these Facebook users to express and emphasise their feelings.

Table 12. Borrowing of English words in French

French	English	translation
1. C'est	fun	It is fun
2. C'est	ok now?	Is it ok now?
3. C'est	me	It's me

Borrowing of English Words

Some 55% of respondents have the habit of mixing French with English. Some of the respondents (70%) usually regard English as their second language. Those expressions that are commonly used are shown in Table 12.

Result and Further Discussion

Some simple English words are mixed with French in some phrases, like 'fun' and 'ok'. They are common and familiar to most second language speakers of English. So, some of the French youth would mix them, as many of them claimed that it is just their personal way of speaking. However a good number of the respondents (40%) said they may not always use this form. Indeed, some of them would use 'c'est bon' to replace 'c'est ok'. On the other hand, they claim that they would not use some complex English words in Facebook. This may be due to their lower levels of competency of English and also, these complex words are too hard and not so common to be well-handled by people with low proficiency levels. As a result, only these kinds of common and simple English words are just used on Facebook.

Inversion of Words 'Verlan'

About 70% of the respondents reported that they often use inversed words, known as verlan, in their Facebooks.

The most common 'verlans' are shown in Table 13.

Result and Further Discussion

The listed inversed words above are borrowed from suburban slang. A number of common words are in verlan, a slang form of French in which syllables are inverted (Laurence, 2005). Concerning the structures of these words, for example in number 2 'a fond', the latter part of the letters in 'fond' are inversed, becoming 'donf'. To give more insight to the number 1 'Béton', the mechanism of change can be explained like this: 'the word 'tomber', divides into 'tom'+'ber'. And swapping the order of these two, we get 'béton' (the spelling of 'verlan' words represents their

Table 13.

Inversed words	French translation	English translation
1. Béton	Tomber	Fall
2. A donf	A fond	Totally
3. Tigen	Gentil	Nice
4. Gro zoubi	Gros bisous	Big kisses
5. cimer	Merci	Thanks

pronunciation, note in this example that the word final sound '-er' of 'tomber' is written -é when it occurs in the middle of a word, but the sound is nevertheless the same).' (Ellis, 2002). Note also that the 'm' becomes 'n'.

The original of verlans were made from the "blurs" (by Arabic youth), but according to Papenfuss (2006), now we have a fast-developing street slang everywhere — schoolyards, newspapers, the Internet, movies, ad campaigns, TV and radio. These verlans have become more common and widespread. The French youth can easily gain exposure to them and thus are prone to using them in CMC tools like Facebook.

Conclusion

From the above analysis of Facebook language used by French youth, the linguistic characteristics in their corpus are divided into different categories. From the collected corpus above, we can conclude that originally, because of a combination of space limitation and heavy charges of SMS in France, simplification and abbreviation became a norm in mobile phone texting. Afterwards, it then spread to other CMC tools like blogs and Facebook.

Generally, the youth use these styles of expressions, firstly, for showing their own creativity and also for keeping abreast with the ever-changing language in the fast-paced world of technology; secondly, as a means to show conformity to the general trends, accommodating their friends, the youth resort to practices like the use of abbreviations, verlans, and code-mixing of French and English.

It can be claimed that the youth don't like lengthy messages, which seems to be a waste of their time. They want to use more innovative ways to express themselves to the extent that these informal platforms allow them to do so. Finally, differences of cultural and linguistic backgrounds also lead to variations in the forms of language. Different people have different ways of writing.

YOUTUBE: THE WORLD'S POPULAR VIDEO-CMC

In this section of the chapter I turn to the discussion of YouTube, the most quintessential of video-CMC.[3] We do a discussion of YouTube and how it can be used for learning, in general, and language learning, in particular. We explore how a relatively new technology like YouTube has changed the way students communicate and learn, the main intention being to get teachers, students, and administrators to think about and hopefully change the way they do things, by providing them with the context for why change is necessary.

Introduction: YouTube Background

With the catchphrase, "Broadcast Yourself", YouTube is a popular free video sharing website (www.youtube.com) created in 2005 which allows users to upload, view, and share video clips. Videos can be rated, and the average rating and the number of times a video has been watched are both published. Anyone who signs up for a free account can upload a movie file in any format. YouTube will compress it and have it available for viewing in a compatible way to all viewers, no matter what kind of computer or how fast their connection is.

YouTube is known for showcasing copy written music videos, clips of popular television shows and is a vehicle for aspiring film makers, musicians and anybody with a point of view to voice their opinion to the world. Little known to the rest of the world, is the fact that there is a growing educational section of the site. The demand for learning and community is greater than ever, and likely to be everlasting. The demand for consuming and sharing video information is now fundamentally unstoppable. (Kaufman, 2006) The technology has grown based on what the cultural and communication needs of our society have demanded.

You Tube and Language Change

In August 2006, Stanford Law School professor and social activist Larry Lessig, declared that text — text, on which most of us were raised and through which most of us communicate — is ... dead, that the written word has become the "'Latin' of our modern times"; that the ordinary language, the "vulgar" or vernacular language, the new language of the street is video and sound and that the software suites that facilitate video and sound editing are now communication tools (Kaufman, 2006). As mentioned in earlier sections of the book, new tools and media of communication demand the creation of new forms and ways of communication. These new forms compete with existing forms and ways of communication, leading to changes in the way we use language in its various forms, including spoken and written forms.

Adkins (2007) believes that, YouTube is unique in its literacy. He argues that the nature and design of YouTube does not require you to be as literate (i.e. in the traditional sense) as other virtual communities might. As long as you have a basic understanding of the language so that you can use the website and respond to messages from other users, you can use the site. Adkins writes that creating videos and verbal communication is much more important than being able to read or write in the YouTube environment. Reading and writing is no longer deemed as important as in the formal classroom. Many teachers utilizing 21st century learning techniques are incorporating informal learning into the traditional classroom setting. As long as you can post a video of something interesting, no one knows or cares if you can read or write (Adkins, 2007).

Pete (2006) compares an inaugural speech given by George Washington to that of Bill Clinton in the 1990s. A comparison of readability statistics shows that George Washington's speech was at a 12th grade reading level, whereas Clinton's was rated at an 8th grade reading level. He made this comparison to show the similarity to technobabble as seen in today's IM sessions. His writing shows that literacy is continuing its evolution. In our speech-heavy classrooms, the "Washingtons" do their best to talk to the "Clintons"; but the gap between the two groups is widening. Based on this premise, he blames this literacy gap on the over 30% student drop-out rate. (Pete, 2006) Responses to his article show general agreement. One proponent of Pete's writing further notes: "the most creative kids I know are not creating with words, many new ideas they engender are being transmitted in other formats. Educators comment on being fixed upon words -writing and reading of them...- as the means of exchanging what they've learned and how they think. Paper is not the medium the new generation uses to record and exchange information and to share culture, beliefs, and history." (Pete, 2006) The article shows that YouTube allows for creativity and transmission of information through visual and oral communication, and no pencils are necessary.

Questions

Some of the questions that this section on YouTube will attempt to answer include: How can teachers incorporate the power of YouTube technology into their pedagogy and curriculum? How are YouTube and Web 2.0 technologies[4] changing the way we share and communicate? What makes YouTube a thriving online community? How does YouTube differ and how is it similar to other communication and learning tools? Will this technology be around in 10 years? How might it evolve our communication as a society?

Defining Important Terms/Background History

According to a Nielsen/NetRatings report in July 2006, 12-17 year olds (also known as Millennials) are 1.5 times more likely to visit YouTube than the average Internet surfer. (Oishi, 2007) It is ranked 7th in the world for internet traffic and shows up to 100 million video clips a day. The Online Publishers Association (OPA) released a study that says video viewing online has reached the point where it is "a routine practice" for many internet users. (Geller, 2006) According to one estimate, almost half of all video online today is user–generated. (Wang, 2006)

In examining these questions, based on reviewing YouTube related articles, we talk about "Millennials". Millennials were born roughly between 1980 and 1994. They've grown up with more choices and more selectivity in the products and services that they use. This group of kids expects interactivity. Whether they're using Facebook, MySpace, iPods, instant messaging or text messaging, Millennials are plugged in. (Sweeney, 2007) Some other traits of Millennials are that they rarely read newspapers or books. They are impatient and goal-oriented. They hate busy work, they learn by doing and they are used to instant feedback. They think it's cool to be smart and they have friends from different ethnic backgrounds. They want flexibility in the classroom and in their lives.

Why Do People YouTube?

A study was done out of Kansas University in 2006, posting a video on the site asking the question "Why Do you YouTube"? The researchers gathered experiential data from immersing themselves in the community and fully participating in the observations. An analysis on the 370 video responses to this question was completed as of Feb. 2007. The 10 researchers on the project, each took 37 responses and put the answers into a spreadsheet to see what the most common answers were (Wesch, 2007). The top ten answers to why people use YouTube are listed below:

1. 61% said they tube to connect with others or to be social
2. 43% said they tube for fun or entertainment
3. 41% said they simply like watching YouTube
4. 33% said they tube to express their opinions
5. 25% said they tube to be creative
6. 19% said they tube because they are bored ("nothing better to do")
7. 17% said they tube because it is more "real" or authentic than commercial productions
8. 16% said they tube hoping they might become famous

9. 15% said they tube to see what other people think of them
10. 12% said they tube because they are addicted

The results were valuable in breaking down some common assumptions. First, most people encounter YouTube through ridiculous and spectacular viral videos. It was surprising to find out that the #1 answer to why people YouTube was to connect with others and be social. There was a strong selection bias, as those who responded to the question were likely going to be the same people trying to connect and be social with others. Some of the answers were non-answers (for fun / like watching), but others show some significant self-reflexivity among Tubers. Some people tube in order to see what other people think of them. Some suggested that they can take the comments they receive on YouTube to better themselves and become more socially adept in their face-to-face encounters. Interestingly, while YouTube would seem to be the most public of spaces imaginable on the entire planet, it seems that some people actually view it as a somewhat private space where few if any of the people they encounter face-to-face will ever find them. (Wesch, 2007)

Who are These YouTubers?

YouTube usage is greatest among young males. Over three-quarters (76%) of 18 to 24 year old males say they have watched a video at YouTube, and 41% visit YouTube frequently. Recent research by Harris Interactive suggests one-third of frequent YouTube users are watching less TV to watch videos online. Over four in 10 (42%) online U.S. adults say they have watched a video at YouTube, and 14 percent say they visit the site frequently. (McNerny, 2007) Of all frequent YouTube users, two-thirds (66%) claimed they are sacrificing other activities when on YouTube. Although their visits to the site were most likely at the expense of visiting other websites (36%), time spent watching TV is next most likely to have taken a hit (32%). YouTube also cut into email and other online social networking (20%), work/homework (19%), playing video games (15%), watching DVD(s) (12%) and even spending time with friends and family in person (12%). Aongus Burke, Senior Research Manager of Harris Interactive Media and Entertainment Practice said, "We know from some of our other data on teens that YouTube is just as popular with them as it is with young adults". Harris polled 2,309 U.S. adults (ages 18 and older), of whom 363 were frequent YouTube viewers, conducted online between December 12 and 18, 2006. (McNerny, 2007, see Figures 12 and 13).

The Harris poll was conducted online within the United States between December 12 and 18, 2006 among 2,309 adults (aged 18 and over), of whom 363 are frequent YouTube viewers. Figures for age, sex, race, education, region and household income were weighted where necessary to bring them into line with their actual proportions

Figure 12: Time spent on YouTube

TIME SPENT ON YOUTUBE

"About how much time do you spend on YouTube?"

Base: U.S. adults having ever watched a video on YouTube

	YouTube Viewers
	%
Uses YouTube Frequently	**33**
More than 2 hours a week	2
1-2 hours per week	7
I'm there frequently, but less than 1 hour per week	24
I've only visited YouTube once or a few times	67

in the population. Propensity score weighting was also used to adjust for respondents' propensity to be online. All surveys were subject to several sources of error. These included: sampling error (because only a sample of a population was interviewed); measurement error due to question wording and/or question order, deliberately or unintentionally inaccurate responses, non-response (including refusals), interviewer effects (when live interviewers are used) and weighting.

Figure 13: Activities people spend less time on because of YouTube

	Frequent YouTube Viewers
	%
Spending Less Time	**66**
Using other websites	36
Watching TV	32
Emailing, chatting online, blogging, etc	20
Working or doing homework	19
Playing video games	15
Spending time in person with friends/family	12
Watching videos on DVD	12
Reading magazines/newspapers	11
Talking to other people on the phone	9
Going to the movies	7
Exercise	1
Other	2
I don't think I'm spending less time doing anything because of my time at YouTube	34

YouTube Barriers

Because most online video databases are not intended for education, they contain material that is inappropriate for minors. Teachers should select clips carefully and supervise use of the sites. In the future, it is likely that indexes of educational material and designated student-safe areas will make access to appropriate material easier. YouTube allows for most content except for pornography. As a community, a member can request to take a posting down.

Users should be careful to avoid copyright violation, because some of the material comes from commercial broadcasts. In other parts of the world, sites like YouTube are banned for content that are controversial. In Thailand, the military backed government in the era 2006-2007 banned a 44 second video that mocked the country's monarch. Thais blamed the government for turning an obscure bit of nonsense into a global free-speech issue. Sites like YouTube face huge problems if they wield editorial power and huge problems if they don't. As soon as they make any decision about editorial issues they will risk accepting responsibility for the site's whole contents. Therefore they leave that power to the users.

There is strong debate on taking down copyrighted materials from YouTube. The company has been sued from to time by a number of media companies. Enforcing the copyright on the majority of the videos that make YouTube a world hub would eliminate the greatest archive of pop culture ever assembled and an incredibly significant historical resource. Elsevier (2002) talks about how the historical and personal value of taking down such an interactive journey into history would be irreplaceable. Where else can kids today see and hear the world that was here before they were, on-demand and a click or two away? The opportunity was unavailable to previous generations, so the benefits for today's learners are huge.

Many teachers are hesitant to integrate Web 2.0 and ICT tools into their curriculum because of fear or unacceptability. Some instructors are not familiar enough with the technology, which causes all sorts of unwanted delays, problems and anxieties. The instructor needs to be organized and take the time to set up the course and facilitate any technology issues, if necessary. Training and help for new YouTube English users is available for free on the web here: http://www.google.com/support/youtube/?hl=en_US

What is a Learning Community and how Does YouTube Fit into this Concept?

There are many books that share the insight that a networked culture is enabling a new form of bottom-up power, as diverse groups of dispersed people pool their expertise and confront problems that are much more complex than they could handle individually. They are able to do so because of the ways that new media platforms support the emergence of temporary social networks that exist only as long as they are needed to face specific challenges or respond to the immediate needs of their members.

Online learning communities have been called "adhocracies" (though some like Conversational Learning Community that I have developed as reported in chapters 10 and 11 have an organizational structure). An adhocracy is a form of social and political organization with few fixed structures or established relationships between players and with minimum hierarchy and maximum diversity. In other words, an adhocracy is the polar opposite of the contemporary university (which preserves often rigid borders between disciplines and departments and even constructs a series of legal obstacles that make it difficult to collaborate even within the same organization).

Online information sharing sites allow users to be producers of content instead of just consumers. This concept is crucial when it comes down to motivating and engaging young learners. When this shift happens, students have more ownership and put more effort into their work. When they are given the power to have their voice heard by many, they see the applied outcomes of their learning automatically. "On demand" and "Just in Time" are phrases that commonly come up with the new generation of learners. Growing up with the internet has given them choice, immediate feedback from other users and freedom to find their own answers when they need to.

In an opinion article by McKinney (2006), he discusses what makes YouTube work. The concept of Peer to Peer file sharing is a key issue here. Peer to Peer file sharing is where individual people are responsible for hosting available resources (as the central server does not have them) and for making its shareable resources available to peers that request it. The dispersing of information on demand to various places and formats is an important concept in YouTube. YouTube provides user with the unique HTML code for each movie so they can place it onto a page or do as they wish with it. A user can take a movie and send a link to her friend, post it on her MySpace page, put it in their blog or search for it on Google within a couple of clicks. The information has become easily accessible. The author states that YouTube has made three main innovations:

1. Making it incredibly easy for users to contribute videos,
2. Creating a central location to find video clips, with lots of good ways to find them (browsing, searching), and
3. Giving the Internet a way to link to videos, and giving television clips a way to exist on the Internet. (McKinney, 2006)

McKinney states that YouTube solves the problem of the difficult process of putting video online, hard to find (doesn't show up in searches), complicated formats intended to deter copying, etc. Videos historically never really worked as part of the web. Online video has always felt forced, tacked on to the web rather than integrated into it.

YouTube around the World

Kaufman compares other new educational video endeavors around the world. In the United Kingdom, the British government is digitizing explicitly for teaching, learning, and research 3,000 hours of television news and cinema newsreels from the ITN Reuters Archive. "Newsfilm Online," that was scheduled to be completed and released in 2007, features downloadable moving image content under licensing schemes allowing users to edit material to suit their own purposes. These resources are being encoded simultaneously as Windows Media, Apple QuickTime, and MPEG–2 files, supported by an extensive and easily searchable — database and supplemented by 450,000 pages of video bulletin scripts. This is happening with the explicit goal of making the use of video as easy and intuitive as it is for the written word. With the support of the William and Flora Hewlett Foundation, Open University launched "OpenLearn" in October 2006 — to make its video and other rich–media resources, many of which were produced in association with the BBC, freely available online alongside learning–support and social–media collaboration tools.

Kaufman writes about other countries embracing these technologies. He talks about the Japanese government supporting far–reaching digitization and preservation initiatives around audiovisual assets at state broadcaster NHK. The Brazilian and Canadian governments and education systems have done likewise with their assets. Tens of millions of people are expected to interact with these assets in the coming years. Scholars have begun to envision creating an integrated, networked cultural record where librarians, curators, archivists, and the private sector have joined forces with the objective of "creating universal access to knowledge anywhere and everywhere." Librarians have begun speaking of building the "global digital library." Museum curators have spoken of "heading toward a kind of digital global museum." (Kaufman, 2006) Cultural and educational institutions are increasingly moving to embrace social media and the power of Web 2.0.

In the United States, universities from Yale to UC Berkeley, Stanford, MIT, Arizona State, Duke, Michigan, Penn State, and the University of Illinois have launched a significant effort to post new video of classes, lectures, and legacy holdings online, free. Public TV producers and stations like WNET and KQED are now launching educational experiments to explore how its assets (programs on history, nature, and international affairs) can be deployed more fluidly and systematically in the classroom. This is important because students, faculty, museum visitors, library users and others will be able to access all of these media, work with it, and contribute to it on an accelerating basis.

Implications for the Classroom

YouTube's "cool factor" with kids has great potential for the classroom. Videos are free and can be viewed immediately without having to download any software. It's easy to incorporate them into a lesson if you have a broadband internet connection. There are lots of ways for teachers to link YouTube to their lessons so they can stimulate bored kids and allow their class to see relevancy in what they're learning. With a digital camera or video camera teachers can post their own unique content within minutes. There is a special educational section of YouTube which now contains 53 videos. (Rapacki, 2007) The sites k12 education group is located here: www.youtube.com/group/k12.

There are instructional videos which overview using blogs, wikis and other social media in the classroom. This safe area for posting and searching is growing, as more teachers and students submit their work. With enough people contributing educational videos, the effect could snowball, creating a multi-disciplinary, international community with vast resources. Digitization and preservation initiatives for cultural materials are taking place in the context of a new cultural expectation: people believe they should have access and a right to work with and manipulate video and text online. (Kaufman, 2006) These cultural expectations directly affect literacy and learning.

For creative educators, there are potentially thousands of ways to utilize YouTube in their curriculum and add to their current teaching pedagogy. These videos don't necessarily provide content, but they can stimulate the interest that makes curriculum relevant. If teachers are unable to find the perfect clip for their lesson, they can create and post their own material. In addition to the k12 user group, creative keyword searches can lead to discussions on greater learning topics. For example, entering "time lapse" brings you videos of flowers blooming and ice melting. These videos can help jumpstart conversations about life cycles in nature. A search for "WW2" introduces the oldest YouTube contributor ever seen, a 94 year old vet relating his firsthand experiences of a battle long past. These videos have a large potential impact on students. One class viewed the Dove "realbeauty" ads to jumpstart a discussion on issues of self image and the manipulative power of advertising.

Comparisons to other ICT Tools

The Newsweek Magazine describes sites such as Flickr, MySpace, Craigslist, Digg and YouTube as putting the "We" in Web. It suggests that these sites represent the next phase in the digital revolution. It's no longer about the technologies per se but about the communities that have grown up around them, popularly known as social networking communities. In addition to YouTube, Google Video and Revver.com provide similar services. Zudeo.com offers a growing collection of higher-quality videos with longer playback times. These ICTs contain several features that fulfill the needs of students. They provide community, almost immediate feedback and gratification, freedom of choice, pacing and privacy, and a space to learn without the fear of failure or being made fun of.

Young people need spaces in which to discover their identities. Web 2.0 technologies like YouTube and MySpace can be ideal tools for students to explore. User profiles allow students to personalize and customize their pages. Young adults spend many hours of the typical day hanging out with their friends. Many kids love spending a large part of their day socializing. As with adults, teens tend to define themselves by the company they keep, and this feature allows teens a variety of ways to experiment with defining themselves through their contacts.

As our culture becomes more participatory, young people are creating their own blogs and podcasts, developing their own profiles on MySpace, and producing their own YouTube videos and Flickr photos. Bulletins, blogs or journaling have for a long time been a way for teens to carry on dialogue with themselves or their network of friends about their lives. The appeal of blogs is in their immediacy. Blogs also have privacy settings that can be applied to individual entries. There are also several other interactive features which allow teens to explore who they are and what interests them. The search engine is attached to places like Amazon.com where they can attach an image of a book they're reading, or upload an MP3 of a song that they currently love.

Much as engineering students learn by taking machines apart and putting them back together, many students can understand how various things work by taking their culture apart and remixing it. Web 2.0 technologies are giving students the resources to solve problems on demand. YouTube and other ICTs allow them the flexibility to create their own learning environments by constructing videos and the ability to search for videos and a variety of subjects.

Predictions for the Future

We should encourage the new generations of learners to acquire skills and experiences as public intellectuals and to share their insights with a larger public from wherever they happen to be situated. A modern learning environment should work not by defining fields of study but by removing obstacles so that knowledge can circulate and be reconfigured in new ways. If YouTube is integrated into the 21st century classroom, the learning implications will be positive. The walls that separate the study of different subjects will be broken down, and full collaboration between students will make combining theory and practice a breeze.

Technologies like YouTube will continue to change education, learning and the workplace over the next 10 years. The future workforce will need to use technology efficiently, and jobs will surely shift as we've seen in the past. A student comments on the effects of technology and integrating into the workplace, "I definitely think we are going to be working more than our parents simply because of the integration of technology and the tools that are required to learn and use in everyday business. Technology forces us to be more productive, so in an eight hour day we are expected to do four, five, and six times as much". (Sweeney, 2007)

Researchers are already describing immersive online game worlds such as Second Life as the beginnings of Web 3.0. These issues reflect changes that cut across culture, commerce, technology, learning, and social organization. Imagine what would happen if academic departments operated more like YouTube, allowing for rapid deployment of scattered expertise and the dynamic reconfiguration of fields. This new form of academic unit would be coined a "YouNiversity". (Jenkins, 2007)

Kaufman gives a range of recommendations for video, education, and open content that could involve the constellation of existing and potential partners from public institutions and private enterprise.

1. Establish a research center for the future of the moving image in education.
2. Launch new, self consciously high – quality educational productions in television, film, video, and radio.
3. Develop a strategy for privileging library, museum, and university – sourced moving images in the online chaos of the YouTube world.
4. Support a research fellows programs, bringing specialists to cultural and educational institutions to work with video in particular.
5. Publish research papers systematically on these topics.
6. Design college level and K – 12 courses around these initiatives.
7. Launch standards and best practices workshops for moving image questions.
8. Explore experimentation around with large data sets or digital material.
9. Support the establishment of the American Archive.
10. Lastly, release moving image and recorded sound material for the public to engage with.

His recommendations interlock and are laid out to be cross–pollinating and thought provoking. He talks about the ripe opportunity at the start of 2007 to embark on a bold strategy of envisioneering in this area. The unifying concept behind all of the initiatives— whether they are test beds, R and D efforts, innovations, laboratories, factories, task forces, commissions, working groups, study groups, projects, institutes, initiatives — is that the future is unknown, and we have to try to understand it. We have to encourage experimentation involving video and recorded sound in the library, in schools, in business, in publishing, in museums.

If this trend continues, then an iPod, or a device its size, will be able to hold a year's worth of video (8,760 hours) by 2012, all the commercial music ever created by 2015, and all the content ever created (in all media) by 2020. (Kaufman, 2006) Today's children will be able to wear all the media ever created around their necks when they reach college. The portability of media archive and display and production devices by that time will represent a radical technology shift.

Conclusion

YouTube not only uses bits of copyrighted material to grab young people's attention, but the site has provided a launching pad to whole new forms of user-generated video that have gained instant popularity. A conceptual shift is happening in schools where students have an urgency and determination to master the "big picture" issues shaping their lives. For example, 21st century learners need to have an understanding of complex systems like the modern workplace, the environment, international relations, social interactions and cultures. To educate such students, we don't so much need a faculty as we need an intellectual network. Students should be encouraged to network broadly and draw on the best thinking about their topic, wherever they can find it. The contemporary educational landscape is blurring the lines between learning consumption and production, between creating our own learning and thinking.

Social networking sites can be powerful tools for self discovery and learning, since they give power as well as responsibility to the user. YouTube can be a valuable resource and is just one example of a potential Web 2.0 tool which in the hands of enlightened educators, can inspire students and support their digital learning style. Young people will always want a voice – it's up to teachers to utilize available resources, to help our students create content that is relevant and engaging. The next section focuses on the third video CMC I have set out to describe and discuss.

MMOS

Having looked closely at Facebook and YouTube, I will now concentrate on analyzing a third type of video-CMC called generically Massively Multiplayer Online (MMO) video games, or simply MMOs for short. We will do a case study of a particular one called World of Warcraft (WoW) that involves a lot of participant interaction mostly through video but of course also through text.[5] This case study of MMOs will try to analyze the linguistic features in MMO language through studying a particular MMO Role-Playing Game (MMORPG).

'Massively Multiplayer Online Games' as a CMC Platform

MMOs, which are identified as Multi-User Virtual Environments (MUVEs) by some scholars (Delwiche, 2006), are video and computer games that require users to connect to a particular server provided by the game producer and to actually play the game. The users, apart from playing the game alone, can also communicate and cooperate with each other in order to achieve different objectives in the game. These MMOs therefore serve as a kind of CMC platform since an adequate level of digital literacy is required in playing MMOs (so as to allow users to play the game and communicate with others without major obstacles).

Compared to other forms of CMC such as MSN conversations and mobile phone messaging, the information being exchanged in MMOs is relatively more complex in nature. Shortenings of words and sentences that are commonly observed in many CMC can also be found in MMO languages. Moreover, the exchange process includes both text-based and graphic-based information, indeed, more image-based communication, hence the designation video-based CMC that I have accorded them here, and topics covered can vary from game-related content discussions to real-life casual conversations. High levels of expressivity and flexibility in MMO languages are generally observed.

Recent studies have suggested that MMOs are worth focusing on since 'they have developed many characteristics of physical communities such as specialized language, political structures, complex social rituals, and shared history' (Steinkuehler, 2004). However, a vast number of studies in the literature focuses on the psychological, sociological and commercial values of MMOs (Delwiche, 2006; Gee, 2004; Hertz, 2002; Yee, 2006); those that try to examine the linguistic aspects in MMOs are comparatively rare. I complement these studies with a close examination of the language used in one of the most popular MMOs nowadays, an MMO which has over 10 million subscribers all over the world – *World of Warcraft* (WoW).

CASE STUDY: WORLD OF WARCRAFT

A Brief Introduction of WoW

World of Warcraft (WoW) is an English-based MMORPG (Asian versions are translated versions). Users in this game (hereafter known as 'gamers') have to create and control their own characters, generally known as 'avatars' in MMOs, and explore the virtual world with planets called Azeroth and Draenor. Upon creation of a character, the gamer will first have to decide which faction the character belongs to, either 'Alliance' or 'Horde', and then the gamer will have to choose the character's 'Race' and 'Class'. Communications between gamers of the two different factions are not possible, for example, a speech produced by an Alliance gamer can only be read by other Alliance gamers, and not by Horde gamers (pseudo-words instead of the actual English words would appear on the screen of the Horde gamers). The main aim of the game requires gamers to fight and kill monsters, some of which can be fought one-on-one, whereas some can only be brought down by a group of 40 gamers with characters hitting the level cap. Apart from leveling, fighting and killing, a wide range of other activities can also be done in WoW. Commercial activities include manufacturing, trading and auctioning virtual items and goods, as well as providing supportive services to other gamers such as 'enchanting' and 'teleporting' services. Social activities include establishing communities, known as 'Guilds', and even organizing and attending ceremonial events such as virtual in-game weddings.

Communication Methods in WoW

In WoW, communication between gamers is usually achieved through text-based synchronous conversations. This activity is usually performed in a rectangular chat-box which is placed at the lower left corner of the screen (default settings). Similar to other CMC platforms like MSN, the name of the speaker and the channel he/she belongs to are displayed automatically once a text-based utterance is produced.

Below is an illustration:

[Chat Channel] [Character's Name]: (text-based utterance)
[2. Trade] [Nozzwolf]: can someone link me fiery weapon plz?...

Having the information displayed in such a way, a single chat-box can therefore be used to display multiple utterances that are being produced simultaneously from different sources. In addition, voice chat functions are also available in some channels, such as the 'Group channels' and the 'Custom chat channels'.

Below are the three main categories of text-based utterances that can appear in WoW:

1. Situational Utterances
2. Group Channels
3. Chat Channels

Situational Utterances

A character can produce text-based utterances anytime during the game. However, the gamer first has to decide the 'nature' of the utterances. There are mainly three types of these utterances.

Utterance Type Effect

Whisper (Generally known as 'Tell' in MMOs)
One-to-one: direct conversation between two gamers
Say One-to-many: short effective range
Yell One-to-many: long effective range

The utterance will be displayed on the screen in the following way:

[Character's Name] (utterance type): (text-based utterance)
e.g. [Tibsy] yells: can a rogue plz pick lock 225 LP minimum

Group Channels

Whenever a gamer partners up with other gamers, all of them are assigned to a group channel compulsorily. There are four different group channels in WoW:

Group Channel Effect

Group One-to-many; utterances can be perceived by all group members regardless of range.
Raid One-to-many; utterances can be perceived by all raid members regardless of range.
Guild One-to-many; utterances can be perceived by all guild members regardless of range.
Battleground One-to-many; utterances can be perceived by all members in a battleground regardless of range.

The utterance will be displayed on the screen in the following way:

[Group Channel] [Character's Name]: (text-based utterance)
e.g. [Guild] [Doomfree]: vex 5d finish ur mc donald la

Chat Channels

Once a character has successfully entered the virtual world, he/she will automatically be assigned to join several 'chat channels' as long as they are accessible. A character can join multiple channels simultaneously as long as he/she is eligible for the channel, and he/she can leave the channels manually whenever he/she wants to. While several channels are official channels, gamers themselves can also create their own custom channels to serve different purposes, such as effective communication between different friends and groups. The official channels in WoW include:

Chat Channel Purpose

General For casual conversations within a particular city Trade
i) For commercial activities
ii) For recruitment of guild members
Local Defense
i) For casual conversations within a non-city region
ii) For defense of a particular region
Looking For Group For battle group formation

The utterance will be displayed on the screen in the following way:

[Chat Channel] [Character's Name]: (text-based utterance)
e.g. [2. Trade] [Nozzwolf]: can someone link me fiery weapon plz?...

Linguistic Features of the WoW (English Version) Language

In WoW, gamers are assigned to different game servers, known as 'Realms'. Gamers in different realms cannot communicate with each other (at least within the framework of the game), therefore it is possible that the use of language in different realms might differ from one another. The realm that this study is investigating is an Oceanic realm called 'Aman'thul'. The dominant language used for communication between gamers in this realm is English. Some general features that have been observed in the communication between users in this realm include:

Table 14.

General Communication and Commercial Activities			
Shortening	**Target expression**	**Shortening**	**Target Expression**
LF	Looking For	WTB	Want To Buy
LFG	Looking For Group	WTS	Want To Sell
LFW	Looking For Work	WTT	Want To Trade
LFM	Looking For More	PST	Please Send Tell
LFxM; x=number	Looking For x More	mt	Mistell
Brb	Be Right Back	gtg	Gotta Go
Cya	See Ya	oic	Oh I See
u(r)	You(r)	btw	By The Way
r	are	4	for

1. The Presence of Shortened Words
2. The Presence of Emoticons
3. The Usage of Capped Letters
4. The Usage of Punctuation Marks
5. Grammatical Reductions

The Presence of Shortened Words

The WoW language, like that of many other CMCs, contains many shortened words. Methods of shortening include both phonetic-based and spelling-based ones. The phonetic-based method includes the use of letter homophones such as 'u' for the pronoun 'you', and the use of number homophones such as '4' for the preposition 'for'. The spelling-based method include initialism such as 'LFM' for 'looking for more' and the formation of acronyms using particular sound segments such as 'LW' for 'leatherworker'. Irregular shortened words can also be observed, such as 'chanter' for 'Enchanter' and 'MgT' for 'Magisters' Terrace'. Tables 14, 15, 16, and 17 show some common shortened words used in the game.

Role-Specific

'Role' refers to the roles of each gamer in a group battle.

There are mainly three roles in WoW (see also Table 18):

i) Tanks (characters that confront monsters and hold them in place)
ii) Damage Dealers (characters that deal major damage to monsters)
iii) Healers (characters that keep the whole group alive)

Table 15.

Expressions			
Shortening	**Target expression**	**Shortening**	**Target Expression**
LOL	Laughing Out Loudly	OMG	Oh My God
ROFL	Rolling On the Floor Laughing	LMAO	Laughing My Ass Off
FTW	For The Win	WTH	What The Hell
FTL	For The Lose	STFU	Shut the F**k Up
Soz	Sorry	Gratz	Congratulations

Table 16.

Location-Specific Cities			
Shortening	**Target expression**	**Shortening**	**Target Expression**
Shat	Shattrath City	SW	Stormwind City
IF	Ironforge	Darn	Darnassus

Profession-Specific

'Profession' refers to the commercial abilities a character possesses in the game (Table 19 shows a few examples).

Talent-Specific

'Talent' refers to the battle skills a character possesses in the game (Table 20 shows a few examples).

Table 17.

Dungeons **(a few examples)**			
Shortening	**Target expression**	**Shortening**	**Target Expression**
UB	Underbog	Arc	Arcatraz
MT	Mana Tombs	MgT	Magisters' Terrace

Table 18.

Shortening	**Target expression**	**Shortening**	**Target Expression**
MT	Main Tank	OT	Off Tank
DD	Damage Dealer	Alt.	Alternate Character

Table 19.

Shortening	Target expression
JC	Jewelcrafter
DE	Disenchant
Chanter	Enchanter
Xmute	Transmute

The Presence of Emoticons

In the WoW language, emoticons are often applied in utterances. However, unlike MSN conservations and SMS messaging, only a very limited amount of emoticons are observed in WoW. Emoticons are mainly used to express the mood of the speaker. Table 21 shows some of the common ones.

The Usage of Capped Letters

In some situations in WoW, a whole string of utterance will be typed in capped letters instead of normal uncapped letters. Such an utterance will usually be regarded as being emphasized, or a 'shout'. It has been observed that gamers are highly aware of utterances in capped letters, and that improper usage of capped utterances by one will generally be regarded as a rude behavior by other gamers. Below are some examples:

Giving Orders:
[Battleground] [Tuffenuff]: DONT STOP AT IBGY
Shouts:
[2.Trade] [Tinng]: AAAAAARRRRRRRRRHHHHHHH

Table 20.

Shortening	Target expression
MD	Misdirection
Rez	Resurrection
SoC	Seed of Corruption
Brez	Resurrection (In Battle)

Table 21.

Emoticon	Mood Expressed	Emoticon	Mood Expressed
=) ; =] ; =D	Smiling	=(	Frowning
='(	Crying	XD	Laughing
O_O ; o_O ; =O	Shocked	=P	Smiling (Tricky)
^_^	Happy	>_<	Sad

Table 22. Punctuation Marks Omitted

Punctuation Mark	Indication
(.) Fullstop	Completion of an utterance
(') Apostrophe	Contraction

The Usage of Punctuation Marks

In the WoW language, some punctuation marks are frequently omitted, whereas some others are often preserved (see Tables 22 and 23)

Grammatical Reductions

In the WoW language, grammatical reductions are found to be quite common. Generally, only the lexical items are often retained in an utterance. Similar to other CMC, loose grammaticality constraints can be observed. Below are some categories that are often omitted in utterances (words in parentheses are those omitted):

Pro-drop:
[1.General] [Lilfighter]: paying 2g for stock run
[1.General] [Lilfighter]: **(I'm)** paying 2g for stock run
Copula / Modal Verb:
[5. jingkirk] [Dranoelzero]: rag u there?

Table 23. Punctuation Marks Preserved

Punctuation Mark	Indication
(!) Exclamation Mark	Emphatic mood
(?) Question Mark	Interrogative Mood

[5. jingkirk] [Dranoelzero]: rag **(are)** u there?
('jingkirk' is a custom channel created by the gamers themselves)
Article:
[2.Trade] [Bankrunner]: WTB dm run 10g pst
[2.Trade] [Bankrunner]: WTB **(a)** dm run 10g pst
Preposition:
[2.Trade] [Bankrunner]: WTB dm run 10g pst
[2.Trade] [Bankrunner]: WTB dm run **(for)** 10g pst

Discussion: Distinctiveness, Economy of Expression and Expressivity

As Bodomo (2008) has suggested, in many CMC environments, the way expressions are constructed are actually induced by a competition between the 'economy of expression' and 'expressivity'. In the WoW language, the 'economy of expression', very often, overrides the importance of 'expressivity', and this can be seen from the frequent usage of shortened words, the usage of emoticons and capped letters for emphases, as well as the omission of some punctuation marks and functional words for the purpose of efficiency. In addition, through examining the linguistic features in the WoW language, the importance of the 'distinctiveness of expression' has been observed. In MMOs, where communication is often under heavy time pressure (due to tasks and objectives in the game-play), gamers would normally try to use the minimum effort in expressing themselves, but at the same time, maintaining the distinctiveness of their expressions. That is to say, gamers will try to use longer expressions if the shortened expression would lead to ambiguity. This tendency can be seen from two phenomena observed in the WoW language.

The first phenomenon is that, though shortening is a common practice in the WoW language, the method of shortening using a combination of letter and sound homophones at the same time is very rare. Common CMC shortening words like 'l8' for 'late' and 'b4' for 'before' are rarely used in the WoW language. One explicit reason is that typing words using keyboards is much easier than using the number pads in mobile phones, therefore the efforts required in typing 'late' and in typing 'l8' using a keyboard would not have a significant difference. Another possible reason that is related to the 'distinctiveness of expression' is that in MMOs, 'l8' might also be interpreted as the number '18' (the letter 'l' and the number '1' has no visual difference in 'Times New Roman' for example) or 'level 8' as well. An expression like 'im l8' can therefore be interpreted as 'I'm late', 'I'm eighteen' or 'I'm at level8'. In order to avoid this kind of ambiguity, gamers would generally prefer not to use a combination of letter and sound homophones for shortening.

The second phenomenon observed in the WoW language is that, though initialism is widely used in creating new shortened words, especially for locations, rarely will there be locations that share the same shortened form. If there are two locations that share the same shortened form, gamers would normally lengthen the expression of either one of the locations. A nice example from the game is that on 25 March 2008, a new dungeon called 'Magisters' Terrace' was introduced. Initially, on the first day, the WoW community in the server Aman'thul referred to the dungeon as 'MT'. However, soon after the release, they realized that using 'MT' to refer to 'Magisters' Terrace' was problematic since the shortened form for 'Mana Tombs', another dungeon, is also 'MT'. In order to avoid ambiguity, new shortened forms including 'MgrT', 'MgT', and 'MagT' were suggested. Due to the 'economy of expression', the shortest form without ambiguity – 'MgT', consequently became the most widely accepted shortened form for 'Magisters' Terrace'.

To state in simple terms, in MMO communications, users have a tendency to use a minimum effort in expressing the most accurate information. Linguistic features that are not significant in the delivery of meaning will very often be left out. It is therefore significant to note that linguistic features such as shortening as seen here with video-based CMC can be markedly different from those in text-based CMC for some of the reasons described here.

A MISCELLANY OF EMERGING MEDIA

We have in previous sub-sections given detailed discussions of three video-based CMC including Facebook, YouTube and MMO. I shall in this section briefly mention other, mostly video-based CMC that are emerging, and which deserve more in-depth analysis beyond the scope of this book.

Skype

This is how Skype introduces itself on its website at: www.skype.com:

Skype was founded in 2003 by Niklas Zennström and Janus Friis. Skype created a little piece of software that makes communicating with people around the world easy and fun. With Skype you can say hello or share a laugh with anyone, anywhere. And if both of you are on Skype, it's free. (Retrieved on November 18, 2008)

Skype has indeed changed the way we communicate through computers since it began in 2003. Before this period, participants in a CMC situation could exchange text and even video but it was not possible to do voice telecommunication directly through the computer with internet connection. One had to incorporate and involve expensive landlines and telephone companies. But all that changed when Skype made it possible for two or more users separated by millions of kilometers to see each other's video photos, communicate with each other through the usual synchronous chatting via text and at the same time actually hear each other and converse on internet phone! Communication through the internet is free for the parties and to landlines but landline users using the internet to communicate with them have to pay. Other services like video-conferencing are possible. Communication has changed so much that it has even come along with a new verb in the English language, "to skype", meaning to communicate through the internet using a video-based computer-mediated communication tool called Skype. The Skype as a CMC medium has grown in popularity since it was acquired by eBay from its two Estonian inventors. It will certainly continue to grow in stature, given its popularity among the youth and as its software gets translated into more and more languages.

Flickr

This is how Flickr introduces itself on its website at: www.flickr.com:

Flickr - almost certainly the best online photo management and sharing application in the world - has two main goals:

1. We want to help people make their content available to the people who matter to them....2. We want to enable new ways of organizing photos and video... (Retrieved November 18, 2008).

Flickr began life in 2004, right on the heels of Skype, and like Skype has grown to be a very popular CMC tool among the youth and other avid users of the internet. This popularity is due to the ease with which it allows CMC players to use it as a repository for their documents, such as photos and videos.

Second Life

Here is how Second Life is introduced on its own website at http://secondlife.com/:

Second Life® is a 3-D virtual world created by its Residents. Since opening to the public in 2003, it has grown explosively and today is inhabited by millions of Residents from around the globe. From the moment you enter the World *you'll discover a vast digital continent, teeming with people, entertainment, experiences and opportunity. Once you've explored a bit, perhaps you'll find a perfect parcel of land to build your house or business. You'll also be surrounded by* the Creations *of your fellow Residents. Because Residents retain intellectual property rights in their digital creations, they can buy, sell and trade with other Residents.* The Marketplace *currently supports millions of US dollars in monthly transactions. This commerce is handled with the inworld unit of trade, the Linden™ dollar, which can be converted to US dollars at several thriving online Linden dollar exchanges.*

Welcome to the Second Life world. We look forward to seeing you inworld. (Retrieved on November 18, 2008)

This three dimensional online video-game that simulates real world life whereby someone can enter that world and begin a new life started in 2002 and has become very popular, even to the point that some events in that virtual world sometimes spill into the real world, such as real world couples divorcing because of relationships by their avatars or imagined characters in the Second Life.

To conclude this sub-section on a miscellany of emerging media, I have just selected three of these exciting video-based CMC media to showcase what the future of video-based CMC would be like. But there are certainly many existing media such as Twitter at twitter.com, Hulu at hulu.com, and new ones yet to be created as technologies get more and more advanced.

But what are the consequences of these for educational issues? Indeed, one thing common with all these video-based CMC media is that they can serve as repositories for learning objects – resources that can be used for educational purposes. YouTube has become an important source of learning objects for teachers to use in their lessons and there are even reports that one teacher has based his entire semester-long lecture on YouTube.

But these media can even have consequences beyond the classroom. Again to take the YouTube as an example, it and other video-based CMC media are being used vigorously by politicians and others who need to get their messages to the public. A famous example of this political and public relations use of the YouTube and the internet as a whole is how Barack Obama based a large part of his campaign for the votes of young people and other internet users to eventually win the US presidential elections of November 2008.

CONCLUSION: FROM TEXT-BASED TO VIDEO-BASED CMC

In this last chapter of the book, I have outlined some emerging themes and media within CMC. I have claimed that CMC as a discipline is at a crucial exciting stage in its development during which we are moving away from text-based CMC to an emphasis on video-based CMC. Many technologies and media that are invented now for the purposes of CMC, especially social networking systems, are incorporating more and more images and video streams, from still images such as photos and emoticons found in Email, SMS to video clips and streams found in YouTube, Online Games, and Skype as discussed in this chapter.

Undoubtedly therefore the main theme of the CMC in the immediate future is the development of more and more video-based CMC, bringing together text and sound to create a lively communication situation around the video object. A further theme is the imitation and simulation of real-life situations. We are seeing more and more systems trying to simulate and thus emulate the complexities of real-life situations and typical examples of these are the MMOs that we have seen but also Web 3 features[6] such as Second Life.

There are theoretical and practical implications for the discipline of Linguistics- and Literacy-Based CMC as a sub-discipline of CMC. Theoretically, ideas and models of language based on the text alone will no longer be enough as theories of language structure and use. We will need integrated approaches to theorizing about the form and function of language that take into consideration text, sound, image and video.

Practically, in terms of the way we present language and literacy in learning situations and in terms of the way we develop language-based technologies, there must be a paradigm shift that involves moving away from mere text-based educational and technological models to multi-media based models in the form of incorporating video to enhance text. Text in the written word has served us for long and may still serve us for a long time to come but certainly video is now the new text for 21st Century communications.

REFERENCES

Adkins, R. (2007). *YouTube and Literacy.* https://webspace.utexas.edu/rra56/RHE330/vcreporthome.html

Ashley, J. (2003). *Synchronous and Asynchronous Communication Tools.* Executive Update Online. Retrieved April 16, 2008, from http://asaecenter.org/PublicationsResources/articledetail.cfm?ItemNumber=13572

Barry, J. A. (1991). *Technobabble.* Cambridge, Mass: MIT Press.

Bodomo, A. (2008). LING2041 Language and Information Technology Lecture Notes.

Bodomo, A., & Lee, C. (2002). Changing forms of language and literacy: technobabble and mobile phone communication. *Literacy and Numeracy Studies: An International Journal in Education and Training of Adults, 12*(1), 23–44.

Chow, H. T. L. (2008). *Massively Multiplayer Online Games (MMOs) as a Platform for Synchronous Information Exchange*. Term Paper for the course Language and Information Technology. HKU

Colman, A. M. (2001). *Homophones. A Dictionary of Psychology*. Retrieved April 17,2008, from http://www.encyclopedia.com/doc/1O87-homophones.html

Crystal, D. (2001). *Language and the Internet*. Cambridge: Cambridge University Press.

Daughmaudytè. Jurga. (2006). Le Langage SMS Dans le Francais. *Kalbotyra, 56*(3). Retrieved April 17, 2008 URL: http://www.ceeol.com/aspx/getdocument.aspx?logid=5&id=787c4e06-64d0-48b5-a1a1-b532a0d84da3

December, J. (1996). Units of analysis for internet communication. *Journal of Computer-Mediated Communication, 1*(4).

Delwiche, A. (2006). Massively multiplayer online games (MMOs) in the new media classroom. *Educational Technology & Society, 9*(3), 160–172.

(Ed.). *Avatars at Work and Play: Collaboration and Interaction in Shared Virtual Environments* (pp. 187-207). London: Springer-Verlag.

Ellis (2002). Argot and Verlan. *Comtemporary France Online*. Retrieved April 16, 2008, from http://www.well.ac.uk/cfol/argot.asp#verlan

Elsevier, R. (2002). YouTube Comes to the Classroom. Reed Business Information 2002. SLJ 53 no1 Jan. 2007.

Fu Wing Man, C. *2008. Language Use in Facebook: A Study of French Youth*. Term Paper for the course Language and Information Technology. HKU

Gee, J. P. (2004). *Learning about Learning from a Video Game: Rise of Nations*. Retrieved 28 April 2008, from http://distlearn.man.ac.uk/download/RiseOfNations.pdf

Geller, M. 2006 *The Rise of Online Video,* Media Connection http://www.imedia-connection.com/content/8888.asp

Gibbons, J. (1987). *Code-mixing and Code Choice: A Hong Kong Case Study*. Clevedon: Multilingual Matters.

Harris Interactive Survey. (2007) http://www.harrisinteractive.com/services/bschools.asp (retrieved March 3, 2009)

Herring, S. C. (1996). Two variants of an electronic message schema. In Susan C. Herring (Ed.), *Computer-mediated communication: Linguistic, social and cross-cultural perspectives* (pp. 81–108). Amsterdam: John Benjamin's.

International, S. I. L. (1997). What is an acronym? *SIL International.* Retrieved April 16, 2008, from http://www.sil.org/lingualinks/LANGUAGELEARNING/OtherResources/GlssryOfLnggLrnngTrms/WhatIsAnAcronym.htm

James, G. (2001). Cantonese particles in Hong Kong students' English emails. *English Today, 17*, 9–16. doi:10.1017/S0266078401003029

Jenkins, H. (2007). From YouTube to YouNiversity. *The Chronicle of Higher Education, F16*, 2007.

Kaufman, P. (2006). *Video, education and open content: Notes toward a new research and action agenda.* First Monday: Peer reviewed Internet Journal.

Laurence, T. (2005). Presentation of self in French teens blogs. *Laboratoire Gestic.* Retrieved April 18, 2008, from http://gesture-lyon2005.ens lsh.fr/article.php3?id_article=254

Lee, C. K.-M. (2002). *Chinese and English computer-mediated communication in the context of New Literacy Studies.* MPhil Thesis, Hong Kong: University of Hong Kong.

Lee, C. K.-M. (2007). Linguistic features of email and ICQ instant messaging in Hong Kong. In Brenda, Danet & Herring, Susan. C (Eds.), *Sociolinguistics.* Oxford University Press.

Li, D. C. S. (1998). The plight of purists. In Pennington, M. C (ED.), *Languagein Hong Kong at Century's End.* Hong Kong: HKU Press. 161-190.

Li, Y. (2000). Linguistic characteristics of ESL writing in task-based e-mail activities. *System, 28*, 229–245. doi:10.1016/S0346-251X(00)00009-9

Liénard, F. (2005). Langage texto et langage contrôle. Description et problèmes. *Linguisticae Investigationes, 28*(1), 49–60. doi:10.1075/li.28.1.06lie

Lung, K. Y. F. (2008). *The use of Facebook and its Linguistic Features Among Studies of the University of Hong Kong.* Term Paper for the course: Language and Information Technology. HKU

Matthews, S., & Yip, V. (1994). *Cantonese: a comprehensive grammar*. London: Routledge

McKinney, J. 2006 Why YouTube Works. *Unsought Input*. http://www.unsoughtinput.com/index.php/2006/08/21/why-youtube-works/

McNerny, T. (2007). One-Third of Frequent YouTube Users are Watching Less TV to Watch Videos Online. *Harris Interactive*. http://www.harrisinteractive.com/news/allnewsbydate.asp?NewsID=1168

Oishi, L. (2007). Did You Just See That? *Technology and Learning, 27*(6).

Papenfuss, M. (2006). French youths speaking their own language. *USA Today*. Retrieved April 15, 2008, from http://www.usatoday.com/news/world/2006-01-05-french-slang_x.htm

Pete, O. (2006). Changing World, Changing Literacy. *Ed Tech Journeys*. http://preilly.wordpress.com/2006/11/07/changing-world-changing-literacy/

Rapacki, S. (2007). *Social Networking Sites: Why Teens Need Places like MySpace*. Young Adult Library Services, Wadsworth Ohio.

Shortis, T. (2001). *The language of ICT*. London: Routledge.

Steinkuehler, C. A. (2004). Learning in massively multiplayer online games. In Kafai, Y. B., Sandoval, W. A., Enyedy, N., Nixon, A. S. & Herrera, F. (Eds.), *Proceedings of the 6th International Conference of the Learning Sciences*, Mahwah, NJ, USA: Erlbaum, 521-528.

Sweeney, R. (2007). How the New Generation of Well-Wired Multitaskers Is Changing Campus Culture. *The Chronicle of Higher Education* 1 Jan 5 2007.

Wang, S. (2006). *The Long Tail: Why Aggregation, Context & Not (Necessarily) Content are King* at http://www.bearstearns.com/bscportal/research/analysts/wang/112706/Slide1.htm

Wesch, P. (2007) Why do You Tube? - Video Response Analysis Digital. *Ethnography*. http://mediatedcultures.net/ksudigg/?p=82

Yee, N. (2006). The Psychology of MMORPGs: Emotional Investment, Motivations, Relationship Formation, and Problematic Usage. In R. Schroeder & A. Axelsson

Yee Wai Man, J. (2008). *Linguistic Features of MSN in specific Hong Kong Contexts*. Term Paper for the course, Language and Information Technology. HKU

Zhou Zhou, T. (2008). *A Closer Look at Mobile Phone Novels: Features and Implications*. Term Paper for the course, Language and Information Technology. HKU

ENDNOTES

1 I thank Frances Lung for collecting the data for this section of the book as part of a term paper (Lung 2008) for my course in Language and Information Technology.

2 I thank Cindy Fu Wing Man for collecting the data for this section of the book as part of a term paper (Fu 2008) for my course in Language and Information Technology.

3 I thank Erin Eng of Santa Clara University, California for mining information and gathering facts and analysis about YouTube for this section of the book as part of a term paper (Eng 2007) for my course in Language Education, and Information Technology in the Spring Semester of 2007.

4 Web 2.0 refers to what is considered as the second generation of the World Wide Web in which social networking communities are rampant. The term became popular after a Media Web conference on "Web 2.0" was held in 2004.

5 I am particularly grateful to my student, friend, and a seasoned video games player, Leonard Ignatius Chow Hin Ting, who spent many hours online playing WoW in order to extract data for me and as part of his term paper (Chow 2008) for my course, Language and Information Technology in the Spring Semester of 2008).

6 The following quotation by Cate Metz explains the various Web versions up to Web 3.0: "Just in case you missed it, the web now has version numbers. [...] a man named Dale Dougherty dreamed up something called Web 2.0, and the idea soon took on a life of its own. In the beginning, it was little more than a rallying cry, a belief that the Internet would rise again. But as Dougherty's Silicon Valley start-ups—and blogs are already abuzz with talk of the Web's next generation. To many, Web 3.0 is something called the Semantic Web, a term coined by Tim Berners-Lee, the man who invented the (first) World Wide Web. In essence, the Semantic Web is a place where machines can read Web pages much as we humans read them, a place where search engines and software agents can better troll the Net and find what we're looking for. "It's a set of standards that turns the Web into one big database," says Nova Spivack, CEO of Radar Networks, one of the leading voices of this new-age Internet." (Retrieved on March 4, 2009: http://www.pcmag.com/article2/0,2817,2102852,00.asp)

About the Author

Adams B. Bodomo is Associate Professor at the School of Humanities, University of Hong Kong. He obtained his PhD from the Norwegian University of Science and Technology after obtaining Bachelors and Masters Degrees at the University of Ghana with specializations in subjects as diverse as Linguistics, Computer Science, French, Swahili, and African Studies. He has taught at the University of Ghana, Stanford University, and Santa Clara University, California, USA. His current teaching and research interests include formal theoretical linguistics, computer-mediated communication, and African Studies, particularly Africa-China relations. He has published more than 10 books and 50 journal articles in major publishing houses such as Stanford University CSLI Publications and IGI Global Publications, and in major academic journals in the fields of Linguistics and Computer-mediated Communication such as *Natural Language and Linguistic Theory*, *Linguistic Inquiry*, *Studia Linguistica*, *International Journal of Technology and Human Interaction*, and the *International Journal of Web-based Learning and Teaching Technologies*.

Index